AF594971

Carbon-Based Polymer Nanocomposites for High-Performance Applications

Carbon-Based Polymer Nanocomposites for High-Performance Applications

Special Issue Editor

Ana María Díez-Pascual

MDPI • Basel • Beijing • Wuhan • Barcelona • Belgrade

Special Issue Editor
Ana María Díez-Pascual
Alcalá de Henares
Spain

Editorial Office
MDPI
St. Alban-Anlage 66
4052 Basel, Switzerland

This is a reprint of articles from the Special Issue published online in the open access journal *Polymers* (ISSN 2073-4360) from 2019 to 2020 (available at: https://www.mdpi.com/journal/polymers/special_issues/carbon_based_polymer_nanocomposites).

For citation purposes, cite each article independently as indicated on the article page online and as indicated below:

LastName, A.A.; LastName, B.B.; LastName, C.C. Article Title. *Journal Name* **Year**, *Article Number*, Page Range.

ISBN 978-3-03928-991-2 (Hbk)
ISBN 978-3-03928-992-9 (PDF)

Contents

About the Special Issue Editor

Ana María Díez-Pascual graduated in Chemistry in 2001 (awarded Extraordinary Prize) at the Complutense University (Madrid, Spain), where she completed her Ph.D. (2002–2005) on dynamic and equilibrium properties of fluid interfaces under the supervision of Prof. Rubio. In 2005, she worked at the Max Planck Institute of Colloids and Interfaces (Germany) with Prof. Miller on the rheological characterization of water-soluble polymers. During 2006–2008, she was a postdoctoral researcher at the Physical Chemistry Institute of the RWTH-Aachen University (Germany), where she worked on the layer-by-layer assembly of polyelectrolyte multilayers onto thermoresponsive microgels. Then, she moved to the Institute of Polymer Science and Technology (Madrid, Spain) and participated in a Canada–Spain joint project to develop carbon nanotube (CNT)-reinforced epoxy and polyetheretherketone composites for transport applications. Currently, she is a postdoctoral researcher at Alcala University (Madrid, Spain) focused on the development of polymer/nanofiller systems for biomedical applications. She has published over 100 SCI articles (97% in Q1 journals). She has an H-index of 41 and more than 3500 total citations. She has published 21 book chapters, 2 monographies, and edited 3 books, and is the first author of an international patent. She has been invited to impart seminars at prestigious international research centers (i.e., Max Planck in Germany, NRC in Canada, School of Materials in Manchester, U.K.). She was awarded the TR35 2012 prize by the Massachusetts Technological Institute (MIT) for her innovative work in the field of nanotechnology.

Editorial

Carbon-Based Polymer Nanocomposites for High-Performance Applications

Ana Maria Díez-Pascual

Department of Analytical Chemistry, Physical Chemistry and Chemical Engineering, Faculty of Sciences, Alcalá University, 28871 Madrid, Spain; am.diez@uah.es; Tel.: +34-918-856-430

Received: 5 April 2020; Accepted: 8 April 2020; Published: 10 April 2020

Carbon-based nanomaterials such as carbon nanotubes, graphene and its derivatives, nanodiamond, fullerenes, and other nanosized carbon allotropes have recently attracted a lot of attention among the scientific community due to their enormous potential for a wide number of applications arising from their large specific surface area, high electrical and thermal conductivity, and good mechanical properties [1,2]. The combination of carbon nanomaterials with polymeric matrices (i.e., thermoplastics, epoxies, conducting polymers, biopolymers, etc.) leads to new nanocomposites with improved structural and functional properties due to synergistic effects [3], with applications in a variety of fields, such as in electronics, energy storage, automobiles, aerospace engineering, biomedicine, and so forth.

In particular, the properties of carbon-based polymer nanocomposites can be easily tuned by carefully controlling the carbon nanomaterial synthesis route and additionally the versatile synergistic interactions amongst the nanomaterials and polymers. In this regard, non-covalent and covalent approaches have been used to modify the surface of carbon nanomaterials with the aim of improving their dispersion and interfacial interactions [4,5]. The non-covalent strategies are based on the intermolecular interaction on the nanomaterial surface via physical adsorption and/or wrapping [6], though the nanomaterial–polymer interfacial interaction is typically weak, and this limits the effective stress transfer. These comprise solution mixing, melt-blending, and in situ polymerization. The solution method requires the dispersion of both the carbon nanomaterial and the polymeric matrix in a suitable solvent [7]. The melt-blending process involves the blending of the carbon nanomaterial into a molten polymer matrix under intense shearing [8]. Nanocomposites can also be prepared via in situ polymerization [9], in which the carbon nanomaterial is initially swollen by the monomer, and upon addition of the initiator, the polymerization begins by light irradiation or heat. The covalent method relies on the formation of a chemical bond between the polymer and the nanomaterial [10], leading to a strong interfacial interaction, though can disrupt the conjugated πsystem of the nanomaterial, hence modifying the properties. Thus, novel surface modifications of carbon nanomaterials are required in order to develop nanocomposites with improved properties compared with conventional composites. This Special Issue, with a collection of 14 original contributions and one review, provides selected examples of the most recent advances in the preparation and characterization of polymer nanocomposites incorporating carbon nanotubes and graphene or its derivatives for a variety of applications.

Graphene oxide (GO), the oxidized form of graphene, has attracted a lot of interest as nanoscale material [5]. It is solution processable, amphiphilic, and biocompatible, hence it can interact with biological cells and tissues [7]. This nanomaterial presents a large number of oxygenated groups, mainly epoxide, hydroxyl, and carbonyl on the basal plane and carboxyl acids on the edges, and can produce stable dispersions in water [11]. Further, it possesses outstanding strength and exceptional optical and electronic properties joint with high stability, flexibility, and optical transparency [12,13]. Nonetheless, it is insoluble in non-polar and polar aprotic solvents, which limits some applications. In this regard, different functionalization approaches have been carried out, like the reaction with organic hexamethylene diisocyanate (HDI) [14]. The resulting HDI-GO was mixed with conductive polymers, including poly(3,4-ethylenedioxythiophene):poly(styrenesulfonate) (PEDOT:PSS) [15], polyaniline (PANI) [16], and polypyrrole-3-carboxylic acid (PPy-COOH) [17], in different weight percentages via a simple solution casting method, and the nanocomposites were analyzed by different techniques to obtain information on the effect of the HDI-GO level of functionalization and percentage on the nanocomposite properties. The best properties were obtained for nanocomposites with HDI-GO contents in the range of 2–5 wt %. These nanocomposites are very suitable to be used in energy storage devices, fuel cells, batteries, supercapacitors, solar cells, touch panels, and so forth.

Other carbon-based nanomaterials, such as carbon nanotubes (CNTs), carbon dots (C dots), and carbon aerogels (CAGs) have also been mixed with different polymers for energy storage applications [18,19]. In particular, CNT-PANI and CNT-PPy have been used as cathodes into Li-ion batteries, leading highly efficient discharge ranges [20]. CNTs are also used as anodes for this type of batteries, though they exhibit very high capacity loss and low initial columbic efficiency. To solve this issue, CNTs can be prelithiated via cyclic voltammetry with a polymer coating such as p-sulfonated poly(allyl phenyl ether) (SPAPE) [21], which eliminates the initial irreversible capacity of the CNT electrodes. This is a simple, fast and inexpensive method to attain high performance flexible anodes for microbatteries.

Numerous approaches for the production of nanocomposites based on CNTs and conductive polymers have been developed [22,23]. A thoughtful issue in the development of these nanocomposites is the CNT tendency to aggregation. To prevent nanotube agglomeration, in situ oxidative polymerization in the presence of the CNTs can be carried out. The use of ultrasound guarantees a homogenous CNT dispersion in the reaction medium and inhibits coagulation during polymerization. In particular, works focused on the oxidative polymerization of aniline in the presence of CNTs have demonstrated that PANI forms a uniform coating on the CNT surface that results in strong π–πinteractions between the quinoid units of PANI and the aromatic rings of the CNTs [24]. This improves the electron transport in the nanocomposite and consequently the electrical performance. Following a similar approach, hybrid nanocomposites were prepared by in situ oxidative polymerization of a conjugated polyacid, diphenylamine-2-carboxylic acid (DPAC) in the presence of SWCNTs in both alkaline and acidic media [25]. The influence of the reaction pH and the SWCNT content on the structure, morphology, thermal stability, and electrical properties of nanocomposites was assessed. The nanocomposites displayed good electrical conductivity and thermal stability, and are suitable for application in supercapacitors, rechargeable batteries, sensors field-emission devices, and dye-sensitized solar cells.

On the other hand, a polyacrylamide-alginate hydrogel was mixed with different amounts of multiwalled carbon nanotubes (MWCNT) to develop a flexible and durable electrode. The electrical conductivity and the mechanical characteristics of the electrode (tensile strength and stretching modulus) gradually increased with increasing MWCNT concentration, with an optimal concentration at 2.79 wt %. Further, the thermal stability of the electrode improved upon raising the nanotube content. These nanocomposite films exhibited an interfacial double-layer capacitance at the highest nanotube content, corroborating their suitability for lightweight flexible energy storage applications [26].

Flexible pressure sensors have also attracted a lot of interest due to their potential in flexible robots, wearing electronic apparatus and electronic skins [27]. Amongst them, piezoresistive sensors are the most preferred due to their easiness of manufacture and simple collecting signal output [28]. A typical method to prepare piezoresistive sensors is to fabricate nanocomposites with an elastomer like polyurethane (PU) as matrix and conductive carbon nanomaterials like reduced graphene oxide (rGO) as a filler [29]. In this regard, an rGO/PU nanocomposite was fabricated by soaking an rGO aqueous solution with a PU foam and subsequently reduced with hydrazine vapors. This manufacturing process is cheap and highly reproducible. The developed sensor exhibited a simple structure, good sensitivity, durability, cycling stability, and fast response, hence it can be applied in piezoresistive devices, leading to outstanding sensing performance. Nonetheless, the properties of this type of nanocomposites are frequently limited by the heterogeneous graphene dispersion within the PU matrix. To solve this issue, an ionic liquid can be used as a dispersing agent [30]. The resulting nanocomposites exhibited a lower electrical percolation threshold, about 2 wt %, than the counterparts with pristine graphene, together with higher conductivity values. Further, the mechanical properties, namely Young's modulus and tensile strength, also improved while the ductility was maintained. Further, electrostatic spraying technique has also been used to improve the dispersion of nanofillers like MWCNTs within a PU coating as matrix [31], and to enhance the antistatic and mechanical properties. Besides the MWCNT distribution, the surface hardness and wear resistance of the nanocomposites were significantly improved. The optimal wear resistance was attained for the nanocomposite with only 0.3 wt % MWCNT.

Thermoelectric generators comprising flexible and lightweight p- and n-type single-walled carbon nanotube (SWCNT)-based composites also display enormous potential for wearable electronic applications [32]. They are able to transform heat into electric energy and vice versa under a temperature gradient without moving parts. These generators are environmentally friendly, noiseless, and display a long lifetime, which makes them highly interesting for power generation in electronic devices. In recent work, the thermoelectric performance of SWCNT flexible films doped with polyethyleneimine (PEI) and Nafion were assessed. The nanocomposites were manufactured via simple solution casting followed by vacuum filtration [33]. The SWCNT/PEI nanocomposites were switched from p- to n-type upon addition of PEI contents higher than 13 wt %. Furthermore, interconnected SWCNTs networks were produced due to an outstanding SWCNT dispersion and film formation. A thermoelectric generator with three thermocouples of p- and n-type SWCNT/PEI nanocomposites yielded an open circuit voltage of 17 mV and a maximum output power of 224 nW at 50 K, which are very encouraging results for wearable autonomous devices.

High-performance hybrid microfibers comprising hyaluronic acid and MWCNTs have been prepared by a wet-spinning method [34]. The matrix acts as a biosurfactant and a crosslinker, thereby improving the MWCNT dispersion. Different factors influencing the nanocomposite properties including the MWCNT content, dispersion time and injection speed have been investigated. The hybrid with 1.4 wt % MWCNT showed outstanding mechanical properties, with a Young's modulus as high as 9 GPa and tensile strength of 130 MPa, combined with superior flexibility and stability due to the excellent mechanical and electrical properties of MWCNTs. Thus, the matrix provides good mechanical support whilst the nanotubes offer a stable conductive path. The developed hybrid microfibers are highly suitable to be used in the fields of energy storage, micro devices, sensors, and intelligent materials. A wet spinning approach has also been used to develop polyacrylonitrile (PAN) grafted-amino-functionalized MWCNT nanocomposite fibers [35]. The nanocomposites were synthesized through in situ polymerization in aqueous solvent. The grafting degree of PAN onto the functionalized MWCNTs was determined as 73%. The nanocomposite fibers displayed a more compact structure and a more homogenous diameter distribution than the pristine fibers. Further, the addition of the grafted MWCNTs enhanced the level of crystallization, crystal size, tensile modulus, and strength of the PAN fibers. High-performance polymeric hybrids can also be developed by other approaches like a combination of in-mold decoration and microcellular injection molding [36], the method that minimizes the bubbles and improves the surface quality.

Self-healing polymeric materials hold the ability of being mended, which extends their self-life. They are based on the capacity of the linking moieties to split and reform after exposure to a stimulus, typically heat or light [37]. A lot of efforts have been devoted to the development of self-healing polymer nanocomposites incorporating nanofillers via in-situ polymerization to improve the nanofiller distribution, strength, modulus, and toughness. Recently, electrically induced self-healed nanocomposites incorporating CNTs have been developed as actuators and for electronic applications [38]. In this case, the healing process takes place by heat generation when an electrical current goes through a conductive matrix filled with a CNT network. The Diels–Alder reversible cycloaddition is one of the most common chemical reactions used for thermal healing. In this regard, electrically conductive self-healing nanocomposites consisting of MWCNTs dispersed into thermally reversible crosslinked polyketones have been developed [39]. The reversible nature relies on both covalent (Diels-Alder) and non-covalent (hydrogen bonding) interactions. The thermomechanical properties of the nanocomposites were easily tuned by modifying the MWCNT, diene-dienophile and hydroxyl moiety fractions. Nanocomposites with 5 wt % MWCNT exhibited up to 10^4 S/m electrical conductivity, and reached temperatures in the range of 120–150 under 20–50 V. This novel method could be used to fabricate thermo-reversible, thermo-adhesive, electrically conductive, and self-repairing systems.

The combination of both inorganic nanoparticles and CNTs in a polymeric matrix is frequently used to obtain multifunctional materials with superior properties [40,41]. The nanoparticles such as TiO_2, SiO_2, ZnO, Ag, and nanoclays aid to enhance the physical and mechanical properties as well as flame retardant activity, thermal stability, and permeability, among others. Nanocomposites of polypropylene filled with both TiO_2 nanoparticles and CNTs have been prepared via melt extrusion, with total nanofiller contents of 1, 5, and 10 wt % [42]. The thermal stability, Young modulus and electrical conductivity of the matrix increased while the level of crystallinity and thermo-oxidative degradation decreased with increasing nanofiller loading. The nanocomposites displayed improved flame retardancy, leading to a significant decrease of the peak heat release rate for nanofiller loadings of 5 and 10 wt %.

High-performance flexible supercapacitors based on a cellulose acetate membrane filled with both CNTs and Ag nanoparticles have been recently developed via a low-temperature, rapid and controlled direct writing method [43]. The Ag nanoparticles were uniformly dispersed within the CNTs, and both embedded within the porous structure of the cellulose membrane. The nanocomposite electrode showed outstanding electrochemical and mechanical electrochemical performance. The synergistic effect of both fillers led to enhanced conductivity, and also improved the specific volumetric capacitance. At an optimal concentration of 5 wt % of both nanofillers, excellent cycling stability was attained, with 75.92% capacitance retention.

Acknowledgments: A.M.D.-P. wishes to acknowledge the MINECO for a "Ramón y Cajal" research fellowship cofinanced by the EU.

Conflicts of Interest: The authors declare no conflict of interest.

References

1. Thostenson, E.T.; Ren, Z.; Chou, T.-W. Advances in the science and technology of carbon nanotubes and their composites: A review. *Compos. Sci. Technol.* **2001**, *61*, 1899–1912. [CrossRef]
2. Diez-Pascual, A.M.; Luceño-Sanchez, J.A.; Peña-Capilla, R.; Garcia-Diaz, P. Recent Developments in Graphene/Polymer Nanocomposites for Application in Polymer Solar Cells. *Polymers* **2018**, *10*, 217. [CrossRef] [PubMed]
3. Diez-Pascual, A.M.; Naffakh, M.; Gonzalez-Dominguez, J.M.; Anson, A.; Martínez-Rubi, Y.; Martínez, M.T.; Simard, B.; Gomez, M.A. High performance PEEK/carbon nanotube composites compatibilized with polysulfones-I. Structure and thermal properties. *Carbon* **2010**, *48*, 3485–3499. [CrossRef]
4. Diez-Pascual, A.M.; Martinez, G.; Gonzalez-Dominguez, J.M.; Anson, A.; Martínez, M.T.; Gomez, M.A. Grafting of a hydroxylated poly(ether ether ketone) to the surface of single-walled carbon nanotubes. *J. Mater. Chem.* **2010**, *20*, 8285–8296. [CrossRef]
5. Dreyer, D.R.; Park, S.; Bielawski, C.W.; Ruoff, R.S. The Chemistry of Graphene Oxide. *Chem. Soc. Rev.* **2010**, *39*, 228–240. [CrossRef] [PubMed]
6. Diez-Pascual, A.M.; Naffakh, M.; Gomez, M.A.; Marco, C.; Ellis, G.; Gonzalez-Dominguez, J.M.; Anson, A.; Martínez, M.T.; Martínez-Rubi, Y.; Simard, B.; et al. Influence of a compatibilizer on the thermal and dynamic mechanical properties of PEEK/carbon nanotube composites. *Nanotechnology* **2009**, *20*, 315707–315720. [CrossRef]
7. Díez-Pascual, A.M.; Díez-Vicente, A.L. Poly(propylene fumarate)/polyethylene glycol-modified graphene oxide biocomposites for tissue engineering. *ACS Appl. Mater. Interfaces* **2016**, *8*, 17902–17914. [CrossRef]

8. Diez-Pascual, A.M.; Naffakh, M.; Marco, C.; Ellis, G. Rheological and tribological properties of carbon nanotube/thermoplastic nanocomposites incorporating inorganic fullerene- like WS2 nanoparticles. *J. Phys. Chem. B* **2012**, *116*, 7959–7969. [CrossRef]
9. Wang, H.L.; Hao, Q.L.; Xia, X.F.; Wang, Z.J.; Tian, J.; Zhu, J.H.; Tang, C.; Wang, X. In situ fabrication of nanoscale graphene oxide/polyaniline composite and its electrochemical properties. *Adv. Mat. Res.* **2010**, *148–149*, 1547–1550. [CrossRef]
10. Diez-Pascual, A.M.; Naffakh, M. Grafting of an aminated poly(phenylene sulphide) derivative to functionalized single-walled carbon nanotubes. *Carbon* **2012**, *50*, 857–868. [CrossRef]
11. Díez-Pascual, A.M.; Hermosa-Ferreira, C.; San Andres, M.P.; Valiente, M.; Vera, S. Effect of graphene and graphene oxide dispersions in poloxamer-407 on the fluorescence of riboflavin: A comparative study. *J. Phys. Chem. C* **2017**, *121*, 830–843. [CrossRef]
12. Zheng, Q.; Kim, J.K. Synthesis, structure and properties of graphene and graphene oxide. In *Graphene for Transparent Conductors. Synthesis, Properties and Applications*; Springer: New York, NY, USA, 2015; pp. 29–94.
13. Salavagione, H.J.; Díez-Pascual, A.M.; Lázaro, E.; Vera, S.; Gómez-Fatou, M.A. Chemical sensors based on polymer composites with carbon nanotubes and graphene: The role of the polymer. *J. Mater. Chem. A* **2014**, *2*, 14289–14328. [CrossRef]
14. Luceño, J.A.; Maties, G.; Gonzalez-Arellano, C.; Díez-Pascual, A.M. Synthesis and characterization of graphene oxide derivatives via functionalization reaction with hexamethylene diisocyanate. *Nanomaterials* **2018**, *8*, 870. [CrossRef] [PubMed]
15. Luceño Sánchez, J.A.; Peña Capilla, R.; Díez-Pascual, A.M. High-Performance PEDOT: PSS/hexamethylene diisocyanate-functionalized graphene oxide nanocomposites: Preparation and properties. *Polymers* **2018**, *10*, 1169. [CrossRef]
16. Luceño Sánchez, J.A.; Díez-Pascual, A.M.; Peña Capilla, R.; García Díaz, P. The effect of hexamethylene diisocyanate-modified graphene oxide as a nanofiller material on the properties of conductive polyaniline. *Polymers* **2019**, *11*, 1032. [CrossRef]
17. Luceño Sánchez, J.A.; Díez-Pascual, A.M. Grafting of polypyrrole-3-carboxylic acid to the surface of hexamethylene diisocyanate-functionalized graphene oxide. *Nanomaterials* **2019**, *9*, 1095.
18. Bose, S.; Kuila, T.; Mishra, A.K.; Rajasekar, R.; Kim, N.H.; Sharma, K. Carbon-based nanostructured materials and their composites as supercapacitor electrodes. *J. Mater. Chem.* **2012**, *22*, 767–784. [CrossRef]
19. Siwal, S.S.; Zhang, Q.; Devi, N.; Thakur, V.K. Carbon-based polymer nanocomposite for high-performance energy storage applications. *Polymers* **2020**, *12*, 505. [CrossRef]
20. Shi, Y.; Peng, L.; Ding, Y.; Zhao, Y.; Yu, G. Nanostructured conductive polymers for advanced energy storage. *Chem. Soc. Rev.* **2015**, *44*, 6684–6696. [CrossRef]
21. Sugiawati, v.A.; Vacandio, F.; Yitzhack, N.; Ein-Eli, J.; Djenizian, T. Direct pre-lithiation of electropolymerized carbon nanotubes for enhanced cycling performance of flexible li-ion micro-batteries. *Polymers* **2020**, *12*, 406. [CrossRef]
22. Wu, T.M.; Lin, Y.W.; Liao, C.S. Preparation and characterization of polyaniline/multiwalled carbon nanotube composites. *Carbon* **2005**, *43*, 734–740. [CrossRef]
23. Wu, T.M.; Lin, Y.W. Synthesis, characterization, and electrical properties of polypyrrole/multiwalled carbon nanotube composites. *J. Polym. Sci. Polym. Chem.* **2006**, *44*, 6449–6457. [CrossRef]
24. Sainz, R.; Benito, A.M.; Martínez, M.T.; Galindo, J.F.; Sotres, J.; Baró, A.M.; Corraze, B.; Chauvet, O.; Maser, W.K. Soluble self-aligned carbon/polyaniline composites. *Adv. Mater.* **2005**, *17*, 278–281. [CrossRef]
25. Ozkan, S.Z.; Karpacheva, G.P.; Kostev, A.I.; Bondarenko, G.N. Formation features of hybrid nanocomposites based on polydiphenylamine-2-carboxylic acid and single-walled carbon nanotubes. *Polymers* **2019**, *11*, 1181. [CrossRef] [PubMed]
26. Thibodeau, J.; Ignaszak, A. Flexible electrode based on mwcnt embedded in a cross-linked acrylamide/alginate blend: Conductivity vs. stretching. *Polymers* **2020**, *12*, 181. [CrossRef] [PubMed]
27. Castro, H.F.; Correia, V.; Pereira, N.; Costab, P.; Oliveiraa, J.; Lanceros-Méndez, S. Printed Wheatstone bridge with embedded polymer based piezoresistive sensors for strain sensing applications. *Addit. Manuf.* **2018**, *20*, 119–125. [CrossRef]
28. Liu, K.; Zhou, Z.; Yan, X.; Meng, X.; Tang, H.; Qu, K.; Gao, Y.; Li, Y.; Yu, J.; Li, L. Polyaniline nanofiber wrapped fabric for high performance flexible pressure sensors. *Polymers* **2019**, *11*, 1120. [CrossRef]

29. Zhong, w.; Ding, X.; Li, W.; Shen, C.; Yadav, A.; Chen, Y.; Bao, M.; Jiang, H.; Wang, D. Facile Fabrication of Conductive Graphene/Polyurethane Foam Composite and Its Application on Flexible Piezo-Resistive Sensors. *Polymers* **2019**, *11*, 1289. [CrossRef]
30. Aranburu, N.; Otaegi, I.; Guerrica-Echevarria, G. Using an ionic liquid to reduce the electrical percolation threshold in biobased thermoplastic polyurethane/graphene nanocomposites. *Polymers* **2019**, *11*, 435. [CrossRef]
31. Wang, F.; Feng, L.; Lu, M. Mechanical Properties of multi-walled carbon nanotube/waterborne polyurethane conductive coatings prepared by electrostatic spraying. *Polymers* **2019**, *11*, 714. [CrossRef]
32. Zhang, Y.H.; Park, S.J. Flexible organic thermoelectric materials and devices for wearable green energy harvesting. *Polymers* **2019**, *11*, 909. [CrossRef] [PubMed]
33. Peng, X.-X.; Qiao, X.; Luo, S.; Yao, J.-A.; Zhang, Y.-F.; Du, F.-P. Modulating carrier type for enhanced thermoelectric performance of single-walled carbon nanotubes/polyethyleneimine composites. *Polymers* **2019**, *11*, 1295. [CrossRef] [PubMed]
34. Zheng, T.; Nuo, X.; Kan, Q.; Li, H.; Lu, C.; Zhang, P.; Li, X.; Zhang, D.; Wang, X. Wet-spinning assembly of continuous, highly stable hyaluronic/multiwalled carbon nanotube hybrid microfibers. *Polymers* **2019**, *11*, 867. [CrossRef] [PubMed]
35. Zhang, H.; Quan, L.; Gao, A.; Tong, Y.; Shi, F.; Xu, L. The structure and properties of polyacrylonitrile nascent composite fibers with grafted multi walled carbon nanotubes prepared by wet spinning method. *Polymers* **2019**, *11*, 422. [CrossRef] [PubMed]
36. Guo, W.; Yang, Q.; Mao, H.; Meng, Z.; Hua, L.; He, B. A combined in-mold decoration and microcellular injection molding method for preparing foamed products with improved surface appearance. *Polymers* **2019**, *11*, 778. [CrossRef]
37. Thakur, V.K.; Kessler, M.R. Self-healing polymer nanocomposite materials: A review. *Polymer* **2015**, *69*, 369–383. [CrossRef]
38. Swait, T.J.; Rauf, A.; Grainger, R.; Bailey, P.B.S.; Laerty, A.D.; Fleet, E.J.; Hand, R.J.; Hayes, S.A. Smart composite materials for self-sensing and self-healing. *Plast. Rubber Compos.* **2012**, *41*, 215–224. [CrossRef]
39. Lima, G.M.R.; Orozco, F.; Picchioni, F.; Moreno-Villoslada, I.; Pucci, A.; Bose, R.K.; Araya-Hermosilla, R. Electrically Self-Healing Thermoset MWCNTs composites based on diels-alder and hydrogen bonds. *Polymers* **2019**, *11*, 1885. [CrossRef]
40. Naffakh, M.; Diez-Pascual, A.M.; Marco, C.; Ellis, G. Morphology and thermal properties of novel poly(phenylene sulfide) hybrid nanocomposites based on single-walled carbon nanotubes and inorganic fullerene-like WS2 nanoparticles. *J. Mater. Chem.* **2012**, *22*, 1418–1425. [CrossRef]
41. Naffakh, M.; Diez-Pascual, A.M.; Gomez-Fatou, M.A. New hybrid nanocomposites containing carbon nanotubes, inorganic fullerene-like WS2 nanoparticles and poly(ether ether ketone) (PEEK). *J. Mater. Chem.* **2011**, *21*, 7425–7433. [CrossRef]
42. Cabello-Alvarado, C.; Reyes-Rodríguez, P.; Andrade-Guel, M.; Cadenas-Pliego, G.; Pérez-Álvarez, M.; Cruz-Delgado, V.J.; Melo-López, L.; Quiñones-Jurado, Z.V.; Avila-Orta, C.A. Melt-mixed thermoplastic nanocomposite containing carbon nanotubes and titanium dioxide for flame retardancy applications. *Polymers* **2019**, *11*, 1204. [CrossRef] [PubMed]
43. Guan, X.; Cao, L.; Huang, Q.; Kong, D.; Zhang, P.; Lin, J.; Li, W.; Lin, z.; Yuan, H. Direct writing supercapacitors using a carbon nanotube/Ag nanoparticle-based ink on cellulose acetate membrane paper. *Polymers* **2019**, *11*, 973. [CrossRef] [PubMed]

Article

The Effect of Hexamethylene Diisocyanate-Modified Graphene Oxide as a Nanofiller Material on the Properties of Conductive Polyaniline

José Antonio Luceño Sánchez [1], Ana Maria Díez-Pascual [1,*], Rafael Peña Capilla [2] and Pilar García Díaz [2]

[1] Department of Analytical Chemistry, Physical Chemistry and Chemical Engineering, Faculty of Sciences, Alcalá University, 28805 Madrid, Spain; jose.luceno@uah.es

[2] Department of Signal Theory and Communication, Polytechnic High School, Alcalá University, 28805 Madrid, Spain; rafa.pena@uah.es (R.P.C.); pilar.garcia@uah.es (P.G.D.)

* Correspondence: am.diez@uah.es; Tel.: +34-918-856-430

Received: 21 May 2019; Accepted: 7 June 2019; Published: 11 June 2019

Abstract: Conducting polymers like polyaniline (PANI) have gained a lot of interest due to their outstanding electrical and optoelectronic properties combined with their low cost and easy synthesis. To further exploit the performance of PANI, carbon-based nanomaterials like graphene, graphene oxide (GO) and their derivatives can be incorporated in a PANI matrix. In this study, hexamethylene diisocyanate-modified GO (HDI-GO) nanosheets with two different functionalization degrees have been used as nanofillers to develop high-performance PANI/HDI-GO nanocomposites via in situ polymerization of aniline in the presence of HDI-GO followed by ultrasonication and solution casting. The influence of the HDI-GO concentration and functionalization degree on the nanocomposite properties has been examined by scanning electron microscopy (SEM), Raman spectroscopy, X-ray diffraction (XRD), thermogravimetric analysis (TGA), tensile tests, zeta potential and four-point probe measurements. SEM analysis demonstrated a homogenous dispersion of the HDI-GO nanosheets that were coated by the matrix particles during the in situ polymerization. Raman spectra revealed the existence of very strong PANI-HDI-GO interactions via π-π stacking, H-bonding, and hydrophobic and electrostatic charge-transfer complexes. A steady enhancement in thermal stability and electrical conductivity was found with increasing nanofiller concentration, the improvements being higher with increasing HDI-GO functionalization level. The nanocomposites showed a very good combination of rigidity, strength, ductility and toughness, and the best equilibrium of properties was attained at 5 wt % HDI-GO. The method developed herein opens up a versatile route to prepare multifunctional graphene-based nanocomposites with conductive polymers for a broad range of applications including flexible electronics and organic solar cells.

Keywords: PANI; graphene oxide; hexamethylene diisocyanate; nanocomposite; thermal stability; mechanical properties

---

1. Introduction

Over the last years, conducting polymers such as polypyrrol, polythiophene and polyaniline (PANI) have attracted a lot of interest both at an academic and industrial level owing to their exceptional electrical and optoelectronic properties arising from their expanded π-conjugated electron system, which make them suitable for a wide range of applications, including thermoelectric devices, batteries, solar cells, sensors, actuators, supercapacitors, memory devices, wastewater treatment, separation ions and so forth [1–3]. Amongst them, PANI, a semi-flexible rod polymer, has been one of the most investigated over the last 20 years due to its unique electrochemical performance, adjustable electrical conductivity, facile synthesis, good chemical, environmental and thermal stability, easy doping chemistry, inexpensiveness and numerous uses [1,4]. It can be prepared with a controllable conductivity by adjusting the pH using both the conventional chemical and electrochemical approaches.

PANI can be polymerized into three different oxidation states [5]: The completely reduced leucoemeraldine form, the fully oxidized pernigraniline form and the emeraldine form, comprising oxidized and reduced alternating repeating units (Scheme 1). Further, each form may occur as either a salt or a base, and they have different colours, stabilities, and conductivities. Leucoemeraldine is a colourless substance that slowly oxidizes in air and is electrically insulating. It can be oxidized in an acidic medium (p-doping) to yield the conducting emeraldine salt. Pernigraniline is a blue insulating compound unstable in water that easily decomposes in air. The green emeraldine salt is formed during protonation of the violet emeraldine base with organic and inorganic acids: The protons interact with the imine nitrogen atoms leading to the formation of polycations [6]. The positive charges onto neighbouring nitrogen atoms rise the energy of the polymeric system, hence electron density undergoes redistribution; as a consequence, "unpairing" of the electron pair of nitrogen atoms occurs and cation radicals appear; these are delocalized over a certain conjugation length and provide the electron conductivity of the polymer, which depends strongly on the degree of protonation [5].

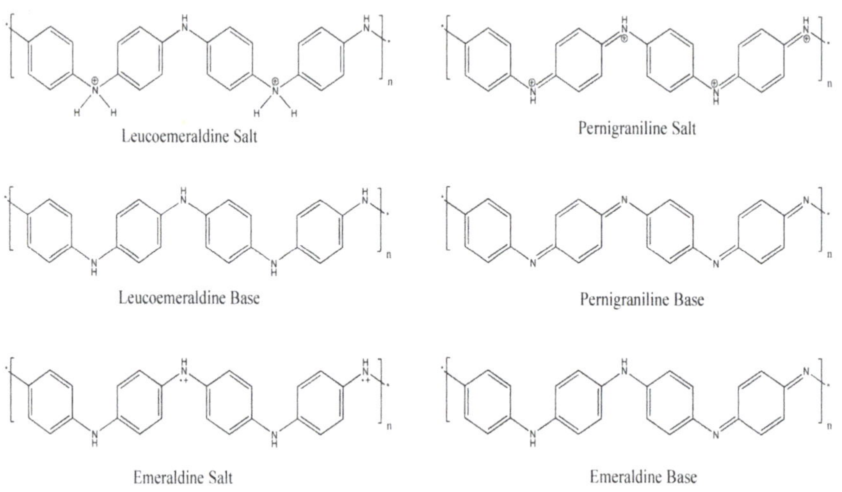

Leucoemeraldine Salt

Pernigraniline Salt

Leucoemeraldine Base

Pernigraniline Base

Emeraldine Salt

Emeraldine Base

Scheme 1. Structure of the different forms of polyaniline (PANI): The completely reduced leucoemeraldine, the completely oxidized pernigraniline and the emeraldine, in the base and salt forms.

Despite the wide spectrum of properties of PANI, it presents several shortcomings for certain applications, namely poor solubility, low mechanical performance and short cycling life [7]. To overcome these issues, PANI can be polymerized in the presence of organic and inorganic nanomaterials. The tailored incorporation of nanoscale materials can lead to nanocomposites with improved properties due to synergistic effects [8]. Thus, various carbon-based materials like mesoporous carbon, carbon nanotubes and graphene have been used as additives to extend the functional uses and improve the performance of PANI. In particular, composites of PANI reinforced with graphite, graphene (G), graphene oxide (GO) and reduced graphene oxide (rGO) have been synthesized via in situ chemical or electrochemical polymerization, covalent and non-covalent functionalization as well as self-assembly strategies [8–12]. The nanomaterial is supposed to increase the specific surface area of the composites, reduce the mechanical deformation of PANI and improve the transportation of the charge carries throughout the matrix due to strong nanomaterial-PANI interfacial adhesion [13]. However, in most of the preceding studies, graphite or graphene aggregates instead of stable graphene dispersions were attained, which reduces the large specific surface area of these nanomaterials and is detrimental for improving the composite electrical properties. In this context, it is crucial to improve the graphene dispersion and its interfacial interaction with the matrix. Consequently, significant effort has been devoted to surface-functionalize graphene and its derivatives as well as to optimize the composite preparation procedures. For example, Xu et al. [14] developed a simple method to prepare hierarchical nanocomposites with aligned PANI nanowires coated with GO nanosheets via dilute polymerization. Wu et al. [15] designed a facile one-step environmentally friendly polymerization method for the preparation of PANI/rGO nanocomposites using dialdehyde starch as reducing and capping agent. Besides, Kumar et al. [16] fabricated PANI-grafted-rGO nanocomposites with high electrical conductivity by functionalizing GO with 4-aminophenol via acyl chemistry. In particular, GO, a graphene derivative that contains epoxides, hydroxyls and carbonyls on the basal planes and carboxylic acids on the edges, has great potential as filler to develop PANI nanocomposites. It presents very large specific surface area, excellent chemical versatility and stability, aqueous processability, amphiphilicity, surface functionalization capability and biocompatibility [17]. More importantly, it can form stable aqueous colloids to enable the assembly of macroscopic structures, which is key for large-scale applications.

In a preceding work [18], we have developed a simple approach to functionalize GO via reaction with organic hexamethylene diisocyanate (HDI) using triethylamine (TEA) as catalyst. The synthesized (HDI)-modified GO nanomaterials, with different functionalization degrees, were found to be more hydrophobic in nature than pristine GO and were easily dispersed in polar aprotic solvents, hence are perfect candidates as fillers of conducting polymers. In the present study, we report a novel method to prepare PANI nanocomposites with homogenously dispersed GO nanosheets via in situ polymerization of aniline in the presence of HDI-GO using ammonium peroxydisulfate as an oxidizing agent in acid medium. HDI-GO can interact with PANI via electrostatic, hydrophobic, π-π stacking and hydrogen bonding interactions, which result in strong polymer-nanomaterial interfacial adhesion. Enhancing the dispersion of GO and its interaction with the conducting polymer matrix is decisive to attain improved electrical, thermal and mechanical properties. The resulting PANI/HDI-GO nanocomposites have been characterized by scanning electron microscopy (SEM), Fourier-transformed infrared spectroscopy (FT-IR), Raman spectroscopy, X-ray diffraction (XRD) analysis, thermogravimetric analysis (TGA), tensile tests and sheet resistance measurements to get information about how the weight fraction and functionalization degree of HDI-GO influence the nanocomposite morphology, structure and properties. This work signifies a great step forward to developing highly conducting PANI-based nanocomposites for application in a variety of organic devices.

2. Materials and Methods

2.1. Reagents

Aniline monomer ($C_6H_5NH_2$, >99%, M_w = 93.13 g/mol, $d_{25°C}$ = 1.02 g/cm^3), ammonium persulfate ($(NH_4)_2S_2O_8$, 98%, M_w = 228.20 g/mol, $d_{25°C}$ = 1.98 g/cm^3), triethylamine (TEA, >98%, M_w = 101.193 g/mol), H_2SO_4, $KMnO_4$, P_2O_5, $K_2S_2O_8$ and H_2O_2 (30 wt % in water) were obtained from Sigma-Aldrich (Madrid, Spain). Graphite powder was purchased from Bay Carbon, Inc. (Bay City, MI, USA). Hexamethylene diisocianate (HDI, >99%, M_w = 168.196 g/mol) was provided by Acros Organics (Madrid, Spain). The organic solvents were HPLC grade and were purchased from Scharlau S.L. (Barcelona, Spain). Deionized water was obtained from a Milli-Q-Water-Purification-System (Millipore, Milford, DE, USA). Aniline was stored in a refrigerator and was distilled under reduced pressure prior to use; the rest of the chemicals were used as received.

2.2. Synthesis of Hexamethylene Diisocyanate-modified Graphene Oxide (HDI-GO)

HDI-GO was prepared according to the procedure described in our preceding works [18,19]. In short, firstly GO was synthesized using a modified Hummers´ method from flake graphite using H_2SO_4, $K_2S_2O_8$, P_2O_5 and $KMnO_4$ as oxidants. Secondly, GO was functionalized by weighing the necessary amount of the synthesized GO powder and placing it into a round-bottom flask, followed by addition of dried toluene under inert atmosphere. In a first trial, GO dispersion was sonicated in an ultrasonic bath for 2 h, and subsequently TEA catalyst and HDI reagent were added dropwise (GO:HDI:TEA weight ratio of 1:1:1); the mixture was then heated to 60 °C and stirred for 12 h under inert atmosphere. The resulting product (HDI-GO 1) exhibited a functionalization degree (FD) of 12.3% as measured by elemental analysis [18]. In a second synthesis, the procedure was similar but the GO dispersion was first ultrasonicated with a tip (3 probe sonication cycles with 5 min of break between cycles at 40% amplitude) followed by the 2 h of bath sonication. The resulting product (HDI-GO 6) had a FD of 18.1 %.

2.3. *Synthesis of PANI/HDI-Modified GO Nanocomposites*

The nanocomposites were prepared via *in situ* polymerization of aniline in acid medium, using $(NH_4)_2S_2O_8$ as oxidizing agent. The structure and nomenclature of the compounds used for the preparation of the PANI/HDI-GO nanocomposites is shown in Scheme 2a.

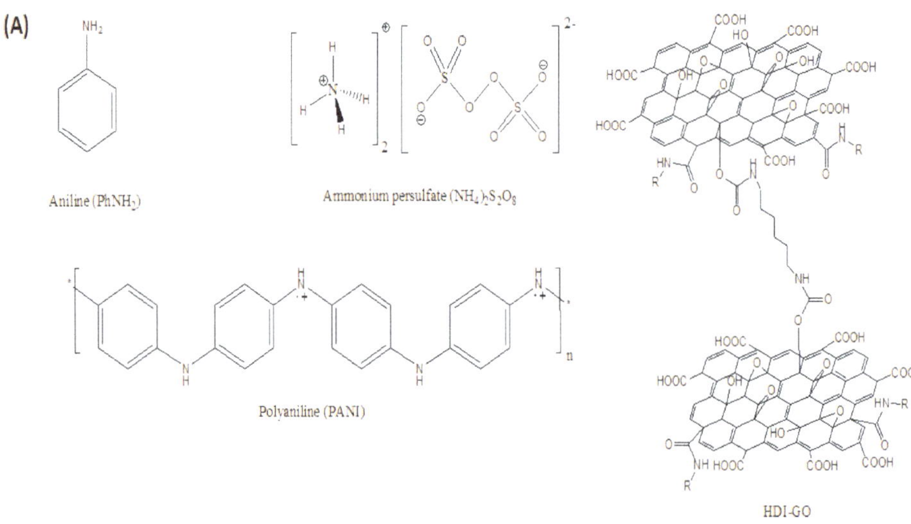

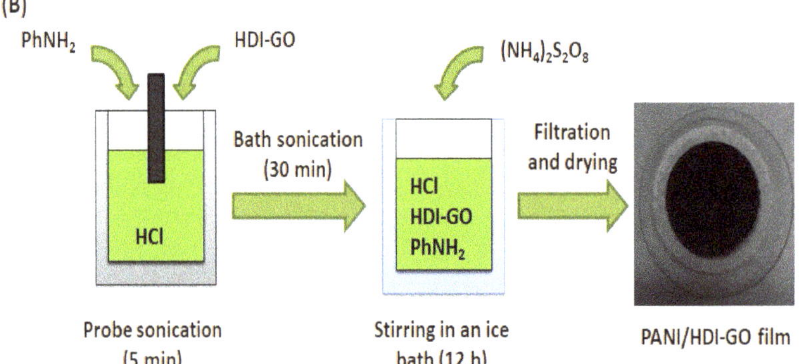

Scheme 2. (**a**) Structure and nomenclature of the compounds used for the preparation of PANI/HDI-GO nanocomposites. (**b**) Diagram of the synthesis procedure of the nanocomposites.

Firstly, 0.1 mL of aniline monomer were placed in a beaker with 25 mL of 1 M HCl, and then the required amount of HDI-GO (0.25, 0.5, 1, 2.5 or 5 mg) was added. The mixture was then sonicated with an ultrasonic tip for 5 min and then in an ultrasonic bath for 30 min. Subsequently, 0.06 g of $(NH_4)_2S_2O_8$ were dissolved in 25 mL of 1 M HCl and then added dropwise to the aniline/HDI-GO mixture under vigorous stirring. Polymerization of aniline started after about 5 min, while the colour of the mixture changed into green. The mixture was then allowed to stir while cooled in an ice bath for 12 h.

The resulting solid precipitates, with HDI-GO weight ratios of 0.5, 1.0, 2.0, 5.0 and 10 wt %, were recovered from the reaction vessel by filtration and repeatedly washed with deionized water, ethanol, and hexane for the elimination of the low molecular weight polymer and oligomers until the filtrates were colourless, and finally dried at 80 °C in an oven for 48 h. A schematic representation of the synthesis procedure of the nanocomposites is shown in Scheme 2b. For comparative purposes, a reference sample containing 5 mg of pristine GO (10 wt % loading) and a reference PANI sample (0 wt % HDI-GO) were prepared in a similar way. Further, to investigate the influence of the level of functionalization of HDI-GO on the nanocomposite properties, two sets of composites were prepared, using HDI-GO 1 and HDI-GO 6, respectively. Photographs of the nanocomposites obtained are shown in Figure S1 in the supplementary material. Homogenous films were also prepared by dropping 5 mL of the liquid nanocomposite mixture onto a glass surface and dried in the oven at 80 °C.

2.4. Intrumentation

A Selecta 3001208 ultrasonic bath (Madrid, Spain) and a 24 kHz Hielscher UP400S ultrasonic tip (Teltow, Germany) with a maximum power output of 400 W, equipped with a titanium sonotrode (Ø = 7 mm, l = 100 mm) were used to prepare the different dispersions.

The morphology of the nanocomposites was analyzed with a SU8000 Hitachi scanning electron microscope (SEM, Hitachi, Ltd., Tokyo, Japan), which operated with an emission current of 10 mA and a 15.0 kV accelerating voltage. Before the experiments, samples were coated with a 5 nm thick Au:Pd overlayer to avoid charge buildup throughout electron irradiation.

Raman spectra were acquired at room temperature with a laser output power of 1 mW using a Renishaw Raman microscope (Gloucestershire, UK) incorporating a He-Ne laser (632.8 nm). To minimize the signal-to-noise ratio, 3 scans were recorded for each composite. Data were then processed with the WiRE 3.3 Renishaw software, and the spectra were normalized to the G band for the sake of comparison.

A TA Instruments Q50 thermobalance (Barcelona, Spain) was used to analyze the thermal stability of the samples via thermogravimetric analysis (TGA) experiments under an inert atmosphere (N_2). Measurements were performed from 100 to 700 °C at a heating rate of 10 °C/min, with a gas flow rate of 60 mL/min. Prior to the tests, samples were dried for 72 h and then located inside aluminum pans.

The zeta potential of aqueous dispersions of the samples was measured at 25 °C on a Zetasizer Nano ZS (Malvern Instruments Ltd, Worcestershire, UK) using disposable plastic cuvettes.

X-ray diffraction (XRD) analysis was carried out with a Bruker D8 Advance diffractometer (Karlsruhe, Germany) using a Cu tube as X-ray source (λ CuKα = 1.54 Å), with a voltage of 40 kV and an intensity of 40 mA.

A standard test method for tensile properties of thin plastic sheeting, ASTM D882, was used for testing the mechanical properties of the composites. Tensile testing was carried out with an Instron 5565 Testing Machine (Norwood, MA, USA). A 1 kN load cell was used and the crosshead speed was 10 mm min^{-1}. The results reported are the mean values for six replicates.

A four-point probe resistivity measurement system (Multiheight Probe station, Leighton Buzzard, UK) was used to determine the sheet resistance (R_s) of the samples. The voltage was measured with a KEITHLEY 2182A nanovoltmeter and the electrical current with a KEITHLEY 6221 current source. R_s was calculated as: $R_s = 4{,}532 \times (V/I)$, being V the test voltage and I the current. To ensure reproducibility, more than 5 measurements were carried out at different points of each sample film, and the results reported correspond to the average values.

3. Results

3.1. Morphology of PANI/HDI-GO Nanocomposites

The morphologies and structures of PANI, GO, HDI-GO and the nanocomposites were studied by SEM, and some representative micrographs are shown in Figure 1. A granular or globular–like morphology is observed for pure PANI, comprising highly aggregated particles (Figure 1b). This is the most common structure found in PANI prepared by precipitation polymerization when using strong oxidants and acidic conditions [20]. Raw GO also appears rather compact and aggregated (Figure 1a), comprising wrinkly and flexible graphene sheets with thicknesses in the range of 10–50 nm. On the other hand, HDI-GO 6 is composed of stacked graphene flakes with a smoother surface topology (Figure 1c), ascribed to the covalent bonding of the HDI chains onto the GO surface. Further, many sheets appear stiffer, likely due to the wrapping of the methylene groups that cover the wrinkles, hence come out thicker, showing flake thicknesses ranging from 20 to 80 nm. Nonetheless, it should be noted that the sample appears quite heterogeneous, which is reasonable considering that it is a mixture of raw GO sheets and HDI modified sheets. Similar morphology was found for HDI-GO 1 (Figure 1e), albeit with thinner and less rigid sheets, suggesting that the flake stiffness and thickness rise with increasing FD.

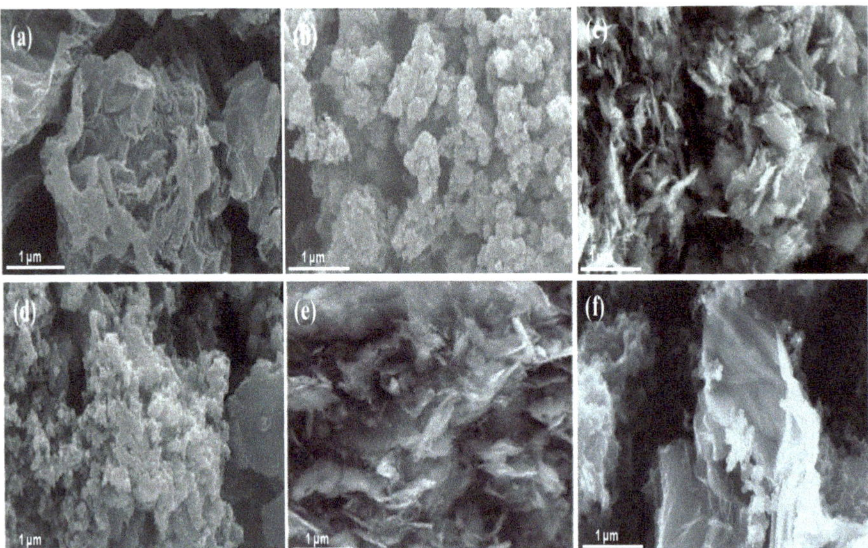

Figure 1. Representative SEM micrographs of raw GO (**a**), neat PANI (**b**), HDI-GO 6 (**c**), PANI/HDI-GO 6 (10 wt %) (**d**), HDI-GO 1 (**e**), and PANI/HDI-GO 1 (10 wt %) (**f**).

Focusing on the nanocomposite with 10 wt % HDI-GO 6 (Figure 1d), it can be observed that the flat and rigid graphene nanosheets are fully embedded within the PANI nanoparticles, forming a dense and interpenetrating network with the matrix particles partly attached onto the HDI-GO surface. It has been reported [21] that when HCl is used as a dopant, aniline monomer adsorbs onto the surface of graphene-based nanomaterials via electrostatic attraction, resulting in the formation of weak charge-transfer complexes between the polymer monomers and the graphitic structure of graphene. In our study PANI should adhere more strongly to the graphene flakes since some of the residual carboxylic groups of HDI-GO are negatively charged due to resonance effects, hence are prone to strongly interact with the positively charged emeraldine salt form of PANI via electrostatic interactions; further, the aromatic moieties of the polymer can interact with the benzene rings of HDI-GO via π-π stacking as well as with the methylene chains of the HDI through hydrophobic interactions. Besides, hydrogen bonding can occur between the amine groups of PANI and the oxygenated functional moieties of HDI-GO (see Scheme 3). Thus, the combination of all these interactions results in strong PANI-HDI-GO interfacial adhesion.

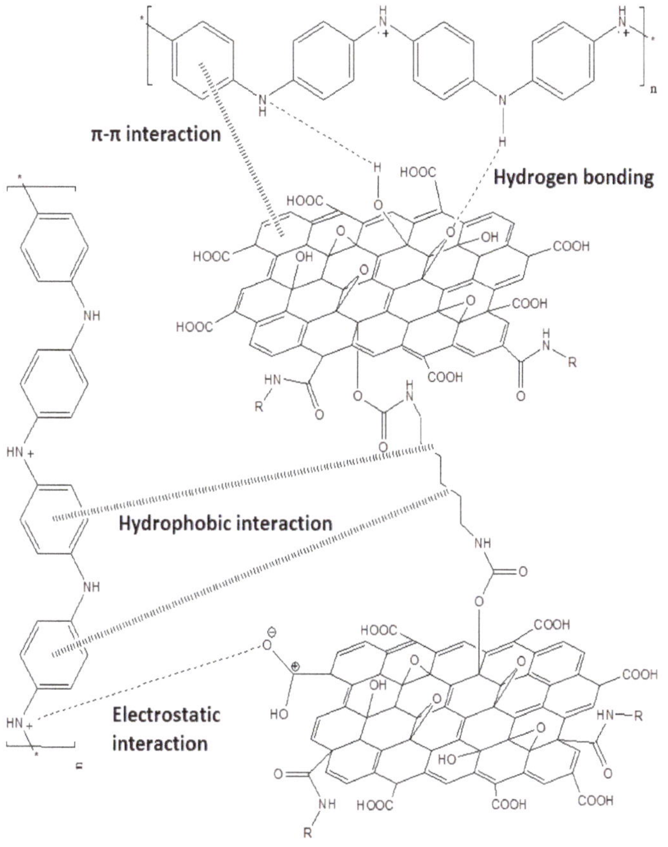

Scheme 3. Representation of the different types of interactions between PANI and HDI-GO.

As a result of the absorption process, the HDI-GO nanosheets were coated by the matrix particles during the in situ polymerization reaction in acid medium. This allowed to prepare homogeneous composites in which PANI and the modified GO are likely intercalated with each other instead of individually being in an agglomerated state and phase separated as typically observed previously [22,23]. The lessening in the strength of the hydrogen bonding interactions among the GO nanosheets upon functionalization with HDI makes the HDI-GO surface more hydrophobic than that of raw GO [18], henceforth it was well dispersed in HCl by the combination of probe and bath sonication used herein. Thus, PANI chains were able to diffuse within the stacked structure of HDI-GO and formed a thin coating onto the graphene nanosheets. This morphology was retained for all the composites reinforced with HDI-GO 6, which showed a relatively smooth and plain surface, and the degree of interpenetration increased with increasing the nanomaterial loading. Another plausible explanation for the observed morphology could be the formation of intimate charge-transfer complexes in the solid state due to very strong donor-acceptor interactions between the polymer and the nanofiller, as reported previously for PANI/rGO nanocomposites [24]. Thus, HDI-GO could act as a stable counterion of the doped state of PANI emeraldine salt, stabilizing PANI in an intermediate oxidation state between the leucoemeraldine and the emeraldine forms. In addition, this charge-transfer in PANI/HDI-GO nanocomposites together with the numerous graphene bundles randomly distributed throughout the matrix would be beneficial for the formation of homogenous conductive paths, and would account for the improved nanocomposite properties including improved electrical conductivity and superior thermal stability, as will be discussed in the following sections.

Conversely, in the nanocomposite with 10 wt % HDI-GO 1 (Figure 1f) there is lack of such interpenetrating polymer-nanofiller network, and both composite components appear much more separated. During the in situ polymerization process, the PANI nanoparticles located within the GO layers and acted as spacers to form gaps between neighboring GO sheets, thus resulting in flakes with a large surface area and high level of exfoliation that retain their bendability and flexibility. Nonetheless, some flakes also display a random decoration of PANI nanoparticles over the wrinkled surface. Similar morphology was found for the rest of nanocomposites incorporating HDI-GO 1, with increasing number of decorated nanosheets with increasing nanofiller loading.

3.2. XRD Analysis of the Nanocomposites

The developed samples were analyzed by XRD measurements, and typical patterns of PANI, GO, HDI-GO 6, HDI-GO 1 and the corresponding nanocomposites with 10 wt % loading are compared in Figure 2. Analogous diffractograms were found for the rest of PANI/HDI-GO nanocomposites, and the data obtained from the diffraction patters are collected in Table S1 (supplementary material). Neat PANI shows a broad peak centered at $2\theta = 25.3^{\circ}$, indicative of its amorphous character and the absence of order arrangement in the polymer chains. This wide peak can also be observed in the diffractograms of the nanocomposites, although appears at lower 2θ values; further, its intensity decreases with increasing nanofiller loading, in agreement with the results reported previously for PANI/GO hidrogels [25]. The presence of GO reduces the percentage of PANI in the nanocomposites and hence weakens the diffraction peaks.

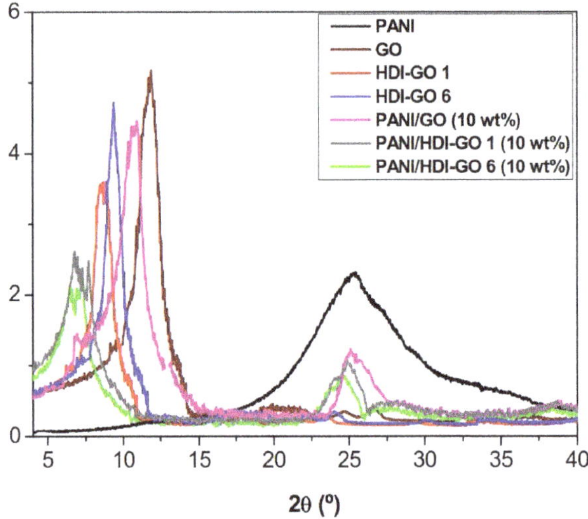

Figure 2. XRD patterns of neat PANI, raw graphene oxide (GO), hexamethylene diisocyanate-modified GO (HDI-GO) 1, HDI-GO 6, and the corresponding nanocomposites with 10 wt % nanofiller content.

The downshift in 2θ is more pronounced for nanocomposites reinforced with HDI-GO compared to that with GO (see Table S1), in particular for those with HDI-GO 6, the utmost decrease being 0.79°

for PANI/HDI-GO 6 with 10 wt % loading. This shift in 2θ towards lower values has been previously reported for PANI/rGO nanocomposites [26], indicative of an expansion of the interlayer distance, and can be explained by the adsorption and intercalation of PANI on the surface and between the HDI-GO sheets. Thus, as the functionalization degree of HDI-GO increases, the stronger the interactions with the matrix chains, the stronger the adsorption of PANI, the more pronounced the shift in the peak maximum, in agreement with the observations from SEM study (Figure 1). Therefore, the absorption and intercalation of PANI on the GO nanosheets seems to be responsible for the nanocomposites structure and morphology.

Regarding pristine GO, a characteristic peak can be observed at 2θ = 11.8° related to the (002) diffraction of the GO sheet [27], from which the interlayer *d* spacing value was calculated to be 0.748 nm according to the Bragg's equation [28]. In the case of HDI-GO 1 and HDI-GO 6, the peak maximum appears at lower 2θ (9.2 and 8.8°, corresponding to interlayer distances of 0.961 and 1.003 nm, respectively) and shows reduced intensity. This increase in the *d* spacing has been previously observed for other nanocomposites comprising polymer chains between GO nanosheets [29,30], and further corroborates the presence of the HDI segments intercalated between the nanofiller layers.

Regarding the nanocomposites, further downshift of this (002) peak is observed, suggesting an additional rise in the interlayer spacing of the carbon nanomaterial compared to either GO or HDI-GO. In the case of PANI/GO (10 wt %), the downshift is relatively small, about 1° compared to raw GO, resulting in a *d* spacing of ~0.8 nm. In contrast, nanocomposites reinforced with 10 wt % HDI-GO 1 and HDI-GO 6 show stronger shifts, and exhibit *d* values of 1.18 and 1.33 nm, respectively (Table S1). These clear increases in the interlayer spacing are indicative of the strong interaction between the PANI segments and the GO nanosheets, and the intercalation of the polymeric chains in between the nanomaterial layers. Once more, the higher the FD of HDI-GO, the larger the interlayer distance in the nanocomposites. The broadening of the (002) peak is also evident for all the PANI/HDI-GO compositions compared to either HDI-GO 1 or HDI-GO 6. Thus, the width at half maximum increased from about 1.4° for HDI-GO 6 to 2.9° for the nanocomposite with 10 wt % loading. This widening is yet another indication of the strong PANI-HDI-GO 6 interactions. The combination of the substantial shift of the (002) peak during the formation of PANI/HDI-GO and the reduction in the GO peak height intensity further corroborate the formation of hybrid intercalated nanocomposites, as observed previously [31]. On the whole, the XRD patterns confirm the efficiency of the in situ polymerization process used in this study to facilitate the diffusion of the PANI chains within the interlayer spacing of the HDI-GO nanosheets.

3.3. *Raman Spectra*

Raman spectroscopy was used to further characterize PANI and the synthesized nanocomposites, and the spectra obtained for neat PANI, GO, HDI-GO, and the nanocomposites with 10 wt % HDI-GO or GO are shown in Figure 3. Data derived from the spectra of all the nanocomposites are collected in Table S1 (supplementary material).

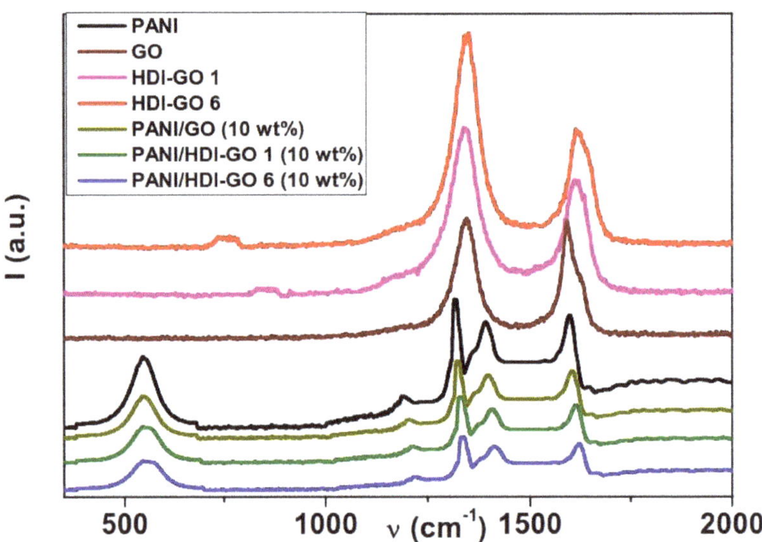

Figure 3. Raman spectra of neat PANI, raw GO, HDI-GO 1, HDI-GO 6, and the corresponding nanocomposites with 10 wt % nanofiller content.

Characteristic Raman peaks of PANI can be observed within the wavelength range 400–1800 cm^{-1}. The C=N stretching vibration in the quinonoid units of PANI is observed at 1396 cm^{-1} [32]. The band at 1191 cm^{-1} corresponds to the C–N stretching vibrations of benzenoid, quinonoid, and polaronic forms of PANI in the emeraldine state, and the presence of C–N^+ vibrations are evident at 1390 cm^{-1} [32,33]. The band at 1597 cm^{-1} corresponds to the C=C stretching of the benzenoid ring vibrations, and that at 555 cm^{-1} is attributed to C–H in plane bending vibrations [33]. Regarding GO, the most prominent features in the spectrum are the disorder induced D band at 1345 cm^{-1} that indicates the level of structural disorder, and the tangencial G band at ~1595 cm^{-1} related to in-plane displacements of the graphene nanosheets [34]. Analogous spectrum is observed for the HDI-GO samples, although they show an enlargement and upshift of the G band, attributed to a change in the electronic structure of GO in the presence of electron-acceptor groups [18]. This upshift can also be associated to a raise in defect density, given that the position of the G band depends on the concentration of defects in the graphene layers [35]. Further, the D to G band intensity ratio (I_D/I_G) offers quantitative information about the amount of defects in graphene sheets: The higher the ratio, the larger the disorder [35]. This ratio is 1 for neat GO, and increases up to 1.55 and 1.73 for HDI-GO 1 and HDI-GO 6, respectively, which corroborates the drop in structural order upon anchoring the HDI chains onto the GO surface.

Focusing on the nanocomposites, the spectra are analogous to that of neat PANI, albeit they exhibit an upshift in the position of the peaks as well as a drop in the intensity of the bands. In the case of PANI/GO (10 wt %), the shift in the band positions is relatively small, indicative of weak interactions between the PANI chains and GO (Table S1). However, PANI/HDI-GO 1 (10 wt %) and PANI/HDI-GO 6 (10 wt %) show strong shifts in the position of the bands, by up to 15 and 26 cm^{-1} in the C=C stretching of the benzenoid ring. This behavior is yet another confirmation of the strong adsorption of the PANI chains onto the GO or HDI-GO surface via π-π stacking, H-bonding, hydrophobic and electrostatic interactions, leading to the formation of a tightly coated PANI layer on the carbon nanomaterial surface, as discussed previously. Thus, the combination of all these interactions results in strong PANI-HDI-GO interfacial adhesion. Moreover, the higher the FD of HDI-GO, the stronger the interactions with the matrix chains, hence the larger the change in the position of the peaks. Thus, the upshifts in the position of the peaks are systematically larger for composites with HDI-GO 6 compared to those with the same loading of HDI-GO 1. An upshift in the position of PANI peaks corresponding to C=N and

C-C stretching modes has also been reported for PANI nanofibers coated with graphene nanosheets, indicating the very intense interaction between the nanofibers and the carbon nanomaterial in the composite [13].

3.4. Thermal Stability of the Nanocomposites

A high thermal stability of nanocomposite films is a desirable factor for certain applications like electromagnetic shielding (EMI) at high temperatures. TGA analysis upon heating under a nitrogen atmosphere was carried to examine the thermal stability of PANI and the PANI/GO nanocomposites (Figure 4). The results derived from all the nanocomposites are collected in Table S2 (supplementary material). Regarding pure PANI, the weight loss below 150 °C can be assigned to the loss of water molecules and other volatile impurities. The first main decomposition stage between 150 and 250 °C is attributed to the deprotonation of PANI through removal of dopant HCl molecules [33], while the second from 340 to 480 °C can be assigned to the exothermic thermal decomposition of PANI backbone. In the case of pristine GO, a one-step degradation process is found, with a main weight loss below 250 °C attributed to the decomposition of surface epoxide, hydroxyl, and carboxylic acid functional groups [18]. In addition, a small weight loss is observed above 250 °C due to the elimination of further functional groups. On the other hand, the HDI-GO samples display two decomposition steps, the first due to the removal of remaining oxygenated surface groups, and the second to the degradation of the HDI chains linked to the GO surface. With increasing FD, the degree of crosslinking between the GO layers increases, hence the thermal stability of HDI-GO improves [18].

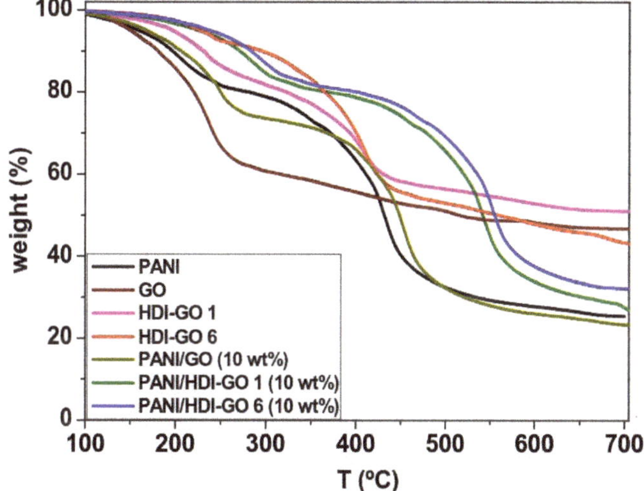

Figure 4. Thermogravimetric analysis (TGA) curves under inert atmosphere of neat PANI, GO, HDI-GO 1, HDI-GO 6, and the nanocomposites with 10 wt % nanofiller loading.

The nanocomposites also exhibit a two-step degradation process, similar to that of neat PANI, albeit shifted to higher temperatures, indicating a thermal stabilization effect induced by the presence of the nanofillers. Nonetheless, the shift depends on the type of nanofiller and its concentration. Thus, upon addition of 10 wt % GO, the initial degradation temperature (T_i) at 2% weight loss increased only slightly (about 12 °C), whereas by addition of the same amount of HDI-GO 1 and HDI-GO 6, it increased significantly, by 53 and 64 °C , respectively. Similar trend is found for the temperature of 10% weight loss (T_{10}) and the temperatures of maximum rate of weight loss of both stages ($T_{maxI,II}$), which show maximum augments of 67 and 90 °C for the nanocomposite with 10 wt % HDI-GO 1 and increments of up to 81 and 109 °C for that with 10 wt % HDI-GO 6, respectively (Table S2). In addition, the weight residue of PANI/GO (10 wt %) is slightly lower than that of neat PANI, whilst that of nanocomposites with 10 wt % HDI-GO 1 and HDI-GO 6 are about 10 and 27% higher, respectively. The extraordinary thermal stability improvements found in the nanocomposites reinforced with HDI-GO can be explained considering the homogenous dispersion of the nanosheets within the PANI chains, as revealed by SEM images (Figure 1d,f) combined with the strong PANI-HDI-GO interfacial adhesion via π-π stacking, H-bonding, hydrophobic and electrostatic donor-acceptor interactions that lead to the formation of intimate charge-transfer complexes in the solid state [24], as discussed earlier. The crosslinked HDI-GO layers, which are randomly dispersed within the polymer matrix, can behave as a barrier and a thermal protecting material that shield the PANI chains from the heat and delay the diffusion of the degradation products from the interior of the nanocomposite to the gas phase through the formation of a tortuous path.

Similar behavior of thermal stability improvement has been reported for PANI nanocomposites reinforced with graphene nanoplatelets via in situ polymerization, attributed to the formation of 3D conducting interpenetrating networks between the polymer and the nanofillers [36]. Thus, in the nanocomposites comprising HDI-GO 6, the stiff graphene nanosheets are fully embedded within the PANI nanoparticles, forming a dense and interpenetrating network with the matrix particles partly attached onto the HDI-GO surface, which can account for the superior thermal stability enhancement observed. Further, the strong PANI:HDI-GO interactions likely confine the rotational movement of the polymeric chains, and the motion restriction of the polymeric chains at the PANI: HDI interface would also be reflected in better thermal stability, as reported for PANI nanofiber-coated polystyrene/GO nanocomposites [37]. The comparison of the degradation temperatures of nanocomposites reinforced increasing amounts of HDI-GO 1 or HDI-GO 6 (Table S2) reveals that all the degradation temperatures systematically increase with increasing HDI-GO loading, likely due to the stronger barrier effect imposed by the nanofillers (i.e., an average increase of 40 °C when the nanofiller loading is increased from 0.5 to 10 wt %). TGA data reveal that the addition of HDI-GO with a high FD strongly improves the thermal stability of PANI, which is an important result from a practical viewpoint including aerospace, electronic and photovoltaic applications.

3.5. Sheet Resistance of the Nanocomposites

Figure 5 shows the sheet resistance (R_s) data of neat PANI and the nanocomposites reinforced with different amounts of HDI-GO 1 and HDI-GO 6. The neat PANI film shows an R_s value close to 135 Ω/sq., which drops steadily with increasing HDI-GO loading, leading to minimum values of 11 and 13 Ω/sq for the nanocomposites reinforced with 10 wt % HDI-GO 6 and HDI-GO 1, respectively. However, the reference PANI/GO sample (blue star in the plot) showed similar sheet resistance to that of PANI. Taking into account that the HDI-GO samples display higher R_s than neat PANI [19], the reduction in sheet resistance found in PANI/HDI-GO nanocomposites is a surprising behavior that could be explained considering the different factors that influence the charge hopping conduction mechanism in this type of conductive polymer, including grain size, level of crystallinity, doping and screening effects as well as conformational changes of the polymeric chains [38].

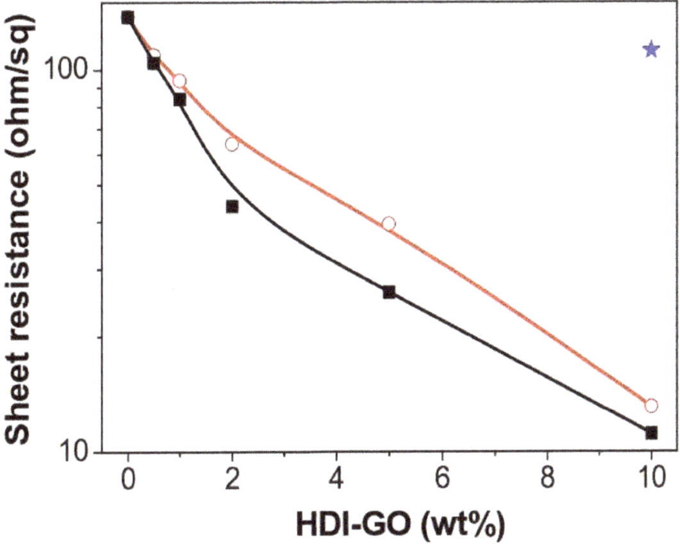

Figure 5. Sheet resistance of PANI nanocomposites reinforced with HDI-GO 6 (solid squares) and HDI-GO 1 (open circles) as a function of the HDI-GO content. Solid lines are guides for the eyes. The blue star corresponds to the PANI/GO (10 wt %) reference composite.

The conduction mechanism is believed to involve polaronic carriers. When the emeraldine base is doped in an acid environment, a polaron or bipolaron can be developed via consecutive formation of positive species. Bipolaron structures are thermodynamically more stable and conductive, and are responsible for the electrical conduction via a jump mechanism [2]. The adjacent nitrogen electron (neutral) moves to a vacant spot and neutralizes it. Consequently, this spot moves, thereby creating new spaces in the nitrogen structure and in the polarons structures, resulting in electron transportation and, thus, electrical conductivity along the chain [38].

The structure, morphology and synthesis method of the nanocomposites has also a profound effect on the electrical conductivity of the materials [39]. Thus, the presence of GO poorly dispersed within PANI has been reported to decrease the conductivity of the polymer [40], while that of composites prepared via in situ emulsion polymerization was significantly improved [41], ascribed to the very homogeneous nanofiller dispersion inside the polymer matrix that made both composite components come close to each other to form conductive paths. Further, a noticeable electrical conductivity improvement was previously reported for PANI/graphene nanoplatelets developed via in situ polymerization, attributed to the formation of 3D conducting interpenetrating networks [36].

Further, our Raman spectra suggest the presence of strong π-π stacking, H-bonding, hydrophobic and electrostatic interactions, in particular for the composites with HDI-GO 6, leading to the formation of a tightly coated PANI layer on the carbon nanomaterial surface. Besides, as mentioned earlier, the HDI-GO could behave as a counterion of the emeraldine salt form of PANI, stabilizing it in an intermediate oxidation state between the leucoemeraldine and the emeraldine structures. This would result in the formation of tight PANI-HDI-GO charge-transfer complexes via strong donor-acceptor interactions, hence better charge carrier transport, and consequently reduced sheet resistance for the nanocomposites. To further corroborate the HDI-GO coverage by the PANI chains, the zeta potential (ζ) of aqueous dispersions of GO, PANI, HDI-GO 1, HDI-GO 6 and their composites with 10 wt % was measured, and the results are included in Table S2 (Supplementary Information). The emeraldine form of PANI carries positive charges, which make it dispersible in aqueous solution, leading to a positive zeta potential of 40 mV, consistent with the results reported previously [42]. ζ of GO is highly negative, due to the large number of surface oxygenated functional groups on the graphene sheets. However, ζ of HDI-GO is significantly smaller (in absolute value), albeit it is still negative, indicative of the presence of remaining oxygen containing groups on the graphene sheets, in agreement with the results from elemental analysis [18]. As expected, with increasing FD, ζ value decreases. Moreover, as compared to the negatively charged GO, all the PANI/GO nanocomposites are positively charged, which indicates coverage of PANI on the graphene surface. The electrostatic interaction between the positively charged PANI and the negatively charged nanofillers is stronger in composites with HDI-GO (particularly those with HDI-GO 6) compared to the reference with GO, hence leading to a lower ζ. Further, with increasing HDI-GO content, the interaction becomes more pronounced, thus ζ decreases.

It is noteworthy that the PANI/HDI-GO samples prepared in this work display higher conductivity than that previously reported for other PANI-GO nanocomposites [40,43,44], which demonstrates the effectiveness of the strategy developed in this study to improve the thermoelectrical properties of conductive polymer/graphene nanocomposites. Further, the values obtained are also higher than those found for PANI/multiwalled carbon nanotube (MWCNT) nanocomposites, likely due to the larger surface area of HDI-GO nanosheets compared to the nanotubes [45]. For the same nanofiller content, R_s is systematically lower for nanocomposites filled with HDI-GO 6 compared to those with HDI-GO 1. This can be rationalized considering that HDI-GO 6, with a higher FD, is more homogeneously dispersed within the matrix, and interacts more strongly with the PANI chains, thus favoring the formation of a more interpenetrated network, and consequently conductive paths, hence lower sheet resistance. It is interesting to note that the nanocomposites with high HDI-GO content exhibit R_s values close to those reported for ITO films coated onto plastic substrates, such as polyethylene terephthalate (PET) [46], hence, they are good candidates to be used as transparent conductive electrodes in conventional panel displays, solar cells, touch panels, and so forth.

3.6. Mechanical Properties

The mechanical properties of PANI/HDI-GO nanocomposites were investigated by tensile tests, and the stress-strain curves of PANI and some PANI-based nanocomposites are shown in Figure S2 in the supplementary material; the results derived from the tensile curves of all the samples tested are depicted in Figure 6. Neat PANI exhibits an elastic modulus close to 2 GPa and a tensile strength value of about 32 MPa, in very good agreement with the data reported earlier [47]. The addition of HDI-GO 6 causes significant enhancements in both parameters (Figure 6a,b), by up to 240 and 258%, respectively, at the highest loading tested. A similar tendency, although with smaller increases, can be observed for the nanocomposites comprising HDI-GO 1. For both types of nanocomposites, the rise is more pronounced up to 2 wt % loading, whilst it is less marked at higher concentrations, and appears to level off at HDI-GO contents > 10 wt %. At low nanofiller weight fractions, there would be larger number of individual HDI-GO nanosheets, hence the reinforcing capability would be stronger than that found at higher loadings, since the HDI-GO nanosheets are more stacked. The modulus and strength improvements attained herein are higher than those reported previously for PANI/graphene composites, corroborating the high reinforcing efficiency of HDI-GO, in particular that with the highest FD, probably due to the combination of a random and very homogenous nanomaterial dispersion within the matrix and a very strong PANI-HDI-GO interfacial adhesion attained via hydrogen bonding, electrostatic, hydrophobic, and π-π stacking interactions, as mentioned above (Scheme 3), together with the high modulus of GO (~207 GPa [48]). The enhancements are also greater than those reported for PANI/MWCNT composites [49], despite the higher modulus of MWCNTs compared to GO, suggesting improved nanofiller-matrix stress transfer in the nanocomposites with HDI-GO nanosheets. The improvements in modulus and strength for the reference PANI/GO (10 wt %) nanocomposite were smaller, around 115 and 85%, respectively, likely due to the presence of aggregates (Figure 1a) that reduce the GO-PANI interfacial area and restrict the load transfer efficiency, combined with the weaker GO-PANI interactions. Good nanofiller dispersion and strong interfacial adhesion to the PANI matrix is the key point to improve the mechanical properties.

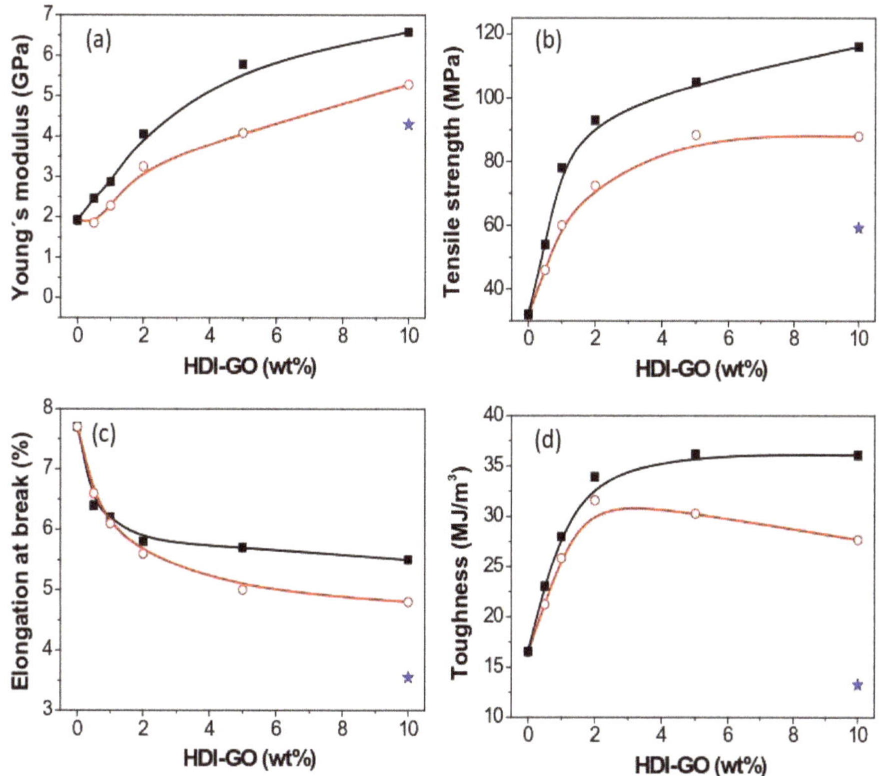

Figure 6. Results derived from tensile tests of PANI nanocomposites reinforced with HDI-GO 6 (solid squares) and HDI-GO 1 (open circles) as a function of the HDI-GO content. (**a**) Young's modulus; (**b**) tensile strength; (**c**) elongation at break; (**d**) toughness. Solid lines are guides for the eyes. The blue star corresponds to the PANI/GO (10 wt %) reference composite.

Regarding the elongation at the break (Figure 6c), PANI presents a value close to 7.8%, which decreases with increasing HDI-GO content, the drop being more pronounced at lower loadings. The maximum drop is found to be 36% for the composite with 10 wt % HDI-GO 1. This is the typical trend found in nanofiller-reinforced polymer nanocomposites, since the fillers restrict the ductile flow of the polymer segments. Further, the strong PANI-HDI-GO interfacial adhesion attained by hydrogen bonding, electrostatic, hydrophobic and π-π interactions contributes to the reduced plasticity. The drop in ductility is in general more pronounced for the nanocomposites with HDI-GO 1 compared to those filled with HDI-GO 6, likely due to the presence of some nanofiller aggregates that restrict more strongly the mobility of the polymeric chains. On the other hand, the reference PANI/GO (10 wt %) nanocomposite shows the most drastic reduction in the elongation at break, about 55% compared to neat PANI, attributed to the larger number of aggregates. Nonetheless, the ductility of the developed nanocomposites is still reasonably good, and the combination of high strength and flexibility is interesting for applications in flexible electronics.

Regarding the toughness of the nanocomposites (the total energy required to break the composite estimated by the area under the stress-strain curve, Figure 6d), the trend observed is quite similar to that of the strength. The nanocomposite with 5 wt % HDI-GO 6 content shows the highest toughness, about 112% higher than that of neat PANI (16.9 MJ/m^3). Beyond this concentration, the toughness remained roughly constant for nanocomposites comprising HDI-GO 6 while dropped slightly for those filled with HDI-GO 1. This trend differs from that found in the reference PANI/GO (10 wt %) nanocomposite that shows a toughness fall of about 28% compared to neat PANI, in agreement with the behavior typically observed in polymer/GO nanocomposites [4], in which the toughness is significantly reduced at high GO contents. The enhanced behavior found herein could be ascribed to the more homogeneous HDI-GO dispersion that avoids the formation of stress concentration points or crack initiators under applied loads, together with the stronger PANI-HDI-GO interfacial adhesion, that would result in a more effective barrier for the propagation of cracks. Further, according to SEM images (Figure 1), PANI and HDI-GO are intercalated with each other instead of individually being in an agglomerated state and phase separated as typically observed for other PANI/GO nanocomposites [22,23], which leads to a decrement in toughness. The simultaneous improvement in ductility and toughness is interesting from an application viewpoint, in particular for the development of flexible electronic devices.

4. Conclusions

Multifunctional high-performance PANI-HDI-GO nanocomposites have been prepared via in situ polymerization of aniline in acid medium in the presence of HDI-GO as nanofillers and $(NH_4)_2S_2O_8$ as oxidizing agent. The morphology, structure, thermal stability, sheet resistance, zeta potential and mechanical properties of the nanocomposites have been characterized to elucidate the influence of the HDI-GO content and level of modification on the final properties. According to SEM observations, the flat and rigid HDI-GO nanosheets are fully embedded within the PANI nanoparticles, forming a dense and interpenetrating network with the matrix particles partly attached onto the nanomaterial surface. PANI can strongly interact with HDI-GO by means of hydrogen bonding, hydrophobic, electrostatic, and π-π stacking interactions, as confirmed by Raman spectroscopy and XRD analysis. TGA and four-point probe measurements reveal that the thermal stability and electrical conductivity of PANI, respectively, gradually rise with increasing HDI-GO loading, the improvement being more significant with increasing HDI-GO functionalization degree. The nanocomposites comprising 5 wt % HDI-GO with the highest modification level display an optimal balance of rigidity, strength, ductility and toughness. These results open novel opportunities for the development of multifunctional nanocomposites based on graphene derivatives and intrinsically conducting polymers to be used in a wide range of applications such as supercapacitors, sensing platforms, solar cells, fuel cells, electrochromic devices, lithium ion batteries, flexible plastics, wearable electronics and so forth.

Supplementary Materials: The following are available online at http://www.mdpi.com/2073-4360/11/6/1032/s1, Figure S1: Photographs of PANI-based nanocomposites incorporating different GO or HDI-GO 6 contents. Figure S2: Stress-strain curves of PANI and some PANI-based nanocomposites. Table S1: Data obtained from the Raman spectra and XRD patterns of PANI/HDI-GO nanocomposites. Table S2: TGA and zeta potential data of PANI/HDI-GO nanocomposites.

Author Contributions: J.A.L.S. performed the experiments; A.M.D.-P. designed the experiments and wrote the paper; R.P.C., P.G.D. and A.M.D.-P. analyzed and discussed the data.

Funding: This research was funded by University of Alcalá, via project CCG2018/EXP-01.

Acknowledgments: J.A.L.S. wishes to acknowledge the University of Alcalá for a "Formación de Personal Investigador (FPI)" PhD fellowship and A.M.D.-P. wishes to acknowledge the MINECO for a "Ramón y Cajal" Potsdoctoral Fellowship cofinanced by the EU.

Conflicts of Interest: The authors declare no conflict of interest.

References

1. Li, D.; Huang, J.; Kaner, R.B. Polyaniline Nanofibers: A Unique Polymer Nanostructure for Versatile Applications. *Acc. Chem. Res.* **2009**, *42*, 135–145. [CrossRef] [PubMed]
2. Bhadra, S.; Khastgir, D.; Singha, N.K.; Lee, J.H. Progress in preparation, processing and applications of polyaniline. *Prog. Polym. Sci.* **2009**, *34*, 783–810. [CrossRef]

3. Díez-Pascual, A.M.; Luceño Sanchez, J.A.; Peña Capilla, R.; García Díaz, P. Recent Developments in Graphene/Polymer Nanocomposites for Application in Polymer Solar Cells. *Polymers* **2018**, *10*, 217. [CrossRef] [PubMed]
4. Wang, L.; Lu, X.; Lei, S.; Song, Y. Graphene-based polyaniline nanocomposites: Preparation, properties and applications. *J. Mater. Chem. A* **2014**, *2*, 4491–4509. [CrossRef]
5. Boeva, Z.; Sergeyev, V. Polyaniline: Synthesis, properties, and application. *Polym. Sci. Ser. C* **2014**, *56*, 144–153. [CrossRef]
6. Focke, W.W.; Wnek, G.E.; Wei, Y. Influence of oxidation state, pH, and counterion on the conductivity of polyaniline. *J. Phys. Chem.* **1987**, *91*, 5813–5818. [CrossRef]
7. Huang, J.; Kaner, R.B. Nanofiber Formation in the Chemical Polymerization of Aniline: A Mechanistic Study. *Angew. Chem. Int. Edit.* **2004**, *43*, 5817–5821. [CrossRef]
8. Peng, C.; Zhang, S.; Jewell, D.; Chen, G.Z. Carbon nanotube and conducting polymer composites for supercapacitors. *Prog. Nat. Sci.* **2008**, *18*, 777–788. [CrossRef]
9. Yan, J.; Wei, T.; Shao, B.; Fan, Z.; Qian, W.; Zhang, M.; Wei, F. Preparation of a graphene nanosheet/polyaniline composite with high specific capacitance. *Carbon* **2010**, *48*, 487–493. [CrossRef]
10. Liu, Y.; Zhou, J.; Zhang, X.; Liu, Z.; Wan, X.; Tian, J.; Wang, T.; Chen, Y. Synthesis, characterization and optical limiting property of covalently oligothiophene-functionalized graphene material. *Carbon* **2009**, *47*, 3113–3121. [CrossRef]
11. Bai, H.; Xu, Y.; Zhao, L.; Li, C.; Shi, G. Non-covalent functionalization of graphene sheets by sulfonated polyaniline. *Chem. Commun.* **2009**, 1667–1669. [CrossRef] [PubMed]
12. Wu, Q.; Xu, Y.; Yao, Z.; Liu, A.; Shi, G. Supercapacitors Based on Flexible Graphene/Polyaniline Nanofiber Composite Films. *ACS Nano* **2010**, *4*, 1963–1970. [CrossRef] [PubMed]
13. Gedela, V.R.; Srikanth, V.V.S.S. Electrochemically active polyaniline nanofibers (PANi NFs) coated graphene nanosheets/PANi NFs composite coated on different flexible substrates. *Synth. Met.* **2014**, *193*, 71–76. [CrossRef]
14. Xu, J.; Wang, K.; Zu, S.; Han, B.; Wei, Z. Hierarchical Nanocomposites of Polyaniline Nanowire Arrays on Graphene Oxide Sheets with Synergistic Effect for Energy Storage. *ACS Nano* **2010**, *4*, 5019–5026. [CrossRef] [PubMed]
15. Wu, W.; Li, Y.; Yang, L.; Ma, Y.; Pan, D.; Li, Y. A Facile One-Pot Preparation of Dialdehyde Starch Reduced Graphene Oxide/Polyaniline Composite for Supercapacitors. *Electrochim. Acta* **2014**, *139*, 117–126. [CrossRef]
16. Kumar, N.A.; Choi, H.; Shin, Y.R.; Chang, D.W.; Dai, L.; Baek, J. Polyaniline-Grafted Reduced Graphene Oxide for Efficient Electrochemical Supercapacitors. *ACS Nano* **2012**, *6*, 1715–1723. [CrossRef] [PubMed]
17. Díez-Pascual, A.M.; Díez-Vicente, A.L. Poly(propylene fumarate)/Polyethylene Glycol-Modified Graphene Oxide Nanocomposites for Tissue Engineering. *ACS Appl. Mater. Interfaces* **2016**, *8*, 17902–17914. [CrossRef] [PubMed]
18. Luceño-Sánchez, J.A.; Maties, G.; Gonzalez-Arellano, C.; Diez-Pascual, A.M. Synthesis and Characterization of Graphene Oxide Derivatives via Functionalization Reaction with Hexamethylene Diisocyanate. *Nanomaterials* **2018**, *8*, 870. [CrossRef]
19. Luceño Sánchez, J.A.; Peña Capilla, R.; Díez-Pascual, A.M. High-Performance PEDOT:PSS/Hexamethylene Diisocyanate-Functionalized Graphene Oxide Nanocomposites: Preparation and Properties. *Polymers* **2018**, *10*, 1169. [CrossRef]
20. Stejskal, J.; Sapurina, I.; Trchová, M. Polyaniline nanostructures and the role of aniline oligomers in their formation. *Prog. Polym. Sci.* **2010**, *35*, 1420–1481. [CrossRef]
21. Du, X.S.; Xiao, M.; Meng, Y.Z. Facile synthesis of highly conductive polyaniline/graphite nanocomposites. *Eur. Polym. J.* **2004**, *40*, 1489–1493. [CrossRef]
22. Male, U.; Srinivasan, P.; Singu, B.S. Incorporation of polyaniline nanofibres on graphene oxide by interfacial polymerization pathway for supercapacitor. *Int. Nano Lett.* **2015**, *5*, 231–240. [CrossRef]
23. Thomas, S.; Zaikov, G.E.; Valsaraj, S.V. *Recent Advances in Polymer Nanocomposites*, 1st ed.; CRC Press: Baton Rouge, LA, USA, 2009; p. 110. ISBN 9789004167261.
24. Vallés, C.; Jiménez, P.; Muñoz, E.; Benito, A.M.; Maser, W.K. Simultaneous Reduction of Graphene Oxide and Polyaniline: Doping-Assisted Formation of a Solid-State Charge-Transfer Complex. *J. Phys. Chem. C* **2011**, *115*, 10468–10474. [CrossRef]

25. Muralikrishna, S.; Nagaraju, D.H.; Balakrishna, R.G.; Surareungchai, W.; Ramakrishnappa, T.; Shivanandareddy, A.B. Hydrogels of polyaniline with graphene oxide for highly sensitive electrochemical determination of lead ions. *Anal. Chim. Acta* **2017**, *990*, 67–77. [CrossRef] [PubMed]
26. Solonaru, A.M.; Grigoras, M. Water-soluble polyaniline/graphene composites as materials for energy storage applications. *Express Polym. Lett.* **2017**, *11*, 127–139. [CrossRef]
27. Kou, L.; He, H.; Gao, C. Click chemistry approach to functionalize two-dimensional macromolecules of graphene oxide nanosheets. *Nano-Micro Lett.* **2010**, *2*, 177–183. [CrossRef]
28. Alexander, L.E. *X-ray Diffraction Methods in Polymer Science*; Wiley-Interscience: New York, NY, USA, 1969; ISBN 9780471021834.
29. Díez-Pascual, A.M.; Ferreira, C.H.; San Andrés, M.P.; Valiente, M.; Vera, S. Effect of Graphene and Graphene Oxide Dispersions in Poloxamer-407 on the Fluorescence of Riboflavin: A Comparative Study. *J. Phys. Chem. C* **2017**, *121*, 830–843. [CrossRef]
30. Wang, C.; Feng, L.; Yang, H.; Xin, G.; Li, W.; Zheng, J.; Tian, W.; Li, X. Graphene oxide stabilized polyethylene glycol for heat storage. *Phys. Chem. Chem. Phys.* **2012**, *14*, 13233–13238. [CrossRef]
31. Vargas, L.R.; Poli, A.K.; Dutra, R.D.C.L.; Souza, C.B.D.; Baldan, M.R.; Gonçalves, E.S. Formation of Composite Polyaniline and Graphene Oxide by Physical Mixture Method. *J. Aerosp. Technol. Manag.* **2017**, *9*, 29–38. [CrossRef]
32. Grzeszczuk, M.; Grańska, A.; Szostak, R. Raman spectroelectrochemistry of polyaniline synthesized using different electrolytic regimes-multivariate analysis. *Int. J. Electrochem. Sci.* **2013**, *8*, 8951–8965.
33. Šeděnková, I.; Trchová, M.; Stejskal, J. Thermal degradation of polyaniline films prepared in solutions of strong and weak acids and in water—FTIR and Raman spectroscopic studies. *Polym. Degrad. Stab.* **2008**, *93*, 2147–2157. [CrossRef]
34. Díez-Pascual, A.M.; Vallés, C.; Mateos, R.; Vera-López, S.; Kinloch, I.A.; San Andrés, M.P. San Influence of surfactants of different nature and chain length on the morphology, thermal stability and sheet resistance of graphene. *Soft Matter* **2018**, *14*, 6013–6023. [CrossRef] [PubMed]
35. Dresselhaus, M.S.; Jorio, A.; Souza Filho, A.G.; Saito, R. Saito Defect characterization in graphene and carbon nanotubes using Raman spectroscopy. *Philos. Trans. A Math. Phys. Eng. Sci.* **2010**, *368*, 5355–5377. [CrossRef] [PubMed]
36. Khasim, S. Polyaniline-Graphene nanoplatelet composite films with improved conductivity for high performance X-band microwave shielding applications. *Results Phys.* **2019**, *12*, 1073–1081. [CrossRef]
37. Grinou, A.; Yun, Y.; Jin, H. Polyaniline nanofiber-coated polystyrene/graphene oxide core-shell microsphere composites. *Macromol. Res.* **2012**, *20*, 84–92. [CrossRef]
38. Bockris, J.O.; Reddy, A.K.N. *Modern Electrochemistry*, 2nd ed.; Kluwer Academic Publishers: New York, NY, USA, 2000; ISBN 978-0-306-46166-8.
39. Colonna, S.; Monticelli, O.; Gomez, J.; Novara, C.; Saracco, G.; Fina, A. Effect of morphology and defectiveness of graphene-related materials on the electrical and thermal conductivity of their polymer nanocomposites. *Polymer* **2016**, *102*, 292–300. [CrossRef]
40. Lu, X.; Dou, H.; Yang, S.; Hao, L.; Zhang, L.; Shen, L.; Zhang, F.; Zhang, X. Fabrication and electrochemical capacitance of hierarchical graphene/polyaniline/carbon nanotube ternary composite film. *Electrochim. Acta* **2011**, *56*, 9224–9232. [CrossRef]
41. Imran, S.M.; Kim, Y.; Shao, G.N.; Hussain, M.; Choa, Y.; Kim, H.T. Enhancement of electroconductivity of polyaniline/graphene oxide nanocomposites through in situ emulsion polymerization. *J. Mater. Sci.* **2014**, *49*, 1328–1335. [CrossRef]
42. Sanches, E.A.; Soares, J.C.; Iost, R.M.; Marangoni, V.S.; Trovati, G.; Batista, T.; Mafud, A.C.; Zucolotto, V.; Mascarenhas, Y.P. Structural Characterization of Emeraldine-Salt Polyaniline/Gold Nanoparticles Complexes. *J. Nanomater.* **2011**, *2011*, 697071. [CrossRef]
43. Yu, L.; Zhang, Y.; Tong, W.; Shang, J.; Lv, F.; Chu, P.K.; Guo, W. Hierarchical composites of conductivity controllable polyaniline layers on the exfoliated graphite for dielectric application. *Compos. Part A Appl. Sci. Manuf.* **2012**, *43*, 2039–2045. [CrossRef]
44. Zhao, Y.; Tang, G.; Yu, Z.; Qi, J. The effect of graphite oxide on the thermoelectric properties of polyaniline. *Carbon* **2012**, *50*, 3064–3073. [CrossRef]
45. Elnaggar, E.M.; Kabel, K.I.; Farag, A.A.; Al-Gamal, A. Comparative study on doping of polyaniline with graphene and multi-walled carbon nanotubes. *J. Nanostruct. Chem.* **2017**, *7*, 75–83. [CrossRef]

46. Lee, J.R.; Lee, D.Y.; Kim, D.G.; Lee, G.H.; Kim, Y.D.; Song, P.K. Characteristics of ITO films deposited on a PET substrate under various deposition conditions. *Met. Mater. Int.* **2008**, *14*, 745. [CrossRef]
47. Husin, M.R.; Arsad, A.; Al-Othman, O. Effect of Graphene Loading on Mechanical and Morphological Properties of Recycled Polypropylene/Polyaniline Nanocomposites. *MATEC Web Conf.* **2015**, *26*, 01008. [CrossRef]
48. Suk, J.W.; Piner, R.D.; An, J.; Ruoff, R.S. Mechanical Properties of Monolayer Graphene Oxide. *ACS Nano* **2010**, *4*, 6557–6564. [CrossRef] [PubMed]
49. Saadattalab, V.; Shakeri, A.; Gholami, H. Effect of CNTs and nano ZnO on physical and mechanical properties of polyaniline composites applicable in energy devices. *Prog. Nat. Sci. Mater. Int.* **2016**, *26*, 517–522. [CrossRef]

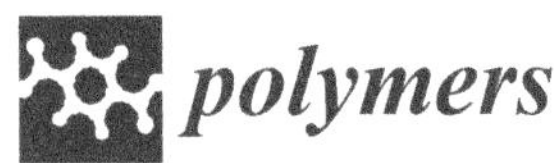

Review

Carbon-Based Polymer Nanocomposite for High-Performance Energy Storage Applications

Samarjeet Singh Siwal [1], Qibo Zhang [1,2,*], Nishu Devi [3,*] and Vijay Kumar Thakur [4,5,*]

1 Key Laboratory of Ionic Liquids Metallurgy, Faculty of Metallurgical and Energy Engineering, Kunming University of Science and Technology, Kunming 650093, China; samarjeet6j1@gmail.com
2 State Key Laboratory of Complex Nonferrous Metal Resources Cleaning Utilization in Yunnan Province, Kunming 650093, China
3 Department of Chemistry, University of Johannesburg, P.O. Box: 524, Auckland Park 2006, South Africa
4 Enhanced Composites and Structures Center, School of Aerospace, Transport and Manufacturing, Cranfield University, Bedfordshire MK43 0AL, UK
5 Department of Mechanical Engineering, School of Engineering, Shiv Nadar University, Uttar Pradesh 201314, India
* Correspondence: qibozhang@kust.edu.cn (Q.Z.); nishu.hooda29@gmail.com (N.D.); vijay.kumar@cranfield.ac.uk (V.K.T.)

Received: 27 January 2020; Accepted: 24 February 2020; Published: 26 February 2020

Abstract: In recent years, numerous discoveries and investigations have been remarked for the development of carbon-based polymer nanocomposites. Carbon-based materials and their composites hold encouraging employment in a broad array of fields, for example, energy storage devices, fuel cells, membranes sensors, actuators, and electromagnetic shielding. Carbon and its derivatives exhibit some remarkable features such as high conductivity, high surface area, excellent chemical endurance, and good mechanical durability. On the other hand, characteristics such as docility, lower price, and high environmental resistance are some of the unique properties of conducting polymers (CPs). To enhance the properties and performance, polymeric electrode materials can be modified suitably by metal oxides and carbon materials resulting in a composite that helps in the collection and accumulation of charges due to large surface area. The carbon-polymer nanocomposites assist in overcoming the difficulties arising in achieving the high performance of polymeric compounds and deliver high-performance composites that can be used in electrochemical energy storage devices. Carbon-based polymer nanocomposites have both advantages and disadvantages, so in this review, attempts are made to understand their synergistic behavior and resulting performance. The three electrochemical energy storage systems and the type of electrode materials used for them have been studied here in this article and some aspects for example morphology, exterior area, temperature, and approaches have been observed to influence the activity of electrochemical methods. This review article evaluates and compiles reported data to present a significant and extensive summary of the state of the art.

Keywords: carbon-based polymer nanocomposite; energy storage; fuel cell; electrochemical devices

1. Introduction

Renewable sources—for example, solar and wind energy—can satisfy the world's power needs, but substitutes for petroleum-derived substances demand a root of carbon fragments [1]. As renewable sources are not spontaneous sources of energy, therefore, storage of that energy generated from renewable sources is a prerequisite for its later use. Also, the innovation of new and compact portable electronic devices has drawn the attention towards efficient energy storage. In the prior two eras, transportable tools/portable devices such as smartphones, laptops, smart health devices, have reduced suggestively in volume, and besides their abilities, storage potentials continue to rise intensely. Not surprisingly, the energy markets of these tools are rather abundant, which has led to a massive improvement in the level of study towards high-performance applications. In current times, millions of dollars have been funded for the investigations that are focused on the evolution of more effective energy storage systems, including fuel cells [2–8], supercapacitors [9], and batteries [10]. Among these applications, battery technology has been one of the most established systems for energy storage as it is useful for high energy requirements owing to their high energy capabilities. Though, despite the dramatic performance with time, there is yet notable room for the development.

Carbon-based polymer nanocomposites (CPNCs) have various applications in the energy accumulation, energy storage, packing, aerospace, and automotive areas [11,12]. The important characteristics of these nanostructured substances are the comfort of processing, configuration adaptability, lightweight, and flexibility to requirements. For energy storage, fuel cell and supercapacitors are supposed to be crucial components in updating the prospect of renewable energy schemes [13,14]. The demand for high energy and power density devices at a low-cost leads to the discovery of novel nanocomposite materials for automotive and electric energy storage applications. Insulating polymers loaded by high-aspect-ratio conductive nanofillers—for example, carbon nanotube (CNT) [15,16] as well as graphene nanoplatelets (GNP) [17,18]—have been proved to be potential dielectric ingredients [19,20]. Scattered nanocomposites within an insulating pattern performed as nanocapacitor electrodes that drive significant improvement in space charge polarization [21,22]. The nanofiller is assumed to become highly conductive to improve the interfacial polarization on the nanofiller/polymer boundary and high exterior area to develop the interfacial polarization density. As a polymer nanocomposite, the polymer matrix, nanofiller, and polymer/nanofiller interface may be polarized. It is observed that this polarization in nanocomposites is plausible, particularly at quite large frequencies, though the polarization of the conductive nanofiller and the interfacial region may happen frequencies under 1 MHz [23,24].

The energy storage methods require unique, paramount, and authentic approaches towards the storage of electric power by alternate renewable origins for assuring appropriate and dependable devices that can store a sufficient quantity of energy and later can be used for transport, electronic gadgets, electric-powered carriers, and distinct purposes. Electrochemical energy storage operations comprise supercapacitors, various kinds of batteries, and fuel cells [25]. Another energy storage method, for example, thermal, compressed air, also flywheel energy aggregate nonelectrochemical storage operations [26–28]. The electrochemical storage methods usually contain the probes, electrolyte media, and power receiver. These ingredients are built from carbon-based nanomaterials, CPs, metal oxides (MO_x), as well as conducting substances. These substances are not well established and there are yet numerous hurdles that need to designate and overcome for the further extensive commercialization of the energy storage applications. Recently, excellent new substances called CPNCs have been examined for electrochemical applications [29].

In the present article, the formerly published data is explained by giving a rationalized review of the state of the system and setting up a wide series of electrochemical energy storage (EES) technologies. EES technology helps in justifying the inconsistency of renewable resources, energy production from them, and storage of that energy. This article includes the reflection on CPNCs for energy storage devices and their high-performance applications. In the remainder of this review article, we briefly presented the characteristics and applications of several kinds of carbon-based polymer nanocomposites towards its use as ingredients of high-performance energy storage substances. An up-to-date explanation of the most advanced progress in the growth of CPNC ingredients intended for their uses as an electrode material probe and electrolyte substantial for fuel cell, supercapacitor (SC), and battery.

2. Electrochemical Energy Storage Applications

One of the paramount scientific and societal tasks in front of us is to provide secure energy for future generations. The prerequisite for storing energy is giving us ideas to explore more and more inexpensive techniques for energy storage. Although researchers and scientists are heading towards renewable sources of energy but to store that renewable energy for future use, we require energy storage devices. One provision is storing energy electrochemically using electrochemical energy storage devices like fuel cells, batteries, and supercapacitors (Figure 1) having a different mechanism of energy storage but have electrochemical resemblances.

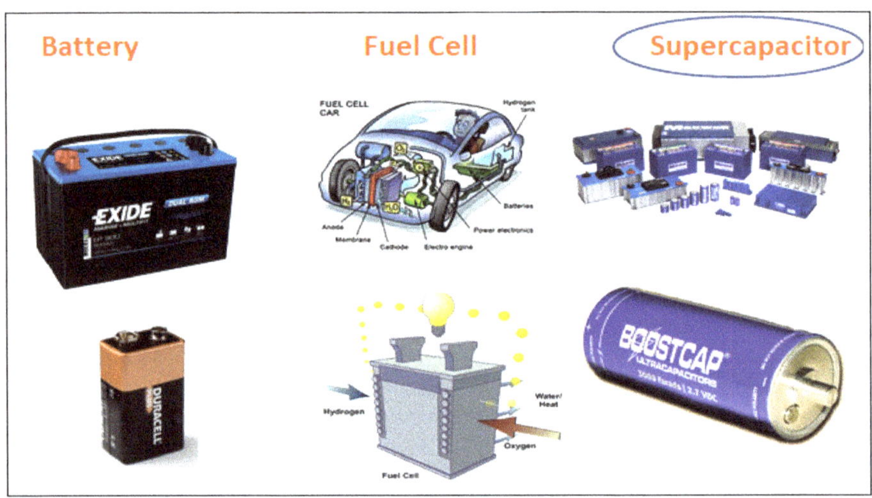

Figure 1. Types of electrochemical energy storage devices.

Energy exists in various forms in nature like radiation, chemical, gravitational, potential, electricity, latent heat, kinetic energy, etc. Energy storage is the capture of produced energy so that it can be used later. Battery and supercapacitors are most commonly used energy storage devices, but fuel cell does not store energy instead produces energy directly from the fuel and oxygen via electrochemical reactions in the cell. The basic operating mechanism and a brief background of these three electrochemical systems is explained in the next few sections.

2.1. Fuel Cells

The first fuel cell (FC) was invented in 1838 by Sir William Grove. The first commercial use of fuel cells was in NASA space programs. FC is electrochemical equipment that harvests electricity directly from fuel instead of accumulating. In addition, they require combustible fuels in stock and a continuous supply of oxygen from the atmosphere. It renovates the chemical energy of the fuel directly into electricity by redox reactions in the cell. They are different from the battery in the location of energy storing and alteration. Fuel cells are open systems, in case of fuel cell, fuels that provide energy are located outside, e.g., oxygen in the air and hydrogen outside in the fuel storage tank which are supplied continuously to the reaction chamber (Figure 2, fuels are supplied from outside into the reaction chamber) while battery is a closed system through the positive and negative electrode as the charge transfer intermediate [30]. Fuel cells have very high energy density even higher than batteries but the lowest power density due to which they are not widely employed towards energy storage and renovation. FCs are still in the developing phase and in search of an effective application that can penetrate the energy market but still be competitive against batteries and supercapacitors.

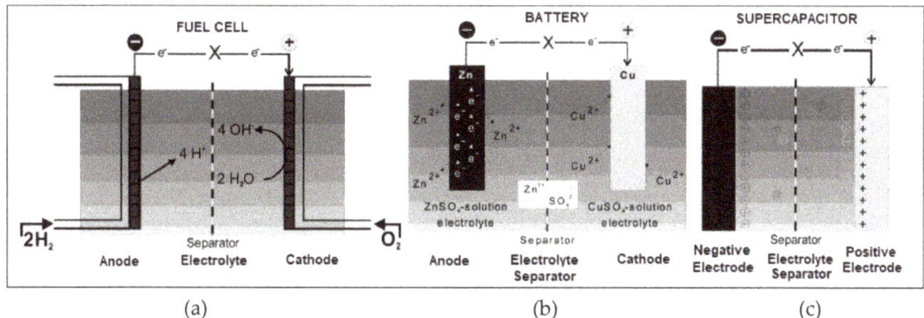

(a) (b) (c)

Figure 2. Representation of a fuel cell (**a**) shows the constant source of ingredients and redox reactions in the cell, a battery (**b**) shows the salient features of battery operation and a supercapacitor (**c**) illustrating the energy storage at probe-conducting solution interface. Reproduced with permission from [30].

2.2. Battery

Battery stores energy via electrochemical reactions. In comparison to fuel cells and supercapacitors, batteries have established the energy storage market till now and most commonly used in mobile phones, laptops, automobiles, homes, offices, portable, mobile, and stationary applications. Batteries vary in size and capacity from an aspirin tablet 1–10 mAh used under air range aids to large buildings with 40 MWh towards energy storage and backup use [30]. Various battery systems like lead–acid, Ni–Cd, Ni–Zn, rechargeable alkaline, Li-ion, and Li-ion polymers are in use.

A simple Ragone plot (Figure 3) shows that FCs are high energy structures whereas SCs are high energy arrangements, batteries are somewhere in between the two but overlapping a part of SC and fuel cells (FCs) [31]. That is why SC and FCs can replace some battery applications. Utilizing high ED of batteries and high PD of SC, a hybrid system [32] may be used to harvest essential properties of both the systems which can be used in practical applications [33]. This combination of both systems can meet the requirements of future power applications as hybrid systems [32].

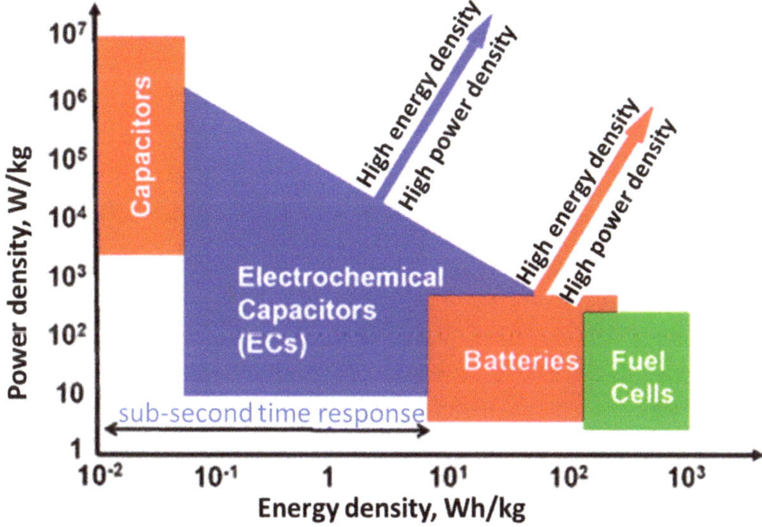

Figure 3. Ragone plot representing the distribution of ED and PD of electrochemical energy storage devices. Reproduced with permission from [31].

2.3. Supercapacitor

Renewable energy schemes have noticeably been painstaking as an influential armament to employ in the environmental revolution and combat global warming [34–36]. In recent years, wind and solar energy have seen unexpected development with enormous cost drops, which has made them more attractive relative to fossil fuels. Though, owing to the worries in power production and demand, energy storage strategies have dynamic appeal in making flexible and dependable energy structures.

Supercapacitors or ultracapacitors having capacitance much higher than usual capacitors are electrochemical energy storage tools which store energy either electrochemically at the exterior of the electrode–electrolyte interface (Figure 2), also known as an electrochemical double-layer capacitor (EDLC) or by redox responses occurring on the surface of electrode recognized as pseudocapacitors. In 1950, General Electric Engineers started experiments using porous carbon electrodes. H. Becker established a low voltage capacitor using the same porous carbon electrodes in 1957 [37]. Recently, Li-ion capacitors came into knowledge and still researchers are aiming to enhance or upgrade specific energy, specific power, cycle stability, and trying to diminish the cost to replace a battery in the energy market [38]. SCs are EDLC, pseudocapacitors, and hybrid capacitors that are pivotal in the mechanism of charge storage and ingredients employed [39].

However, still, SCs are facing material stability challenges, and we need research these issues, so that supercapacitors can also be used as current devices which require high capacity for brief periods. The performance of SC depends upon many factors including electrochemical characteristics of probe material, type of electrolyte in addition to the voltage assortment. Two energy storage mechanisms, EDLC and pseudocapacitive, have been studied depending upon the type of material carbonaceous, metal oxide, or polymers [40,41]. Where two or more above stated materials are used in combination with each other to increase the performance of SC, both the EDLC and redox mechanisms take place in the system [42]. In asymmetric SCs, a combination of EDLC and pseudocapacitor in which one electrode is an electrical double layer electrode that stores energy via the adsorption/desorption of solution ions at the interface, whereas the other electrode is a faradic or pseudo capacitor electrode at which energy is stored by the transfer of electrons during redox reactions. A comparison between EDLC and pseudocapacitive mechanisms are shown in Table 1. Metal oxides, metal sulfides in a combination of carbon materials, have been used for asymmetric or hybrid SCs.

Table 1. Comparison between electric double layer capacitors and pseudocapacitor

EDLC	Pseudocapacitor
Good cyclic stability	Greater specific capacitance
Good power performance	Greater energy density
Formation of the double layer at the interface	No such layer formation
No redox reaction i.e., non-faradaic	Redox reactions, i.e., Faradaic
No mechanism failure	Depends upon redox reactions

3. Carbon-Based Polymer Nanocomposites

Carbon nanomaterials, for example, activated carbon (AC) [43], graphene [44], CNTs [45], carbon dots (C dots) [46], and carbon aerogels (CAGs) [47] can act as potential filler for polymeric materials and help in enhancing the properties of the polymer. The dimensions of nanocomposite have been found to be less than 100 nm. Adding carbon nano-fillers to polymer background results in the high exterior to volume proportion and significantly improves the macroscopic possessions of the polymer including improved mechanical properties and high flexibility.

3.1. Carbon Materials

Carbon has been found to exist in various allotropic forms and sizes from insulator diamond to layered semiconductor graphite to fullerenes and is a versatile element of the periodic table that exhibits outstanding traits. Figure 4 shows the derived nanostructures of carbon in different dimensions (0D, 1D, 2D, and 3D) [48]. Carbon nanomaterials show exceptionally high surface areas, high electronic conductivity, amended capacitance, and reprocessing rate ability on the traditional engineering probe materials, for example, AC. Carbon-based substances, as well as AC, CNTs, along with graphene (GN), have been utilized into EDL SCs owing to their extraordinary chemical in addition to physical assets [49].

Carbon nanostructures, particularly CNTs and GN, that have been used as electrodes for supercapacitors and batteries follow the EDLC mechanism of charge storage at the electrode/solution boundary. There are some limitations of using carbon materials as electrode materials because the storage of charge is limited only by adsorption–desorption of ions at the interface of electrode/solution forming a double layer, and their aggregation during synthesis and processing results in inferior device performance. Their nanocomposites with conducting polymers have high performance and improved properties [50,51]. Carbon nanomaterials—such as CNT [52], GN [53], GN foams [54], and carbon nanofibers [45,55]—with their 3D and 2D architectures have been broadly described owing to their ideal morphological and physical possessions for energy storage applications.

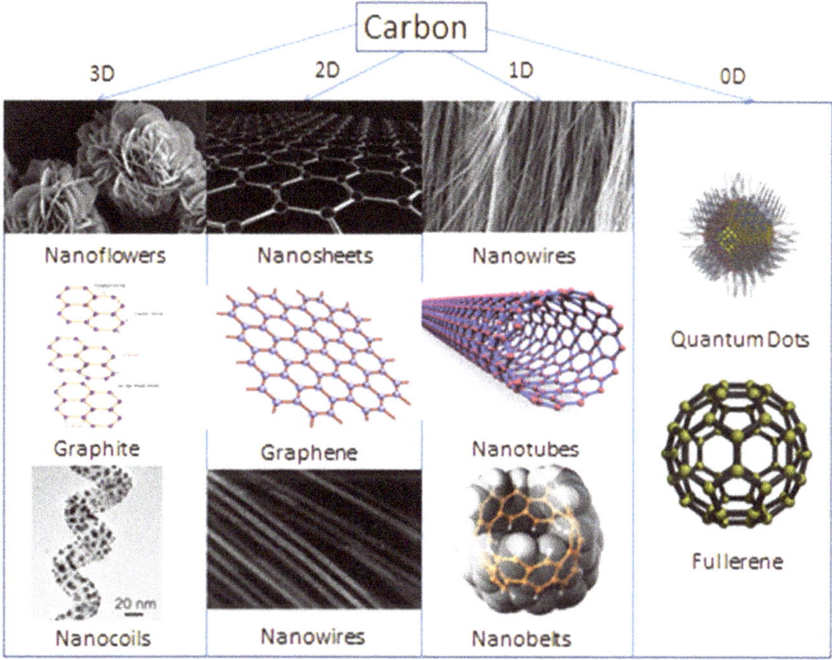

Figure 4. Carbon nanostructures in all three dimensions. Reproduced with permission from [48].

One of the most widely used carbon materials is activated carbon. The AC of high exterior area 1401 $m^2 \cdot g^{-1}$ by a typical pore dimension of 2.2 nm was employed for in situ growth of Bi_2O_3 nanoparticles and used as an asymmetric supercapacitor material [56]. The AC material they used as the electrode was 80 wt % carbon powder, 10 wt % extended graphite, and 10 wt % Teflon (dry) suspension. They have the advantages of high exterior area, cost-effectiveness, accessibility, recognized probe production, and machinery and follow EDLC charge storage mechanism and their capacitance mostly depends on the attainable exterior area towards electrolyte ions. Some other constraints include specific surface area (SSA), pore size, assembly, pore size delivery, surface functionality, as well as electrical behavior to alter their performance. Carbon-based ingredients—for example, AC [57], CAGs [58], CNTs [59], and GN [60]—help high ion dispersion also donate to a higher SC of carbon-based composite substances for conducting polymers (CPs)/MO_x with excellent cyclic stability [61].

AC is of great exterior area, excellent electrical possessions, reasonable price due to physical/chemical activation, SA up to 3000 m^2/g, broad pore size distribution furnishes specific capacitance 100–300 $F{\cdot}g^{-1}$ in aqueous and <150 $F{\cdot}g^{-1}$ organic solution. The shape of the CV curve of carbonaceous materials tends to be nearly rectangular (Figure 5) due to the EDLC mechanism (an ideal EDLC should have a rectangular CV and CD curves are straight lines) [62].

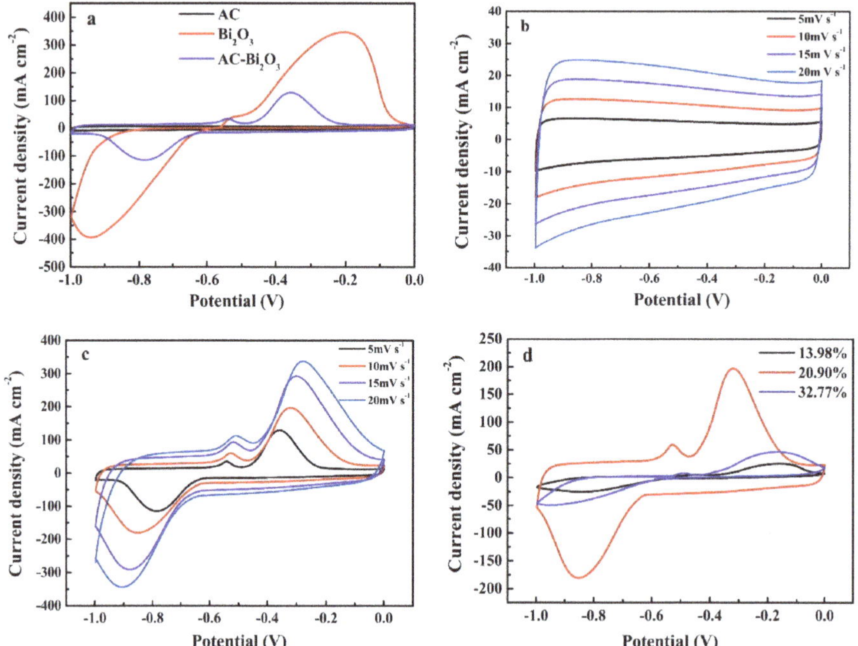

Figure 5. Cyclic voltammogram curves of (**a**) CVs of AC, Bi_2O_3 and AC-Bi_2O_3; (**b**) CVs of AC at different sweep rates; (**c**) CVs of AC-Bi_2O_3 at different sweep rates. (**d**) CVs of different contents AC-Bi_2O_3 composite. Reproduced with permission from ref [62].

Another material is CNTs having exceptional pore construction, virtuous mechanical also thermal steadiness, greater electrical assets, and consistent mesopores that permit a constant charge circulation. CNT has inferior ESR than AC since electrolyte ions may dissociate through the mesoporous system but low SSA (<500 m^2/g) less than AC which leads to low ED. CNT can be single-walled (SW) or multiwalled (MWCNT) and can be chemically activated with KOH resulting towards improvement within the specific capacitance (C_s) [63].

Apart from AC and CNTs, GN—an outstanding carbonaceous material—is a one particle dense film 2D assembly has good electrical performance, chemical durability, and a wide exterior area (nearly 2630 m^2/g) [64]. If the whole exterior area of GN is utilized, it can give C_s up to 550 $F{\cdot}g^{-1}$. GN has a high surface area as together key exteriors of the graphene sheet are eagerly available for electrolyte ions. Reduced graphene sheets showed ED 85.6 Wh/kg on ambient temperature 136 Wh/kg at 1 $A{\cdot}g^{-1}$ [50] which is equivalent to Ni metal hydride battery. There are different methods have been reported by scientists to synthesize GN comprises chemical vapor deposition, arc discharge method, micromechanical exfoliation, unzipping of CNT, electrochemical, and intercalation approaches; but, in all methods, agglomeration and restacking of GN is a problem for mass production which is the only concern for its commercial use in many applications.

Also, carbon black (CB) is the most commonly used conductive additive because of its great electrical conductivity and small price, but it has a drawback that the carbon particles are not able to form carbon to carbon connected conductive networks sometimes. It results in a reduction in activity over a few thousand cycles. Therefore, other carbon materials—such as acetylene black, AC, GN, and MWCNTs—can be used instead of CB to increase stability [65].

3.1.1. Carbon Nanotubes

CNT (Figure 6), a carbon-based material, due to their greater physical properties—for example, high chemical steadiness, feature proportion, mechanical assets and activated surface area in addition to extraordinary electrical assets, and nano-dimensions—has obtained exceptional attention owing to its exciting features. The important purpose of all the fascination nanotechnology has gained is mainly owing to the considerable variation in the feature of material while associated with its atomic and massive state [48]. An exceptional pore structure for charge storing but relatively high materials cost and limitation to additional expanding the active exterior area of the CNTs restrict the commercial application of CNTs based SCs. The high aspect ratio of CNTs helps to create CNT fiber that gives about 10 GPa highly tensile continuous fiber [66]. Highly oxidized SWCNTs of electrical conductivity in the range 100–1000 S·cm^{-1} have been reported to have high performance in supercapacitors and Li-ion batteries as electrochemical conductors [67]. More excellent conductivity and improved charge transmission cylinders of CNTs deliver the greatest encouraging element toward energy-saving purposes. In modern generations, CNTs have been considered as valuable power conductor substances owing to their enhanced electrical performance in addition to the highly-available exterior area.

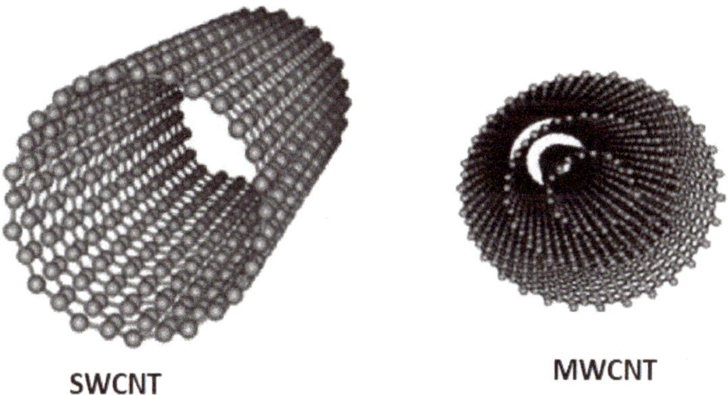

Figure 6. Structure of CNT (SWCNT and MWCNT) carbon-based nanomaterials.

3.1.2. Graphene, Graphene Oxide, and Reduced Graphene Oxide

Graphene (GN) is a carbon-based material (2D form of graphite) having a high exterior area (2675 m^2·g^{-1}) great mechanical strength, greater electrical conductivity (6000 S·cm^{-1}), chemical steadiness was first discovered in 2004, and since then has gained significant attention and use in a variety of applications. It has had great influence under the area of science and technology and is a potential candidate for EDLC electrodes [68]. It comprises a single film, while graphite contains a few films of GN loaded composed through van der Waals interaction and π–π stacking. Agglomeration and stacking of layers reduce its effective exterior area.

Graphene oxide (GO) is a solitary film of sp^2 hybridized carbon particles, arranged in a honeycomb matrix structure, having oxygen as functional assemblies on its basal planes, boundaries and may be developed through exfoliation of GN. GO has been found to be promising carbon material in forming functional nanocomposites due to oxygen functional groups and GO's compatibility with polymers [69].

From a sustainability viewpoint, the hunt of renewable carbon reservoir and research of economical yet straightforward assembly methods are of selective interest in producing GN and their carbon-based analogs within existing power storage purposes. Although, the standard economic policies acquired so far for GN and GN-like substances construction yet cannot compete with the generation of AC [70]. Chemical assembly of GN, GO, and reduced graphene oxide (rGO) are exposed in Figure 7 [71].

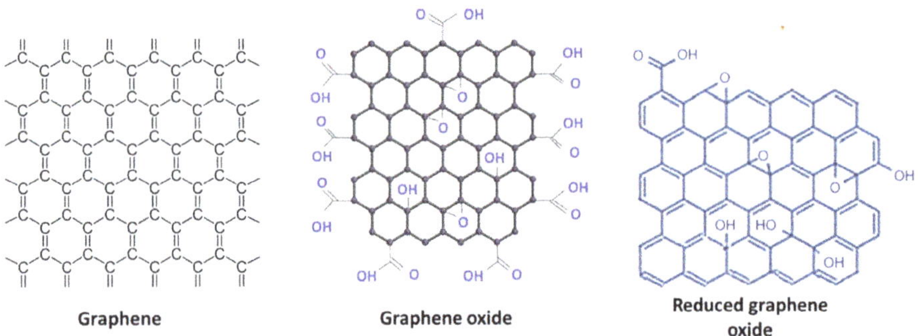

Figure 7. Chemical assembly of GN, GO, and rGO.

Usually, incorporation of GN through chemical oxidation generates GO by entering oxygenated functional assemblies (carboxyl, hydroxyl, and epoxy) in among the carbon films of GN employing oxidants through Hummer's protocol [72–74]. Therefore, the exterior of GO tolerates negative controls owing to his oxygenated practical combinations. The oxygenated efficient assemblies of GO are useful into combining GO by another active kind such these practical clusters may work as an effective fastening position. In contrast to GN, GO is non-conductive owing to the sp^3 hybridization of carbon particles, in addition, the functionalization of GO creates holes at the carbon basal extension owing to the change into the sp^2 hybridized carbon arrangement of the GN layers. To recover the conductivity of GO, deoxygenating or conversion of GO should be brought out.

GN may be modified by chemical reactions plus functionalization towards improving the dispersibility and method ability to improve the synergy through organic polymers owing to the hydrophilic oxygenated functional combinations existing in GO. A quasi-1D single-layer GN nanoribbon (120 nm in width and ~100 μm in length) microelectrode made-up through mechanical exfoliation of graphite, trailed with electron beam lithography method, and oxygen plasma engraving behavior has been investigated as the consequence of chemical functional groups and physical internment at the electrochemical activity [75]. During an electrochemical reaction, the arrangement and localization of electrolyte ions and structuring close to the GN interface can affect the properties at the interface [76]. Application of thermally flaked graphene nanosheets as supercapacitor probe substances has been described to provide the highest C_s of 117 $F{\cdot}g^{-1}$ in acidic media. Toward supercapacitors composed of chemically transformed GN, a C_s of 135 $F{\cdot}g^{-1}$ in alkaline media has been described. A novel 3D ultrathin porous carbon nanosheet has been prepared from biomass by hydrothermal method followed by carbonization and tested for SC application that showed ED (79.4 Wh/kg) and PD (5.1 kW/kg), and imposing cyclic constancy by 94.6% after 5000 charge–discharge (CD) methods [77].

Reduced graphene oxide (rGO) is an extremely conducting kind of GN may be found either by the chemical reduction of GO by some reducing agent like $NaBH_4$, hydrazine, NaOH, Na_2CO_3, and L-ascorbic acid or by electrochemical reduction employing CV or chronoamperometry methods on negative voltage. During the reduction GO revives the graphitic arrangement of carbon particles and separates the oxygenated functional combinations at the exterior and sides of GO, so improving the electrical performance of rGO [78].

3.1.3. Other Carbon Materials

Carbon quantum dots are sp^2 graphitic structure and nano-sized carbon particles with oxidized functionalities and semiconducting properties. The main attraction of using carbon dots in nanocomposites can increase the transport of ions in the CD method, which results in profligate redox reactions and high C_s [79]. CAGs having a high electrochemical exterior area carbon fiber materials are entrenched into an incessant system of CAG towards an arrangement of a clear but porous column suggestively improved the exterior area of carbon fiber materials, also later the improved electrochemical activity has been reported [80].

3.2. *Polymer Materials*

Polymers can be natural or synthetic and made of many repeating monomers units united together to form a long chain. To meet the demands of everyday life, synthetic polymers have been playing an important role from the past century. Natural polymers include proteins, enzymes, DNA, in addition, few cases of synthesized polymers are polyethylene, polyester (PE), polyvinyl chloride (PVC), etc., within the arrangement of plastic fibers and the products with exceptional properties and substituting metal counterparts in every industry with cost-effectiveness [81].

Conducting polymers (CPs) are organic materials where the redox method is applied to stock and release charge, therefore it shows an example of pseudocapacitance [82]. If oxidation occurs ions are moved toward the polymer backbone. During reduction, ions are unrestricted posterior within the electrolyte media. These redox reactions in CPs cause mechanical stress results in limiting stability as of many charge discharges cycles which hinders their application as SC materials. Examples of CPs include PANI [83], polypyrrole (PPy) [84], polythiophene (PTh) [85], etc. PANI has better conductivity, facile synthesis, outstanding capacity, and lower cost. However, owing to repetitive series (CD procedure), this causes puffiness and shrinkage that lead to quick degradation.

Polymers having conducting behavior are of prime interest for electrochemical applications such as solar cells [86], sensors [87], energy storage devices [88], and electronics equipment. The electrochemical energy storage devices like supercapacitors, Li-polymer batteries have been using CPs as high-efficiency electrodes that show reversible redox reactions in electrolytes. Insertion of carbon materials with CPs results in an increase in the stability of the nanostructured materials. Furthermore, in contrast to inorganic resources, the outstanding elasticity of CP also aids in building a supple probe, as well as the flexible tools for wearable electric arrangements [89]. CPs have gained considerable interest and attention due to their conductivity, redox behavior to investigate their necessary substances for exceptional purposes, for example, energy-storing and transformation tools, electrochromic arrays, batteries, membranes, anti-destructive films, etc. [90].

CPs, for example, poly(3,4-ethylene dioxythiophene) (PEDOT) [91], PANI [92], PPy [93], and PTh [94]—owing to their outstanding characteristics such as facile construction, affordability, good electrical performance, chemical endurance, excellent environmental resistance, mechanical adaptability, redox activity, and biocompatibility—are attracting considerable interest and are valuable for numerous purposes, with electrochemical energy storage tools [88], actuators, and neural boundaries. These CPs have delocalized π-electron arrangement, available redox conditions, and manageable physical assets that make these conjugated polymers perfect applicants for developing energy storage materials and device fabrications. CPs can accumulation charges by the method of doping. There are two kinds of doping: p-doping (oxidative), which includes the discharge of π-electrons; and n-doping, which gives clear negative charges. PEDOT may support mutually p-plus n-doping, whereas PANI and PPy support p-doping being their n-doping voltage is significantly less comparable to the standard electrolyte reduction potential.

To build effective CP electrodes/probe, it is usually advisable to store a compact CP film at the electrode because its activity is hugely reliant upon this layer's depth; for example, the areal ED/PD (Wh/cm^2, W/cm^2) or activating drive of a CP probe is equivalent to its thickness [95]. It has been noticed that the initiation of MWCNTs within the polymeric patterns increases the specific exterior area and recovers the electrical performance and mechanical characteristics [96].

PANI having cost-effective, facile construction, great flexibility, and high C_s plays a significant part in the energy accommodation and regeneration patterns. The only limitation is the poor cycle life or durability owing to growing and shortening throughout the doping and dedoping method which drives to the architectural destruction of PANI electrode results in poor charge/discharge capability that can be improved by the addition of carbonaceous materials and metallic compounds, and the composite proved to be potential electrode material [92]. Therefore, the tremendous mechanical robustness of carbon-based substances—for example, GN—is utilized continuously to succeed the mechanical degeneration of CPs and head to high cycling endurance, which is necessary for SCs [78]. These nanomaterials in PANI may decrease the dissipation time, improve the electroactive zones, and notably improve the capacitive activity of nanocomposites that can result in enhanced stable specific capacitance due to synergistic effects [79]. Including the discovery of conductive polyacetylene (PA) during the 1970s, further intrinsically conducting polymers (ICPs) have drawn notable recognition from both science and engineering associations. ICPs usually involve PA, PANI, PPy, PTh, poly(phenylenevinylene) (PPV), polyfuran (PF), etc. Their constructions are illustrated in Figure 8.

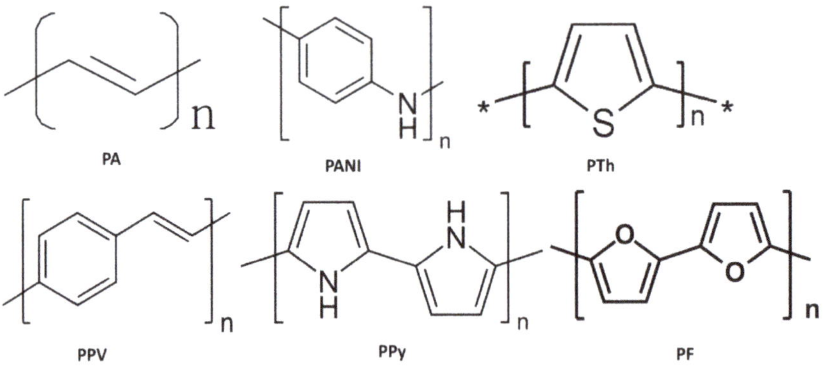

PA PANI PTh

PPV PPy PF

Figure 8. Chemical construction of common ICPs with PA, PANI, PTh, PPV, PPy, and PF.

Investigators are yet concentrating on the advanced substances like carbon-based, metal oxide and CPs or a composite of these. Carbon materials are considered for properties such as great specific area and normal pore size dispersion although their capacitances and energy densities are yet small. With CPs, the challenge is the short lifetime owing to puffiness and dwindling all through CD besides high specific capacitances. Metal oxide possesses high SSA and favor the diffusion of ions. Metal complexes have lower crystallinity and higher SC due to accessibility of chemically active hanging components that can participate in redox cycles [97].

3.3. Carbon Polymer Nanocomposites

Composites materials consists of two or more constituents that may have superior properties than the individual constituents. The reinforcement material typically transmits the distinctive characteristics to the matrix where the matrix and strengthening substantial accompaniment each other for their excellent synergy. If the sizes of materials are in the nanoscale then they are known as 'nanocomposites'. Nanocomposite materials are making a great impact on research and technology globally. The nanofiller reinforced composites possess toughness with stiffness as nanofillers offers an ultrahigh interfacial area with greater thermal and oxidative stability arises from the high surface to volume ratio. Nanocomposites can be classified into three kinds: metal, ceramic, and the polymeric matrix. Our primary interest in this review is polymeric matrix nanocomposites [48].

Polymer composites are composite materials that have been employed towards a diversity of purposes and receiving attention in daily routine. Carbon polymer composites have been formed by using carbonaceous materials as a filler or reinforcement material in the polymer matrix. The incorporation of CP with carbon nanomaterials results in nanocomposite electrodes with high ED and PD simultaneously, owing to the combination of pseudocapacitance and EDLC [89]. To form carbon polymer nanocomposite, proper selection of fillers assists in enhancing the electrical possessions of the polymer nanocomposites where CB, GN, and CNT are conducting fillers whose shape and size and amount of loading change the properties of the nanocomposite [98].

The discovery of CNT and GN has boomed the field and their incorporation inside the polymer matrix results in enhancement of properties and is rapidly expanding. Related to a various array of nanofillers, CNTs have been developed as the common encouraging nanofiller for polymer composites owing to their exceptional mechanical and electrical assets [99].

AC is a widely known electrode material, but transport of ions is slow through the micropores. The porous structure can expedite ion transportation by providing a pathway to ions and therefore a 3D ordered macroporous (3DOM) carbons by a classified porous construction may enable ion conveyance, showing high activity as SC probes have been reported to exhibit. CP-macroporous carbon composite probe substances were developed through deposition of a thin film of PANI at the exterior of 3DOM carbon shows C_s of 1490 F·g^{-1} was perceived at the deposited PANI into the compound probe with excellent rate display and cycle strength were obtained at the complex conductor [100].

Numerous studies have reported on the PPy/carbon compound being employed for probe substances, for example, PPy/AC [101], PPy/MWCNT [102], PPy/carbon foam [103], PPy/ carbon fiber [104], composites. CB/PPy composites are synthesized by the simple chemical oxidative polymerization of pyrrole by the exteriors of the CB nanoparticles—through poly(2-hydroxy-3-(methacryloyloxy) propane-1-sulfonate) (PHMAS), for example—mutually the surfactant and the dopant. The synthesis process of the core–shell CB/PPy nanocomposite shown in Figure 9 [51].

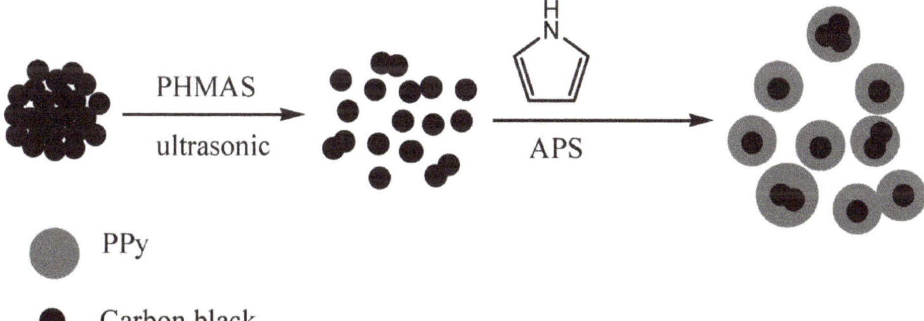

Figure 9. Diagram drawing of the synthesis procedure of the core–shell CB/PPy nanomaterial. Reproduced with permission from [51].

The nanocomposites of CPs including MO_x, CNTs, or GN were incorporated mostly through in situ polymerization within suspensions comprising monomers. Including the extension of an oxidant solution comprising ammonium peroxydisulfate [$(NH_4)_2S_2O_8$], polymerization of aniline at the exterior of CNTs happened to develop CNTs-PANI nanocomposite. GN-PANI nanocomposite coating with layered construction was achieved by the percolation of an aqueous distribution containing positively charged PANI nanofibers and negatively charged chemically changed GN films which produces a steady composite diffusion by electrostatic synergy due to the support of the sonication process [105].

A new polyimide (PI)/ MWCNT nanocomposite by a two-step imidization was synthesized and used as a negative electrode substance for an organic Na-ion battery. A two-step polycondensation reaction made the PI/MWCNT nanocomposite according to the synthetic track presented in Figure 10. Besides its exceptional mechanical resistance, MWCNT works as a conductive material under the nanocomposite and promotes quick electron transmission toward the cathode, and so takes full advantage of the exploitation of the working substance [71].

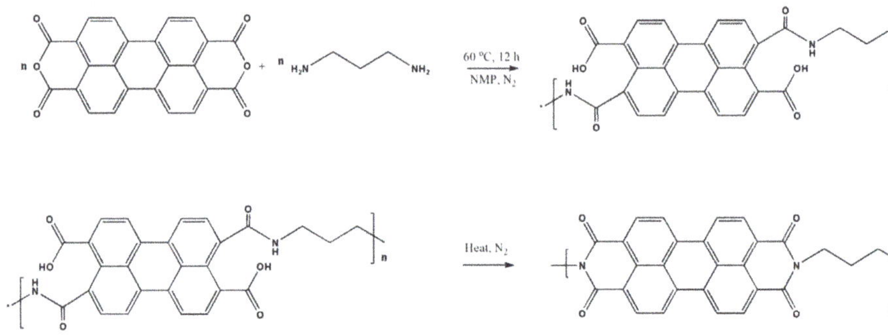

Figure 10. Representation of the synthesis of the PI/MWCNT nanocomposite. Reproduced with permission from [71].

4. Applications of Carbon-based Polymer Nanocomposites

The advances in sustainable and renewable energy storage by renewable energy resources are in great demand to compensate for the energy crisis from conventional energy resources. Electrochemical storage of energy is a medium to store energy for later use that includes electrochemical capacitors, batteries, and FCs that are into urgent demand to develop environmentally friendly energy solutions [106]. The scope of nanoscience and nanotechnology has provided nanocomposites materials with superior properties to be used in industrial and technological development for a wide range of devices. Carbon-based polymer nanocomposites provide a wide spectrum of opportunities to produce novel multifunctional materials individually and on hybridization for applications in thermal materials, electromagnetic interface shielding, sensors, and energy storage [107].

Carbon materials with AC, CNTs, and GN have been widely investigated for SCs applicability even in common they represent smaller ED owing for the quick ions adsorption response, for producing EDLCs. On another side, metal oxides and CPs may give more leading EDs by Faradic reactions including low cyclic durability and PD associated with EDL based SCs. Consequently connecting these substances within their compound construction and applying various charging devices and potential combined impact among each of their ingredients with great aspect proportion, the huge exterior area, and electrical performance is known to be a perfect explanation to enterprise and advance the activity of SCs [106,108].

4.1. Applications of Carbon-Based Polymer Nanocomposites (CPNCs) in Fuel Cells

A FC is crucial energy expertise that has been getting growing recognition to giving substitute environmentally concerned and extremely useful power dynamos during the prior two decades, despite modular in structure by truncated chemical and sound polluting. FCs are electrochemical methods that produce electricity continuously provided the fuel (e.g., H_2) under the appearance of an oxidant, dissimilar batteries holding yields restricted over their sustained chemical power. Furthermore, FC current densities and energy per mass or size are more important compared to different standard power origins. FCs utilise polymer electrolyte membranes (PEMs), for example, PEMFCs and direct methanol fuel cells (DMFCs), such as solution media are the usual encouraging applicants towards low-temperature performances plus mobile and compact applicability [109].

FCs are usually categorized based upon the working circumstances (e.g., temperature), construction (e.g., range of the practice and purpose), and the characteristics of the polymer electrolyte in the FC [30,110]. Table 2 summarizes the working and functional characteristics of the five foremost kinds of FCs.

Table 2. Typical characteristics of various fuel cell systems. Reproduced with permission from [30,110].

Fuel Cells	Anode Feed	Cathode Feed	Working Temp. (°C)	Power Density (mW/cm^2)	Fuel Efficiency
Alkaline fuel cell	Pure H_2	O_2 or air	90–100	100–200	60
PEMFCs	Pure H_2	O_2 or air	50–100	350	60
Phosphoric acid fuel cell	Pure H_2	O_2 or air	150–200	200	40
Molten carbonate fuel cell	H_2 or natural gas	O_2 or air	600–700	100	45–50
Solid oxide fuel cell	Gasoline or natural gas	O_2 or air	700–1000	240	60

4.2. Applications of CPNCs for Li-Ion Battery

Li-ion batteries are among the most encouraging, practical, and virtuous conventional ED methods employed for electrochemical energy storage. The Li-ion battery's efficacy informs its widespread application in electricity storage, as it shows an inherently great ED and configuration versatility. Generally employed in 'nomadic' electrical gadgets as batteries also seems to indicate that they are an essential element in decreasing CO_2 emissions, (i) as a potential energy source for progressive electric carriers and (ii) as a possible buffer energy storing method for handling the alternative renewable energy sources. Therefore, the product of plants and animals' usage of Li-ion batteries is supposed to expanding [111]. Currently, the application of Li-ion batteries in conventional electronic tools—as well as research to obtain further practical and reliable batteries—has grown remarkably. Batteries—with their notable characteristics such as better mechanical features, higher performance, and compactness—are needed for the production of handheld electronics to proceed apace with the rapidly-evolving computing ability, bigger screens, and smaller form factors of these devices. Moreover, there is a growing importance for polymer-based batteries to be combined, including elastic, smooth, and micro-scale electronics. There has been a notable improvement in the problems correlated with these batteries. The exploitation of combustible organic diluents as a conducting solution, growth of Li dendrites, and significant volume variation as an outcome of weak architectural durability are between the obstacles connected by Li-ion batteries [112].

CPs are encouraging ingredients towards organic–inorganic composites for use in Li-ion batteries, owing to their unique properties that include good coulombic proficiency and electrical performance. This encourages the design of a battery's periodical use with slight degeneration. CPs show various benefits, like excellent processability, lower price, acceptable molecular change, and light weight when employed as electrodes. Though poor endurance while driving and low conductivity within degraded state hinder their additional purposes during Li-ion batteries [113]. CPs can be applied as both anodic and cathodic substances, however, they are often employed as cathodes into Li-ion batteries. Various CPs shows considerably diverse EDs, for example, PPy-based probes have EDs of around 10–50 $W{\cdot}h{\cdot}kg^{-1}$ and PDs about 5–25 $kW{\cdot}kg^{-1}$; PANI-based probes give EDs around 50–200 $W{\cdot}h{\cdot}kg^{-1}$ and PDs of 5–50 $kW{\cdot}kg^{-1}$; PTh-based probes give EDs of 20–100 $W\ h{\cdot}kg^{-1}$ and PDs of 5–50 $kW{\cdot}kg^{-1}$ [114,115]. Cheng and co-workers employed directed PANI nanotubes incorporated by

$HClO_4$ amalgamated as anode probe substances showed better activity compared to the commercial PANI particles into Li-ion batteries [116]. The Li-PANI battery achieved a highly efficient discharge range of 75.7 mAh·g^{-1} and managed a 95.5% of the most potent discharge capacity subsequent to 80 runs. However, the PD is comparatively small, the durability of organic substances presents a severe obstacle [117].

Liu and co-workers [118] proposed the development of the design of complex nanomaterials for energy storage. Precisely, exceptional Li-battery chemistries require a standard shift over electrodes which incorporate Li to receive elements based on progress or alloying devices, where the expanded ability is usually followed by extreme volumetric variations, significant bond cleavage, poor electronic/ionic conductivity, and variable electrode/electrolyte interphase.

Among the benefits of high functioning voltage, great storing ability, lowering poisonousness, and continued running period, LIBs have grown to be the most influential and extensively employed rechargeable batteries. The electroactive organic efficient clusters within ICPs have more active redox reaction kinetics compare to common inorganic LIB probe substances. Lately, carbon-based composite probes employing CNTs and GN have been developed that have established the capability and rate recitals of the Li-ion battery. Besides, CPs, that are necessarily PPy and PANI is further used toward the Li-ion battery due to their electrical determination and chemical resistance. Here, we address the purpose of the carbon-based CP composites in the Li-ion battery with their synthesis approaches, structural modification, and electrochemical activity [10]. The CD response of Li-ion batteries depends on the "rocking-chair" notion [119], which is publicized in Figure 11.

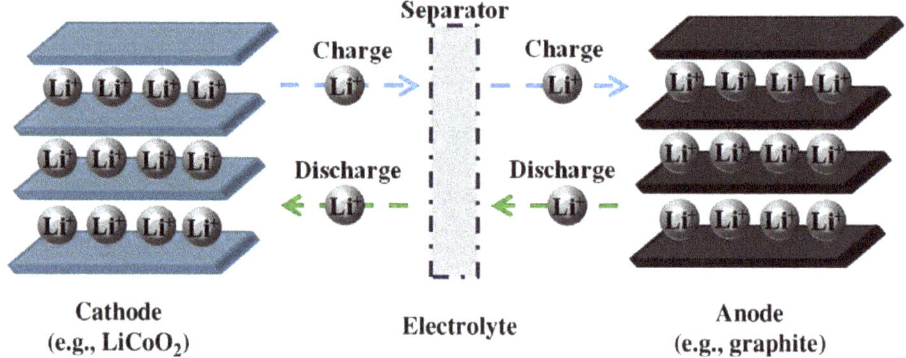

Figure 11. Diagram of the CD procedure in a Li-ion battery. Reproduced with permission from [10].

Sarang et al. [120] examined the n-type redox reaction with poly-fluorene-alt-naphthalene diimide (PFNDI) by way of an organic battery conductor with the maximum capacity noted towards the compound probe being 39.8 mAh/g at 0.5 C, with an n-doping near of 1.6 Li^+ ions/reappearance component. GN@SnS_2 heterojunction nano compound is produced by a microwave-assisted solvothermal method at liquid-phase exfoliated GN (LEGr) [121]. The storing capacity remained 664 mAh·g^{-1} over 200 subsequent runs on 300 mA·g^{-1} current density. The modification procedure and purposes of LEGr@SnS_2 amalgams are shown in Figure 12.

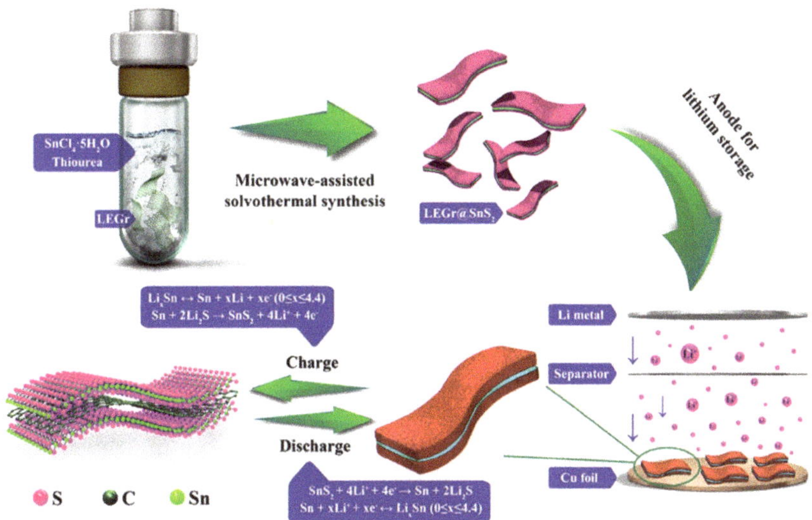

Figure 12. Schematic of the synthesis method and applications of LEGr@SnS_2 heterojunctions. Reproduced with permission from [121].

Based on prior studies, nanotechnology has been giving exceptional resolutions on battery investigation. Among several intricate nanostructure studies, investigators have grown nearer to nailing the enigmas of next-generation battery chemistries. Therefore, it is an excellent opportunity to evaluate development done up until now to extrapolate what might occur in the near future. This report strives to review the critical function of nanotechnology in superior battery methods, highlighting illustrative patterns of Si and Li metal anodes, S cathodes, and composite solid electrolytes. We next address nanomaterials for supercapacitor applications in more detail.

4.3. Applications of CPNCs for Supercapacitors

Due to their high ED and PD, SCs show prodigious latent as high-performance power foundations towards developed machinery. SCs, ultracapacitors or electrochemical capacitors (ECs), are energy-storing tools which store the energy as a charge at the probe exterior or sub-surface film, preferably within the bulk substance as under batteries. As CD transpirs at the facade, it does not cause radical structural reforms against electroactive substances, SCs hold great cycling facility. Owing to those novel characteristics, SCs are perceived as one of the maximum encouraging energy-storing designs [122].

There are two kinds of SCs: EDLCs and pseudocapacitors. In EDLCs, the energy is saved electrostatically on the probe and conducting solution edge into the double layer, although the pseudocapacitor charge storage happens through quick redox reactions at the electrode exterior. Here there are three significant kinds of conductor substances for SCs: carbon-based substances, MO_X/hydroxides, and CPs [123]. EDLCs are only at the exterior part of the carbon-based substances toward a storage charge, hence, they usually show more leading power production and strong cycling capacity. Though, EDLCs have more profound ED values than pseudocapacitors as they include active redox substances to store charge both at the exterior and at the sub-surface film [124]. While carbon-based substances, MO_X/hydroxides, and CPs are the usual electroactive resources for SCs, every kind of matter has its strengths and limitations, such as carbon-based elements having excellent PD and long life cycle, while their short C_s (mostly double layer capacitance) restricts their use in high ED tools.

MO_x/hydroxides maintain pseudocapacitance, as well as double-layer capacitance, and also have a broad charge/discharge voltage scale; however, they have a comparatively low surface area and very poor cycle period. CPs have the benefits of significant capacitance, excellent conductivity, low cost, and efficiency of modification, though they have comparatively short mechanical durability and run time [125]. Joining the unexpected benefits of these nano-scale different capacitive elements to develop nanocomposite electroactive substances is a critical path to regulate, improve, and augment the compositions and characteristics of probe substances for their SC activity. The attributes of nanocomposite electrodes depend not just in the unique ingredients employed but also on the morphology and the interfacial aspects.

Lately, significant works have been allocated to manifest every class of nanocomposite capacitive element; for example, different metal oxides, CPs, CNTs combined with CPs, and GN merged by metal oxides or CPs. Study and invention of nanocomposite electroactive substances for supercapacitors applicability require the attention of several circumstances, for example, material choice, construction techniques, modification method parameters, interfacial properties, electrical performance, nanocrystalline dimension, exterior area, etc. Although an important journey has been completed to improve nanocomposite electroactive elements for SC purposes, there are still many hurdles to be overcome [18]. The stage-wise evolution, research, and development of supercapacitors (and their properties and limitations) are shown in Figure 13 [126].

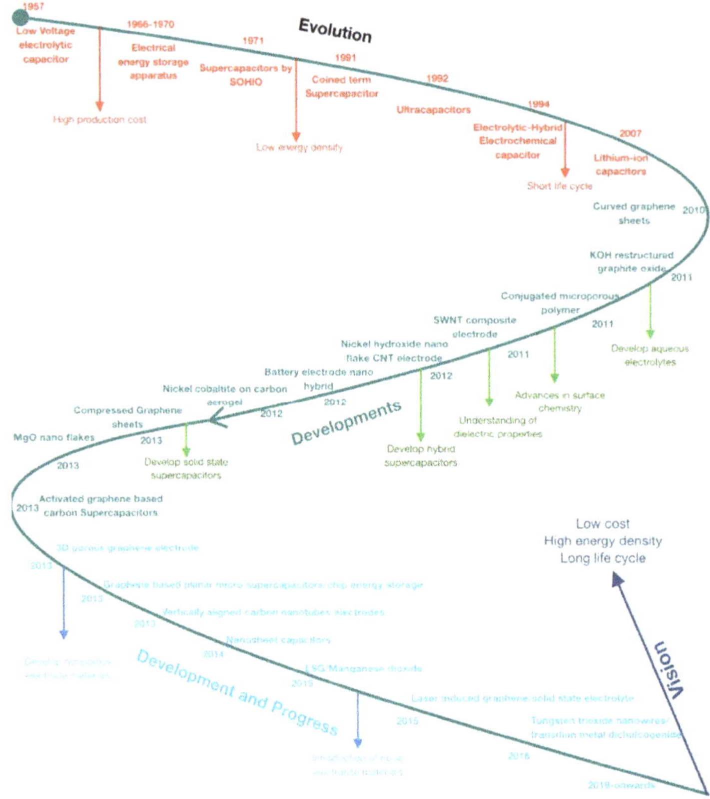

Figure 13. Roadmap of the evolution, progress, and developments of supercapacitors. Reproduced with permission from [126].

Flexible solid-state SCs (FSSCs) head within energy-storing expertise also has drawn widespread recognition due to important novel discoveries toward new wearable microelectronics. Remarkable possible useful purposes—for example, the improvement of piezoelectric, outline, retention, reconstruction, electrochromic, and combined sensor-SCs—are further explained [127].

CPs, for example—PANI, PPy, and PEDOT—are additional kinds of pseudocapacitive substances by the excessive potential to deliver outstanding C_s [128,129]. Xiao et al. [130] developed a distinct rGO/PANI/rGO double-decker composition nanohybrid paper and also investigated its potential as a probe to SSCs. Primarily, the self-supporting GN paper was developed through a print method and sparkling delamination process and demonstrated excellent electrical performance (340 $S{\cdot}cm^{-2}$), light weight, and superior mechanical characteristics. Consequently, PANI was electro polymerised at GN paper by continuous deposition of a slight GN film with a concavity layer to develop a sandwich-structured GN/PANI/GN paper. Interestingly, this novel strategy developed the energy storage ability, the rate execution and cycling durability of the probe. Therefore, the as-achieved SSC showed an exemplary capacitance around 120 $mF{\cdot}cm^{-2}$, which was affirmed at 62% following an improvement in current density on or after 0.1 to 10 $mA{\cdot}cm^{-2}$, including an ED about 5.4 $mW\ h{\cdot}cm^{-3}$.

Current localities are remaining existing at the novel technology that allowed new substances and techniques to energy storage tools. Especially, carbon-based nanomaterials such as GN, carbon nanosheets, CNTs, AC, CAGs, MO_x, CPs, and polymer amalgams have played a vital role within extremely effective supercapacitors [131]. Carbon-based electroactive [132] and PANI nanocomposite material are employed in SC applications [133]. Among the different construction and heteroatom doping, the nitrogen-doped AC matter delivered a maximum capacitance condition around 268 $F{\cdot}g^{-1}$ under symmetric SC agent in the acidic media, in addition, 226 $F{\cdot}g^{-1}$ under the organic environment with 3 V potential windows. The high ED of 60.3 $Wh{\cdot}kg^{-1}$ received within NAC based SC equipment, intimating the excellent potential for technical employment within the energy storage area [134]. GO/PANI nanomaterial, including PANI nanoparticles, consistently covered above GO films has been strongly developed by the support of CO_2. GO/PANI nanomaterial by aniline absorption on 0.1 M shows high C_s (425 $F{\cdot}g^{-1}$) on a current density of 0.2 $A{\cdot}g^{-1}$. The unique electrochemical capacitance and cycle durability owe to the combined impact among the small nanosized PANI nanocomposites and GO by high specific surface area [135].

A paired composite of GN by incorporating iron oxide (rGO/$MeFe_2O_4$) (Me = Mn, Ni) was manufactured employing a novel single-step method by NaOH employing as a coprecipitation and GO proton-rich component. The rGO/$MnFe_2O_4$ compound probe revealed gravimetric capacitance about 147 $F{\cdot}g^{-1}$ including oxidative capacitance around 232 $mF{\cdot}cm^{-2}$ on a sweep rate around 5 $mV{\cdot}s^{-1}$. The ternary GN/metal-doped iron oxide/PPy (rGO/$MnFe_2O_4$/Ppy) compound probe exhibited expressively increased gravimetric capacitance and oxidative capacitance of about 232 $F{\cdot}g^{-1}$ and 395 $mF{\cdot}cm^{-2}$, correspondingly showing the combined effect of PPy additives and the corresponding studies shown in Figure 14 [106].

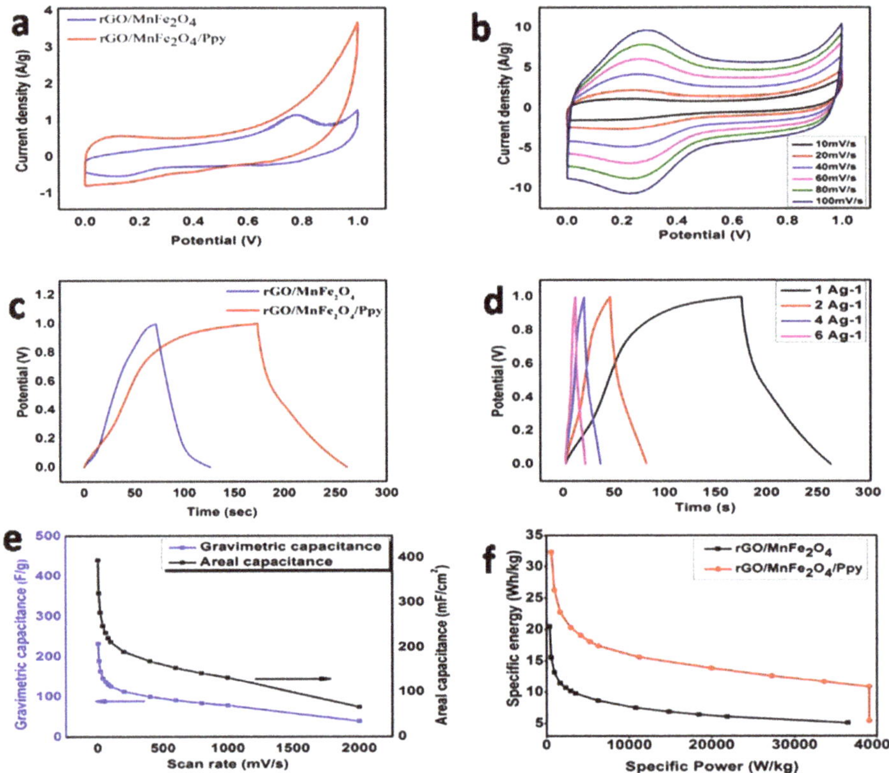

Figure 14. (**a**) CV curves at 5 mV·s^{-1}. (**b**) CV curves for rGO/$MnFe_2O_4$/Ppy nanomaterial on various sweep rates. (**c**) CD curves on 1 A·g^{-1}. (**d**) CD curves for rGO/$MnFe_2O_4$/Ppy on dissimilar current density. (**e**) Gravimetric and areal capacitance of rGO/$MnFe_2O_4$/Ppy on dissimilar sweep rates (5–2000 mV·s^{-1}) (**f**) Ragon designs of rGO/$MnFe_2O_4$ plus rGO/$MnFe_2O_4$/Ppy on dissimilar sweep rates (5–2000 mV·s^{-1}). Reproduced with permission from [106].

Lately, amalgams of polymers plus nanofillers, for example, carbon-based substances have been favorably accepted as probes toward enhancing the activity of SCs applying the high synergistic impact. Biswas et al. [136] integrated GN/PPy composite material presenting gravimetric capacitance about 165 F·g^{-1} on 1 A·g^{-1} current density contained within two conductor's arrangements though applying 1 M NaCl media. Parl et al. [137] adopted graphite/PPy compound towards SC electrodes exhibiting gravimetric capacitance around 400 F·g^{-1} held with three probes arrangement. With the aim of obtaining superior SCs activity, the idea of the ternary operation with connecting these three segments has been introduced [138]. The synthesized ternary PPy/GO/ZnO SC electrodes by calculated its gravimetric capacitance within two probes arrangement to be 94.6 F·g^{-1} on 1 A·g^{-1} by CD arcs. Furthermore, Lim et al. [139] described the ternary PPy/GN/nano MnO_x complex, the gravimetric capacitance of manufactured compound was 320.6 F·g^{-1} on 1 mV·s^{-1} that was abundant advanced compare to the PPy/GN and straight PPy carrying gravimetric capacitance of 255.1 F·g^{-1} and 118.4 F·g^{-1}, respectively. Xiong and co-workers [140] applied three probes method to estimate gravimetric capacitance of ternary cobalt ferrite/GN/PANI nanomaterials, that conferred gravimetric capacitance of about 1133.3 F·g^{-1} on the sweep rate at 1 mV·s^{-1}. These investigations show that the composition of multi-ingredient compound terminals to SCs is a useful and encouraging path that can suggestively advance the activity of SCs.

The combination of PPy/CNT has been employed as an assuring pseudocapacitive cathode toward non-aqueous LIC purposes. The PPy gives high pseudocapacitance through the doping/undoping effect and the CNT incorporation significantly improves electrical performance. The as-developed composite gives exceptional capacities and steadiness (98.7 $mAh{\cdot}g^{-1}$ on 0.1 $A{\cdot}g^{-1}$ plus hold 89.7% following runs on 3 $A{\cdot}g^{-1}$ for 1000 cycles within Li-half cell), that exceeds porous carbon negatives within existing LICs. Moreover, while joined by Fe_3O_4@C positive electrode, the as-developed LICs manifests a greater ED around 101.0 $Wh{\cdot}kg^{-1}$ on 2709 $Wh{\cdot}kg^{-1}$, and yet preserves 70 $Wh{\cdot}kg^{-1}$ through improved PD of 17,186 $W{\cdot}kg^{-1}$ [141].

Recently, carbon nanocomposites (particularly, CNTs plus GN) have been extensively examined as active electrodes in SCs due to their excellent specific exterior area and outstanding electrical and mechanical traits. Current investigation and expansion evidently specify that high-performance SCs can be equipped through electrodes based on vertically-aligned CNTs by unlocked tips, GN sheets with tenable through-thickness π–π stacking connections and/or edge functionalities, and 3D pillared GN-CNT systems. The various materials and their supercapacitor performances are given below in Table 3.

Table 3. Various materials and their supercapacitive performance

Substances	Specific Capacitance (C_s)	Cycling Durability	Ref.
CNT	1.8 $mF{\cdot}cm^{-2}$ on 1 mA	80%; 1000 runs	[142]
MnO_2 nanowire/GN	66.1 $F{\cdot}cm^{-3}$ on 60 $mA{\cdot}cm^{-3}$	96%; 10,000 runs on 0.12 $A{\cdot}cm^{-3}$	[143]
AC cloth	161.2 $mF{\cdot}cm^{-2}$ on 12.5 $mA{\cdot}cm^{-2}$	104%; 30,000 runs on 12.5 $mA{\cdot}cm^{-2}$	[144]
AC	153 $mF{\cdot}cm^{-2}$ on 10 $mV{\cdot}s^{-1}$	93.4%; 1000 runs on 200 $mV{\cdot}s^{-1}$	[145]
PANI hydrogel	430 $F{\cdot}g^{-1}$ on 5 $mV{\cdot}s^{-1}$	86%; 1000 runs on 7.5 $A{\cdot}g^{-1}$	[146]
GN	180.40 $mF{\cdot}cm^{-2}$ on 1 $mA{\cdot}cm^{-2}$	96.8%; 7500 runs on 8 $mA{\cdot}cm^{-2}$	[147]
TiO_2@PANI	775.6 $mF{\cdot}cm^{-3}$ (28.3 $F{\cdot}g^{-1}$) on 10 $mV{\cdot}s^{-1}$	97.2%; 10,000 runs on 100 $mV{\cdot}s^{-1}$	[147]
$NiCo_2O_4$@CNT/CNT	-	95%; 5000 runs on 50 $mV{\cdot}s^{-1}$	[148]
MnO_2/rGO	14 $F{\cdot}cm^{-2}$ (31.8 $F{\cdot}g^{-1}$) on 2 $mV{\cdot}s^{-1}$	100%; 5000 runs on 0.2 $mA{\cdot}cm^{-2}$	[149]
SnS/S doped GN	2.98 $mF{\cdot}cm^{-2}$ on 60 $mA{\cdot}cm^{-2}$	99%; 10,000 runs on 120 $mA{\cdot}m^{-2}$	[150]
Co_3O_4/vertically aligned GN nanosheets	580 $F{\cdot}g^{-1}$ on 1 $A{\cdot}g^{-1}$	86.3%; 20,000 runs on 20 $A{\cdot}g^{-1}$	[151]
MnO_2-CNT-GN	107 $F{\cdot}g^{-1}$	-	[152]
N/O-Enhanced carbon cloth	-	116%; 5000 runs on 5 $mA{\cdot}cm^{-1}$	[153]
MnO_2/CNT	324 $F{\cdot}g^{-1}$ on 0.5 $A{\cdot}g^{-1}$	100%; 5000 runs on 10 $A{\cdot}g^{-1}$	[154]
GN sheets	11.3 $mF{\cdot}cm^{-2}$ on 1 $mV{\cdot}s^{-1}$	-	[155]
PANI-MnO_x	94.73 $mF{\cdot}cm^{-2}$ on 0.1 $mA{\cdot}cm^{-2}$	-	[156]
Graphite/PANI	77.8 $mF{\cdot}cm^{-2}$ on 0.1 $mA{\cdot}cm^{-2}$	83%; 10,000 runs on 1 $mA{\cdot}cm^{-2}$	[157]
KCu_7S_4/GN	-	92%; 5000 runs on 0.8 $mA{\cdot}cm^{-2}$	[158]
Ag/AC	45 $mF{\cdot}cm^{-2}$ on 0.3 $mA{\cdot}cm^{-2}$	86%; 1200 runs on 5 $mA{\cdot}cm^{-2}$	[159]
GN/MWNT	740.9 $mF{\cdot}cm^{-2}$ on 1 $mA{\cdot}cm^{-2}$	85%; 20,000 runs on 15 $mA{\cdot}cm^{-2}$	[160]
SWCNTs	17.5 $F{\cdot}g^{-1}$ on 2 $A{\cdot}g^{-1}$	87.5%; 10,000 runs on 5 $A{\cdot}g^{-1}$	[161]
Au/PANI	26.49 $mF{\cdot}cm^{-2}$ (67.06 $F{\cdot}cm^{-3}$) on 0.5 $mA{\cdot}cm^{-2}$	72.7%; 1000 runs on 200 $mV{\cdot}s^{-1}$	[162]
GO/MOF	250 $mF{\cdot}cm^{-3}$ on 6.4 $mA{\cdot}cm^{-3}$	96.3%; 5000 runs on 50.4 $mA{\cdot}cm^{-3}$	[163]
rGO/PPy	147.9 $F{\cdot}cm^{-3}$ on 5 $A{\cdot}cm^{-3}$	71.7%; 5000 runs on 10 $A{\cdot}cm^{-3}$	[164]
N-Doped rGO	3.4 $mF{\cdot}cm^{-2}$ on 20 $mA{\cdot}cm^{-2}$	98.4%; 2000 runs on 100 $mA{\cdot}cm^{-2}$	[165]

Table 3. *Cont.*

Substances	Specific Capacitance (C_s)	Cycling Durability	Ref.
rGO/MoO_3	404 $F{\cdot}g^{-1}$ on 0.5 $A{\cdot}g^{-1}$	80%; 5000 runs on 2 $A{\cdot}g^{-1}$	[166]
GN	56.5 $F{\cdot}cm^{-3}$ on 0.06 $A{\cdot}cm^{-3}$	-	[167]
b-$Ni(OH)_2$/graphene	2570 $mF{\cdot}cm^{-2}$ on 0.2 $A{\cdot}m^{-1}$	98.2%; 2000 runs on 0.1 $A{\cdot}m^{-2}$	[168]
GN/carbon black nanoparticle	144.5 $F{\cdot}g^{-1}$ on the current density of 0.5 $A{\cdot}g^{-1}$	-	[169]
GN fibers/MnO_2 fibers	42.02 $mF{\cdot}cm^{-2}$ on 0.01 $V{\cdot}s^{-1}$	92%; 1000 runs on 1 $mA{\cdot}cm^{-2}$	[170]
GO	130 $F{\cdot}g^{-1}$ on 5 $mV{\cdot}s^{-1}$	-	[171]
$Cu(OH)_2$//AC	26.4 $F{\cdot}g^{-1}$ on 4 $A{\cdot}g^{-1}$	90%; 5000 runs	[172]
$ZnCo_2O_4$//carbon nanofibers	139.2 $F{\cdot}g^{-1}$ on 2 $mV{\cdot}s^{-1}$	90%; 3000 runs on 50 $mV{\cdot}s^{-1}$	[173]
$CuMnO_2$-gCN	817.85 $F{\cdot}g^{-1}$ at 0.025 $A{\cdot}g^{-1}$	91% up to 1000 cycles	[174]

5. Conclusions and Future Prospects

This review article compiles the state-of-the-art and novel hurdles in carbon-based polymer nanocomposites regarding high-performance energy storage applications. Focusing on the possible studies to improve the electrocatalytic behavior of these combined multiphase nanocomposites and principally at the connection among the composition and processing of carbon-based polymer nanocomposites in terms of exploiting the electrochemical efficiency are described.

The fast consumption of fossil fuels and raised contamination has led to the expansion of energy transformation and storage tools. Fuel cells, Li-ion batteries, and supercapacitors are the possible candidates in these applications. Here, in the event of a fuel cell film, the addition of nanomaterials over polymer membranes improves the physical characteristics as well the proton conductivity, and activity of the membranes means that the nanomaterial does not act as a block to proton movement of the anode to the cathode. This can be accomplished by employing proper incorporated and developing the degree of distribution. Notwithstanding lots of studies which have previously been conducted herein area, up to now, no invention has been completed, and are much more necessary to be fulfilled before the edge use.

Currently, CNTs are used as a substitute to graphite-based anodes. CNT-based anodes with tiny advances in Li storage potential and cyclability associated with graphite-based anodes make it challenging to practice them for substituting graphite-based materials in Li-ion substance. It seems that the numerous encouraging efforts toward developing different energy storage devices may originate from the incorporation of CNTs with another composite with intricate nanostructure schemes. CPNCs as an electrode substance have also been studied. Additional investigations are needed to develop their long-term durability. Mostly, carbon-based materials are being accepted as for supercapacitor electrodes.

GN-based substances are admittedly one of the impressive materials with high potential within the active area of energy storage devices. It has shown great specific capacitance together with polymeric nanocomposite. In practice, GN has been considered the perfect supercapacitor electrode substance due to its vast exterior area, notably high electrical conductivity, and high mechanical robustness. Traditionally, however, extensive works are yet required to apply this potential material into an extremely effective element.

Although the progress has significantly delivered more reliable storage abilities and achievement, tailoring, and optimization of substances are yet in a developmental manner where cost-effective and great execution are of the important interests in functional applicability. To generate a succession of high PD and EDs, hybrid technology is one of the exciting features. For that, we require a much perception of surface chemistry among electrode substances and electrolytes to enhance the interfacial synergies beneficial for improved charge transfer. Nanoarchitectures play an important part in building a structure feature synergy among separate elements, for example, CPNCs. Therefore, integrative directed elements require to be intensively examined to achieve an optimized high-performance energy storage materials that can work with the energy crunch we are suffering now.

Author Contributions: Writing—original draft, S.S.S., and N.D.; Writing—review and editing, S.S.S., N.D., Q.Z., and V.K.T.; Funding acquisition, S.S.S., and Q.Z. All authors have read and agreed to the published version of the manuscript.

Funding: This research work was funded by financial support from the National Natural Science Foundation of China (21962008, 51464028), Candidate Talents Training Fund of Yunnan Province (2017PY269SQ, 2018HB007), and Yunnan Ten Thousand Talents Plan Young & Elite Talents Project (YNWR-QNBJ-2018-346).

Acknowledgments: We are grateful for the financial support from the National Natural Science Foundation of China, Candidate Talents Training Fund of Yunnan Province and Yunnan Ten Thousand Talents Plan Young & Elite Talents Project.

Conflicts of Interest: There is no conflict to declare.

References

1. Davis, S.E.; Ide, M.S.; Davis, R.J.; Bowman, S. Selective oxidation of alcohols and aldehydes over supported metal nanoparticles. *Green Chem.* **2013**, *15*, 17–45. [CrossRef]
2. Siwal, S.; Devi, N.; Perla, V.K.; Ghosh, S.K.; Mallick, K. Promotional role of gold in electrochemical methanol oxidation. *Catal. Struct. React.* **2019**, *5*, 1–9. [CrossRef]
3. Siwal, S.; Devi, N.; Perla, V.; Barik, R.; Ghosh, S.; Mallick, K. The influencing role of oxophilicity and surface area of the catalyst for electrochemical methanol oxidation reaction: A case study. *Mater. Res. Innov.* **2018**, *23*, 440–447. [CrossRef]
4. Samarjeet, S.; Sarit, G.; Debkumar, N.; Nishu, D.; Venkata, K.P.; Rasmita, B.; Malick, K. Synergistic effect of graphene oxide on the methanol oxidation for fuel cell application. *Mater. Res. Exp.* **2017**, *4*, 095306. [CrossRef]
5. Siwal, S.; Matseke, S.; Mpelane, S.; Hooda, N.; Nandi, D.; Mallick, K. Palladium-polymer nanocomposite: An anode catalyst for the electrochemical oxidation of methanol. *Int. J. Hydrogen Energy* **2017**, *42*, 23599–23605. [CrossRef]
6. Siwal, S.S.; Choudhary, M.; Mpelane, S.; Brink, R.; Mallick, K. Single step synthesis of a polymer supported palladium composite: A potential anode catalyst for the application of methanol oxidation. *RSC Adv.* **2016**, *6*, 47212–47219. [CrossRef]
7. Lei, H.; Li, X.; Sun, C.; Zeng, J.; Siwal, S.S.; Zhang, Q. Galvanic Replacement–Mediated Synthesis of Ni-Supported Pd Nanoparticles with Strong Metal–Support Interaction for Methanol Electro-oxidation. *Small* **2019**, *15*, 1804722. [CrossRef]
8. Siwal, S.S.; Thakur, S.; Zhang, Q.; Thakur, V.K. Electrocatalysts for electrooxidation of direct alcohol fuel cell: Chemistry and applications. *Mater. Today Chem.* **2019**, *14*, 100182. [CrossRef]
9. Choudhary, M.; Siwal, S.; Nandi, D.; Mallick, K. Charge storage ability of the gold nanoparticles: Towards the performance of a supercapacitor. *Appl. Surf. Sci.* **2017**, *424*, 151–156. [CrossRef]
10. Li, Y.; Ye, D. Carbon-Based Polymer Nanocomposite for Lithium-Ion Batteries. In *Carbon-Based Polymer Nanocomposites for Environmental and Energy Applications*; Elsevier: Amsterdam, The Netherlands, 2018; pp. 537–557.
11. Abbasi, H.; Antunes, M.; Velasco, J.I. Recent advances in carbon-based polymer nanocomposites for electromagnetic interference shielding. *Prog. Mater. Sci.* **2019**, *103*, 319–373. [CrossRef]
12. Mohan, V.B.; Lau, K.T.; Hui, D.; Bhattacharyya, D. Graphene-based materials and their composites: A review on production, applications and product limitations. *Compos. Part B Eng.* **2018**, *142*, 200–220. [CrossRef]
13. Dang, Z.-M.; Yuan, J.; Yao, S.-H.; Liao, R.-J. Flexible Nanodielectric Materials with High Permittivity for Power Energy Storage. *Adv. Mater.* **2013**, *25*, 6334–6365. [CrossRef] [PubMed]
14. Devi, N.; Ghosh, S.K.; Perla, V.K.; Pal, T.; Mallick, K. Laboratory based synthesis of the pure form of gananite (BiF3) nanoparticles: A potential material for electrochemical supercapacitor application. *New J. Chem.* **2019**, *43*, 18369–18376. [CrossRef]
15. Al-Saleh, M. Electrically conductive carbon nanotube/polypropylene nanocomposite with improved mechanical properties. *Mater. Des.* **2015**, *85*, 76–81. [CrossRef]

16. Tang, H.; Chen, G.-X.; Li, Q. Epoxy-based high-k composites with low dielectric loss caused by reactive core-shell-structured carbon nanotube hybrids. *Mater. Lett.* **2016**, *184*, 143–147. [CrossRef]
17. Romasanta, J.L.; Santana, M.H.; López-Manchado, M.A.; Verdejo, R. Functionalised graphene sheets as effective high dielectric constant fillers. *Nanoscale Res. Lett.* **2011**, *6*, 508. [CrossRef]
18. Wang, F.; Wang, H.; Mao, J. Aligned-graphene composites: A review. *J. Mater. Sci.* **2018**, *54*, 36–61. [CrossRef]
19. Baek, J.E.; Kim, J.Y.; Jin, H.M.; Kim, B.H.; Lee, K.E.; Kim, S.O. Single-step self-assembly of multilayer graphene based dielectric nanostructures. *FlatChem* **2017**, *4*, 61–67. [CrossRef]
20. Lin, B.; Li, Z.-T.; Yang, Y.; Li, Y.; Lin, J.-C.; Zheng, X.-M.; He, F.-A.; Lam, K.H. Enhanced dielectric permittivity in surface-modified graphene/PVDF composites prepared by an electrospinning-hot pressing method. *Compos. Sci. Technol.* **2019**, *172*, 58–65. [CrossRef]
21. Mohanapriya, M.K.; Deshmukh, K.; Chidambaram, K.; Ahamed, M.B.; Sadasivuni, K.K.; Ponnamma, D.; AlMaadeed, M.A.-A.; Deshmukh, R.R.; Pasha, S.K.K. Polyvinyl alcohol (PVA)/polystyrene sulfonic acid (PSSA)/carbon black nanocomposite for flexible energy storage device applications. *J. Mater. Sci. Mater. Electron.* **2017**, *28*, 6099–6111. [CrossRef]
22. Fan, Z.; Wang, D.; Yuan, Y.; Wang, Y.; Cheng, Z.; Liu, Y.; Xie, Z. A lightweight and conductive MXene/graphene hybrid foam for superior electromagnetic interference shielding. *Chem. Eng. J.* **2020**, *381*, 122696. [CrossRef]
23. Al-Saleh, M. Carbon-based polymer nanocomposites as dielectric energy storage materials. *Nanotechnology* **2018**, *30*, 062001. [CrossRef] [PubMed]
24. Cai, C.; Liu, L.; Fu, Y. Processable conductive and mechanically reinforced polylactide/graphene bionanocomposites through interfacial compatibilizer. *Polym. Compos.* **2017**, *40*, 389–400. [CrossRef]
25. Beaudin, M.; Zareipour, H.; Schellenberglabe, A.; Rosehart, W. Energy storage for mitigating the variability of renewable electricity sources: An updated review. *Energy Sustain. Dev.* **2010**, *14*, 302–314. [CrossRef]
26. Arani, A.A.K.; Karami, H.; Soleymani, S.; Hejazi, M. Review of Flywheel Energy Storage Systems structures and applications in power systems and microgrids. *Renew. Sustain. Energy Rev.* **2017**, *69*, 9–18. [CrossRef]
27. Elliman, R.; Gould, C.; Al-Tai, M. Review of current and future electrical energy storage devices. In Proceedings of the 2015 50th International Universities Power Engineering Conference (UPEC), Stoke on Trent, UK, 1–4 September 2015; pp. 1–5.
28. Hannan, M.; Hoque, M.; Mohamed, A.; Ayob, A. Review of energy storage systems for electric vehicle applications: Issues and challenges. *Renew. Sustain. Energy Rev.* **2017**, *69*, 771–789. [CrossRef]
29. Hossain, S.; Hoque, M. Polymer nanocomposite materials in energy storage: Properties and applications. In *Polymer-based Nanocomposites for Energy and Environmental Applications*; Elsevier: Amsterdam, The Netherlands, 2018; pp. 239–282.
30. Winter, M.; Brodd, R.J. What Are Batteries, Fuel Cells, and Supercapacitors? *Chem. Rev.* **2004**, *104*, 4245–4270. [CrossRef]
31. Rolison, D.R.; Long, J.W.; Lytle, J.C.; Fischer, A.E.; Rhodes, C.P.; McEvoy, T.M.; Bourg, M.E.; Lubers, A.M. Multifunctional 3D nanoarchitectures for energy storage and conversion. *Chem. Soc. Rev.* **2009**, *38*, 226–252. [CrossRef]
32. Zuo, W.; Li, R.; Zhou, C.; Li, Y.; Xia, J.; Liu, X. Battery-Supercapacitor Hybrid Devices: Recent Progress and Future Prospects. *Adv. Sci.* **2017**, *4*, 1600539. [CrossRef]
33. Xiu, Y.; Cheng, L.; Chunyan, L. Research on Hybrid Energy Storage System of Super-capacitor and Battery Optimal Allocation. *J. Int. Counc. Electr. Eng.* **2014**, *4*, 341–347. [CrossRef]
34. Manzetti, S.; Mariasiu, F. Electric vehicle battery technologies: From present state to future systems. *Renew. Sustain. Energy Rev.* **2015**, *51*, 1004–1012. [CrossRef]
35. Hoffert, M.I.; Caldeira, K.; Benford, G.; Criswell, D.R.; Green, C.; Herzog, H.; Jain, A.K.; Kheshgi, H.S.; Lackner, K.S.; Lewis, J.S.; et al. Advanced Technology Paths to Global Climate Stability: Energy for a Greenhouse Planet. *Science* **2002**, *298*, 981–987. [CrossRef] [PubMed]
36. Chauhan, A.; Saini, R. A review on Integrated Renewable Energy System based power generation for stand-alone applications: Configurations, storage options, sizing methodologies and control. *Renew. Sustain. Energy Rev.* **2014**, *38*, 99–120. [CrossRef]
37. Hossain, A.; Bandyopadhyay, P.; Guin, P.S.; Roy, S. Recent developed different structural nanomaterials and their performance for supercapacitor application. *Appl. Mater. Today* **2017**, *9*, 300–313. [CrossRef]
38. Jagadale, A.; Zhou, X.; Xiong, R.; Dubal, D.; Xu, J.; Yang, S. Lithium ion capacitors (LICs): Development of the materials. *Energy Storage Mater.* **2019**, *19*, 314–329. [CrossRef]

39. Lin, Z.; Goikolea, E.; Balducci, A.; Naoi, K.; Taberna, P.; Salanne, M.; Yushin, G.; Simon, P. Materials for supercapacitors: When Li-ion battery power is not enough. *Mater. Today* **2018**, *21*, 419–436. [CrossRef]
40. Wang, Y.; Fu, X.; Zheng, M.; Zhong, W.-H.; Cao, G. Strategies for Building Robust Traffic Networks in Advanced Energy Storage Devices: A Focus on Composite Electrodes. *Adv. Mater.* **2018**, *31*, 1700322. [CrossRef]
41. Miller, E.E.; Hua, Y.; Tezel, F.H. Materials for energy storage: Review of electrode materials and methods of increasing capacitance for supercapacitors. *J. Energy Storage* **2018**, *20*, 30–40. [CrossRef]
42. Muzaffar, A.; Ahamed, M.B.; Deshmukh, R.; Thirumalai, J. A review on recent advances in hybrid supercapacitors: Design, fabrication and applications. *Renew. Sustain. Energy Rev.* **2019**, *101*, 123–145. [CrossRef]
43. Farzana, R.; Rajarao, R.; Bhat, B.R.; Sahajwalla, V. Performance of an activated carbon supercapacitor electrode synthesised from waste Compact Discs (CDs). *J. Ind. Eng. Chem.* **2018**, *65*, 387–396. [CrossRef]
44. Tian, J.; Wu, S.; Yin, X.; Wu, W. Novel preparation of hydrophilic graphene/graphene oxide nanosheets for supercapacitor electrode. *Appl. Surf. Sci.* **2019**, *496*. [CrossRef]
45. Zhou, Y.; Zhou, X.; Ge, C.; Zhou, W.; Zhu, Y.; Xu, B. Branched carbon nanotube/carbon nanofiber composite for supercapacitor electrodes. *Mater. Lett.* **2019**, *246*, 174–177. [CrossRef]
46. Genç, R.; Alaş, M.Ö.; Harputlu, E.; Repp, S.; Kremer, N.; Castellano, M.; Colak, S.G.; Ocakoglu, K.; Erdem, E. High-Capacitance Hybrid Supercapacitor Based on Multi-Colored Fluorescent Carbon-Dots. *Sci. Rep.* **2017**, *7*, 11222. [CrossRef]
47. Saliger, R.; Fischer, U.; Herta, C.; Fricke, J. High surface area carbon aerogels for supercapacitors. *J. Non-Cryst. Solids* **1998**, *225*, 81–85. [CrossRef]
48. Al Sheheri, S.Z.; Al-Amshany, Z.M.; Al Sulami, Q.A.; Tashkandi, N.Y.; Hussein, M.; El-Shishtawy, R.M. The preparation of carbon nanofillers and their role on the performance of variable polymer nanocomposites. *Des. Monomers Polym.* **2019**, *22*, 8–53. [CrossRef]
49. Chen, X.; Paul, R.; Dai, L. Carbon-based supercapacitors for efficient energy storage. *Natl. Sci. Rev.* **2017**, *4*, 453–489. [CrossRef]
50. Chen, T.; Dai, L. Carbon nanomaterials for high-performance supercapacitors. *Mater. Today* **2013**, *16*, 272–280. [CrossRef]
51. Yang, C.; Liu, P.; Wang, T. Well-Defined Core–Shell Carbon Black/Polypyrrole Nanocomposites for Electrochemical Energy Storage. *ACS Appl. Mater. Interfaces* **2011**, *3*, 1109–1114. [CrossRef]
52. Hahm, M.G.; Reddy, A.L.M.; Cole, D.P.; Rivera, M.; Vento, J.A.; Nam, J.; Jung, H.Y.; Kim, Y.L.; Narayanan, N.T.; Hashim, D.P.; et al. Carbon Nanotube–Nanocup Hybrid Structures for High Power Supercapacitor Applications. *Nano Lett.* **2012**, *12*, 5616–5621. [CrossRef]
53. El-Kady, M.F.; Shao, Y.; Kaner, R.B. Graphene for batteries, supercapacitors and beyond. *Nat. Rev. Mater.* **2016**, *1*, 16033. [CrossRef]
54. Ping, Y.; Gong, Y.; Fu, Q.; Pan, C. Preparation of three-dimensional graphene foam for high performance supercapacitors. *Prog. Nat. Sci.* **2017**, *27*, 177–181. [CrossRef]
55. Yanilmaz, M.; Dirican, M.; Asiri, A.M.; Zhang, X. Flexible polyaniline-carbon nanofiber supercapacitor electrodes. *J. Energy Storage* **2019**, *24*, 100766. [CrossRef]
56. Qu, D.; Wang, L.; Zheng, D.; Xiao, L.; Deng, B.; Qu, D. An asymmetric supercapacitor with highly dispersed nano-Bi2O3 and active carbon electrodes. *J. Power Sources* **2014**, *269*, 129–135. [CrossRef]
57. Godse, L.; Karandikar, P.; Khaladkar, M. Study of Carbon Materials and Effect of its Ball Milling, on Capacitance of Supercapacitor. *Energy Procedia* **2014**, *54*, 302–309. [CrossRef]
58. Crane, M.J.; Lim, M.B.; Zhou, X.; Pauzauskie, P.J. Rapid synthesis of transition metal dichalcogenide-carbon aerogel composites for supercapacitor electrodes. *Microsyst. Nanoeng.* **2017**, *3*. [CrossRef] [PubMed]
59. He, S.; Hu, Y.; Wan, J.; Gao, Q.; Wang, Y.; Xie, S.; Qiu, L.; Wang, C.; Zheng, G.; Wang, B.; et al. Biocompatible carbon nanotube fibers for implantable supercapacitors. *Carbon* **2017**, *122*, 162–167. [CrossRef]
60. Ke, Q.; Wang, J. Graphene-based materials for supercapacitor electrodes—A review. *J. Mater.* **2016**, *2*, 37–54. [CrossRef]
61. Bose, S.; Kuila, T.; Mishra, A.K.; Rajasekar, R.; Kim, N.H.; Sharma, K. Carbon-based nanostructured materials and their composites as supercapacitor electrodes. *J. Mater. Chem.* **2012**, *22*, 767–784. [CrossRef]
62. Wang, S.; Jin, C.; Qian, W. Bi_2O_3 with activated carbon composite as a supercapacitor electrode. *J. Alloy. Compd.* **2014**, *615*, 12–17. [CrossRef]

63. Wang, G.; Zhang, L.; Zhang, J. A review of electrode materials for electrochemical supercapacitors. *Chem. Soc. Rev.* **2012**, *41*, 797–828. [CrossRef]
64. Shao, Y.; El-Kady, M.F.; Wang, L.J.; Zhang, Q.; Li, Y.; Wang, H.; Mousavi, M.F.; Kaner, R.B. Graphene-based materials for flexible supercapacitors. *Chem. Soc. Rev.* **2015**, *44*, 3639–3665. [CrossRef] [PubMed]
65. Nithya, V.; Hanitha, B.; Surendran, S.; Kalpana, D.; Selvan, R.K. Effect of pH on the sonochemical synthesis of $BiPO_4$ nanostructures and its electrochemical properties for pseudocapacitors. *Ultrason. Sonochem.* **2015**, *22*, 300–310. [CrossRef] [PubMed]
66. Liu, Y.; Kumar, S. Polymer/Carbon Nanotube Nano Composite Fibers—A Review. *ACS Appl. Mater. Interfaces* **2014**, *6*, 6069–6087. [CrossRef] [PubMed]
67. Han, J.T.; Cho, J.Y.; Kim, J.H.; Jang, J.I.; Kim, J.S.; Lee, H.J.; Park, J.H.; Chae, J.S.; Roh, H.-K.; Lee, W.; et al. Structural Recovery of Highly Oxidized Single-Walled Carbon Nanotubes Fabricated by Kneading and Electrochemical Applications. *Chem. Mater.* **2019**, *31*, 3468–3475. [CrossRef]
68. Park, J.W.; Park, S.J.; Kwon, O.S.; Lee, C.; Jang, J.; Jang, J.; Park, J.W.; Park, S.J.; Lee, C.; Kwon, O.S. In Situ Synthesis of Graphene/Polyselenophene Nanohybrid Materials as Highly Flexible Energy Storage Electrodes. *Chem. Mater.* **2014**, *26*, 2354–2360. [CrossRef]
69. Krishnamoorthy, M.; Jha, N. Oxygen-Rich Hierarchical Porous Graphene as an Excellent Electrode for Supercapacitors, Aqueous Al-Ion Battery, and Capacitive Deionization. *ACS Sustain. Chem. Eng.* **2019**, *7*, 8475–8489. [CrossRef]
70. Guan, L.; Pan, L.; Peng, T.; Gao, C.; Zhao, W.; Yang, Z.; Hu, H.; Wu, M. Synthesis of Biomass-Derived Nitrogen-Doped Porous Carbon Nanosheests for High-Performance Supercapacitors. *ACS Sustain. Chem. Eng.* **2019**, *7*, 8405–8412. [CrossRef]
71. Manuel, J.; Zhao, X.; Cho, K.-K.; Kim, J.-K.; Ahn, J.-H. Ultralong Life Organic Sodium Ion Batteries Using a Polyimide/Multiwalled Carbon Nanotubes Nanocomposite and Gel Polymer Electrolyte. *ACS Sustain. Chem. Eng.* **2018**, *6*, 8159–8166. [CrossRef]
72. Azman, N.H.N.; Nazir, S.M.M.; Ngee, L.H.; Sulaiman, Y. Graphene-based ternary composites for supercapacitors. *Int. J. Energy Res.* **2018**, *42*, 2104–2116. [CrossRef]
73. Park, J.; Cho, Y.S.; Sung, S.J.; Byeon, M.; Yang, S.J.; Park, C.R. Characteristics tuning of graphene-oxide-based-graphene to various end-uses. *Energy Storage Mater.* **2018**, *14*, 8–21. [CrossRef]
74. Hou, D.; Liu, Q.; Wang, X.; Quan, Y.; Qiao, Z.; Yu, L.; Ding, S. Facile synthesis of graphene via reduction of graphene oxide by artemisinin in ethanol. *J. Mater.* **2018**, *4*, 256–265. [CrossRef]
75. Zaaba, N.; Foo, K.L.; Hashim, U.; Tan, S.; Liu, W.W.; Voon, C. Synthesis of Graphene Oxide using Modified Hummers Method: Solvent Influence. *Procedia Eng.* **2017**, *184*, 469–477. [CrossRef]
76. Qorbani, M.; Esfandiar, A.; Mehdipour, H.; Chaigneau, M.; Zad, A.I.; Moshfegh, A.Z. Shedding Light on Pseudocapacitive Active Edges of Single-Layer Graphene Nanoribbons as High-Capacitance Supercapacitors. *ACS Appl. Energy Mater.* **2019**, *2*, 3665–3675. [CrossRef]
77. Pykal, M.; Langer, M.; Prudilova, B.B.; Banas, P.; Otyepka, M. Ion Interactions across Graphene in Electrolyte Aqueous Solutions. *J. Phys. Chem. C* **2019**, *123*, 9799–9806. [CrossRef]
78. Liu, X.; Lai, C.; Xiao, Z.; Zou, S.; Liu, K.; Yin, Y.; Liang, T.; Wu, Z. Superb Electrolyte Penetration/Absorption of Three-Dimensional Porous Carbon Nanosheets for Multifunctional Supercapacitor. *ACS Appl. Energy Mater.* **2019**, *2*, 3185–3193. [CrossRef]
79. Devadas, B.; Imae, T. Effect of Carbon Dots on Conducting Polymers for Energy Storage Applications. *ACS Sustain. Chem. Eng.* **2017**, *6*, 127–134. [CrossRef]
80. Qian, H.; Kucernak, A.R.; Greenhalgh, E.S.; Bismarck, A.; Shaffer, M.S.P. Multifunctional Structural Supercapacitor Composites Based on Carbon Aerogel Modified High Performance Carbon Fiber Fabric. *ACS Appl. Mater. Interfaces* **2013**, *5*, 6113–6122. [CrossRef] [PubMed]
81. Luo, H.; Zhou, X.; Ellingford, C.; Zhang, Y.; Chen, S.; Zhou, K.; Zhang, D.; Bowen, C.R.; Wan, C. Interface design for high energy density polymer nanocomposites. *Chem. Soc. Rev.* **2019**, *48*, 4424–4465. [CrossRef]
82. Yang, L.; Huang, X.; Gogoll, A.; Strømme, M.; Sjödin, M. Conducting Redox Polymer Based Anode Materials for High Power Electrical Energy Storage. *Electrochim. Acta* **2016**, *204*, 270–275. [CrossRef]
83. Eftekhari, A.; Li, L.; Yang, Y. Polyaniline supercapacitors. *J. Power Sources* **2017**, *347*, 86–107. [CrossRef]
84. Xie, Y.; Wang, D.; Ji, J. Preparation and Supercapacitor Performance of Freestanding Polypyrrole/Polyaniline Coaxial Nanoarrays. *Energy Technol.* **2016**, *4*, 714–721. [CrossRef]

85. Patil, B.H.; Patil, S.J.; Lokhande, C.D. Electrochemical Characterization of Chemically Synthesized Polythiophene Thin Films: Performance of Asymmetric Supercapacitor Device. *Electroanalysis* **2014**, *26*, 2023–2032. [CrossRef]
86. Singh, K.; Kumar, S.; Agarwal, K.; Soni, K.; Gedela, V.R.; Ghosh, K. Three-dimensional Graphene with MoS_2 Nanohybrid as Potential Energy Storage/Transfer Device. *Sci. Rep.* **2017**, *7*, 9458. [CrossRef] [PubMed]
87. Yun, S.; Freitas, J.N.; Nogueira, A.F.; Wang, Y.; Ahmad, S.; Wang, Z.-S. Dye-sensitized solar cells employing polymers. *Prog. Polym. Sci.* **2016**, *59*, 1–40. [CrossRef]
88. Naveen, M.H.; Gurudatt, N.G.; Shim, Y.-B. Applications of conducting polymer composites to electrochemical sensors: A review. *Appl. Mater. Today* **2017**, *9*, 419–433. [CrossRef]
89. Gomez, I.; Leonet, O.; Blázquez, J.A.; Grande, H.-J.; Mecerreyes, D. Poly(anthraquinonyl sulfides): High Capacity Redox Polymers for Energy Storage. *ACS Macro Lett.* **2018**, *7*, 419–424. [CrossRef]
90. Li, S.; Chen, Y.; He, X.; Mao, X.; Zhou, Y.; Xu, J.; Yang, Y. Modifying Reduced Graphene Oxide by Conducting Polymer Through a Hydrothermal Polymerization Method and its Application as Energy Storage Electrodes. *Nanoscale Res. Lett.* **2019**, *14*, 226. [CrossRef]
91. Yang, C.; Guan, L.; Wei, H.; Guo, Z.; Wang, Y.; Yan, X.; Zhang, X. Polymer nanocomposites for energy storage, energy saving, and anticorrosion. *J. Mater. Chem. A* **2015**, *3*, 14929–14941. [CrossRef]
92. Naresh, V.; Elias, L.; Martha, S.K. Poly(3,4-ethylenedioxythiophene) coated lead negative plates for hybrid energy storage systems. *Electrochim. Acta* **2019**, *301*, 183–191. [CrossRef]
93. Wang, H.; Lin, J.; Shen, Z.X. Polyaniline (PANi) based electrode materials for energy storage and conversion. *J. Sci. Adv. Mater. Devices* **2016**, *1*, 225–255. [CrossRef]
94. Dubal, D.; Caban-Huertas, Z.; Holze, R.; Gómez-Romero, P. Growth of polypyrrole nanostructures through reactive templates for energy storage applications. *Electrochim. Acta* **2016**, *191*, 346–354. [CrossRef]
95. Zhao, S.; Chen, H.; Li, J.; Zhang, J. Synthesis of polythiophene/graphite composites and their enhanced electrochemical performance for aluminum ion batteries. *New J. Chem.* **2019**, *43*, 15014–15022. [CrossRef]
96. Kim, K.-G.; Kim, S.Y. Increase in Interfacial Adhesion and Electrochemical Charge Storage Capacity of Polypyrrole on Au Electrodes Using Polyethyleneimine. *Sci. Rep.* **2019**, *9*, 2169. [CrossRef] [PubMed]
97. Dhibar, S.; Bhattacharya, D.-P.; Ghosh, D.; Hatui, G.; Das, C. Graphene–Single-Walled Carbon Nanotubes–Poly(3-methylthiophene) Ternary Nanocomposite for Supercapacitor Electrode Materials. *Ind. Eng. Chem. Res.* **2014**, *53*, 13030–13045. [CrossRef]
98. Thakur, V.K.; Gupta, R. Recent Progress on Ferroelectric Polymer-Based Nanocomposites for High Energy Density Capacitors: Synthesis, Dielectric Properties, and Future Aspects. *Chem. Rev.* **2016**, *116*, 4260–4317. [CrossRef]
99. Moniruzzaman, M.; Winey, K.I. Polymer Nanocomposites Containing Carbon Nanotubes. *Macromolecules* **2006**, *39*, 5194–5205. [CrossRef]
100. Zhang, Y.; Li, S.; Zhang, J.; Guo, P.; Zheng, J.; Zhao, X.S. Enhancement of Electrochemical Performance of Macroporous Carbon by Surface Coating of Polyaniline. *Chem. Mater.* **2010**, *22*, 1195–1202. [CrossRef]
101. Muthulakshmi, B.; Kalpana, D.; Pitchumani, S.; Renganathan, N. Electrochemical deposition of polypyrrole for symmetric supercapacitors. *J. Power Sources* **2006**, *158*, 1533–1537. [CrossRef]
102. Jurewicz, K.; Delpeux, S.; Bertagna, V.; Beguin, F.; Frackowiak, E. Supercapacitors from nanotubes/polypyrrole composites. *Chem. Phys. Lett.* **2001**, *347*, 36–40. [CrossRef]
103. Zhang, Q.-W.; Zhou, X.; Yang, H.-S. Capacitance properties of composite electrodes prepared by electrochemical polymerization of pyrrole on carbon foam in aqueous solution. *J. Power Sources* **2004**, *125*, 141–147. [CrossRef]
104. Kim, J.-H.; Lee, Y.-S.; Sharma, A.K.; Liu, C.G. Polypyrrole/carbon composite electrode for high-power electrochemical capacitors. *Electrochim. Acta* **2006**, *52*, 1727–1732. [CrossRef]
105. Chauhan, N.P.; Solanki, M.S. Approaches and Challenges of Polyaniline–Graphene Nanocomposite for Energy Application. In *Carbon-Based Polymer Nanocomposites for Environmental and Energy Applications*; Elsevier: Amsterdam, The Netherlands, 2018; pp. 415–436.
106. Ishaq, S.; Moussa, M.; Kanwal, F.; Ehsan, M.; Saleem, M.; Van, T.N.; Losic, D. Facile synthesis of ternary graphene nanocomposites with doped metal oxide and conductive polymers as electrode materials for high performance supercapacitors. *Sci. Rep.* **2019**, *9*, 5974. [CrossRef] [PubMed]
107. Idumah, C.; Hassan, A. Emerging trends in graphene carbon based polymer nanocomposites and applications. *Rev. Chem. Eng.* **2016**, *32*, 223. [CrossRef]

108. Wang, Y.; Wei, H.; Lu, Y.; Wei, S.; Wujcik, E.K.; Guo, Z. Multifunctional Carbon Nanostructures for Advanced Energy Storage Applications. *Nanomaterials* **2015**, *5*, 755–777. [CrossRef] [PubMed]
109. Abouzari-Lotf, E.; Etesami, M.; Nasef, M.M. Carbon-Based Nanocomposite Proton Exchange Membranes for Fuel Cells. In *Carbon-Based Polymer Nanocomposites for Environmental and Energy Applications*; Elsevier: Amsterdam, The Netherlands, 2018; pp. 437–461.
110. Peighambardoust, S.; Rowshanzamir, S.; Amjadi, M. Review of the proton exchange membranes for fuel cell applications. *Int. J. Hydrog. Energy* **2010**, *35*, 9349–9384. [CrossRef]
111. Zhang, R.; Chen, Y.; Montazami, R. Ionic Liquid-Doped Gel Polymer Electrolyte for Flexible Lithium-Ion Polymer Batteries. *Materials* **2015**, *8*, 2735–2748. [CrossRef]
112. Cai, C.; Wang, Y. Novel Nanocomposite Materials for Advanced Li-Ion Rechargeable Batteries. *Materials* **2009**, *2*, 1205–1238. [CrossRef]
113. Shi, Y.; Peng, L.; Ding, Y.; Zhao, Y.; Yu, G. Nanostructured conductive polymers for advanced energy storage. *Chem. Soc. Rev.* **2015**, *44*, 6684–6696. [CrossRef]
114. Nyholm, L.; Nyström, G.; Mihranyan, A.; Strømme, M. Toward Flexible Polymer and Paper-Based Energy Storage Devices. *Adv. Mater.* **2011**, *23*, 3751–3769. [CrossRef]
115. Cheng, F.; Tang, W.; Li, C.; Chen, J.; Liu, H.K.; Shen, P.; Dou, S.X. Conducting Poly(aniline) Nanotubes and Nanofibers: Controlled Synthesis and Application in Lithium/Poly(aniline) Rechargeable Batteries. *Chem. A Eur. J.* **2006**, *12*, 3082–3088. [CrossRef]
116. Deng, W.; Liang, X.; Wu, X.; Qian, J.; Cao, Y.; Ai, X.; Feng, J.; Yang, H. A low cost, all-organic Na-ion Battery Based on Polymeric Cathode and Anode. *Sci. Rep.* **2013**, *3*, 2671. [CrossRef] [PubMed]
117. Magu, T.O.; Agobi, A.U.; Hitler, L.; Dass, P.M. A Review on Conducting Polymers-Based Composites for Energy Storage Application. *J. Chem. Rev.* **2019**, *1*, 19–34.
118. Liu, Y.; Zhou, G.; Liu, K.; Cui, Y. Design of Complex Nanomaterials for Energy Storage: Past Success and Future Opportunity. *Acc. Chem. Res.* **2017**, *50*, 2895–2905. [CrossRef] [PubMed]
119. Ji, L.; Lin, Z.; Alcoutlabi, M.; Zhang, X. Recent developments in nanostructured anode materials for rechargeable lithium-ion batteries. *Energy Environ. Sci.* **2011**, *4*, 2682–2699. [CrossRef]
120. Sarang, K.T.; Miranda, A.; An, H.; Oh, E.-S.; Verduzco, R.; Lutkenhaus, J.L. Poly(fluorene-alt-naphthalene diimide) as n-Type Polymer Electrodes for Energy Storage. *ACS Appl. Polym. Mater.* **2019**, *1*, 1155–1164. [CrossRef]
121. Li, J.; Han, S.; Zhang, C.; Wei, W.; Gu, M.; Meng, L. High-Performance and Reactivation Characteristics of High-Quality, Graphene-Supported SnS_2 Heterojunctions for a Lithium-Ion Battery Anode. *ACS Appl. Mater. Interfaces* **2019**, *11*, 22314–22322. [CrossRef]
122. Brousse, T.; Crosnier, O.; Bélanger, D.; Long, J.W. Capacitive and Pseudocapacitive Electrodes for Electrochemical Capacitors and Hybrid Devices. In *Metal Oxides in Supercapacitors*; Elsevier: Amsterdam, The Netherlands, 2017; pp. 1–24.
123. Han, H.; Lee, S.W.; Moon, K.H.; Cho, S. Fabrication of Solid-State Asymmetric Supercapacitors Based on Aniline Oligomers and Graphene Electrodes with Enhanced Electrochemical Performances. *ACS Omega* **2019**, *4*, 1244–1253. [CrossRef]
124. Oschatz, M.; Borchardt, L.; Hippauf, F.; Nickel, W.; Kaskel, S.; Brunner, E. Chapter Four—Interactions Between Electrolytes and Carbon-Based Materials—NMR Studies on Electrical Double-Layer Capacitors, Lithium-Ion Batteries, and Fuel Cells. In *Annual Reports on NMR Spectroscopy*; Webb, G.A., Ed.; Academic Press: Cambridge, MA, USA, 2016; pp. 237–318.
125. Yang, Q.; Li, Z.; Zhang, R.; Zhou, L.; Shao, M.; Wei, M. Carbon modified transition metal oxides/hydroxides nanoarrays toward high-performance flexible all-solid-state supercapacitors. *Nano Energy* **2017**, *41*, 408–416. [CrossRef]
126. Pan, Y.; Xu, K.; Wu, C. Recent progress in supercapacitors based on the advanced carbon electrodes. *Nanotechnol. Rev.* **2019**, *8*, 299–314. [CrossRef]
127. Dubal, D.; Chodankar, N.; Kim, D.-H.; Gómez-Romero, P. Towards flexible solid-state supercapacitors for smart and wearable electronics. *Chem. Soc. Rev.* **2018**, *47*, 2065–2129. [CrossRef]
128. Soni, R.; Anothumakkool, B.; Kurungot, S. 1D Alignment of PEDOT in a Buckypaper for High-Performance Solid Supercapacitors. *ChemElectroChem* **2016**, *3*, 1329–1336. [CrossRef]

129. Yang, C.; Zhang, L.; Hu, N.; Yang, Z.; Su, Y.; Xu, S.; Li, M.; Yao, L.; Hong, M.; Zhang, Y. Rational design of sandwiched polyaniline nanotube/layered graphene/polyaniline nanotube papers for high-volumetric supercapacitors. *Chem. Eng. J.* **2017**, *309*, 89–97. [CrossRef]
130. Xiao, F.; Yang, S.; Zhang, Z.; Liu, H.; Xiao, J.; Wan, L.; Luo, J.; Wang, S.; Liu, Y. Scalable Synthesis of Freestanding Sandwich-structured Graphene/Polyaniline/Graphene Nanocomposite Paper for Flexible All-Solid-State Supercapacitor. *Sci. Rep.* **2015**, *5*, 9359. [CrossRef] [PubMed]
131. Dubey, R.; Guruviah, V. Review of carbon-based electrode materials for supercapacitor energy storage. *Ionics* **2019**, *25*, 1419–1445. [CrossRef]
132. Bashid, H.A.; Lim, H.N.; Hafiz, S.M.; Andou, Y.; Altarawneh, M.; Jiang, Z.T.; Huang, N.M. Modification of Carbon-Based Electroactive Materials for Supercapacitor Applications. In *Carbon-Based Polymer Nanocomposites for Environmental and Energy Applications*; Elsevier: Amsterdam, The Netherlands, 2018; pp. 393–413.
133. Li, R.; He, C.; Han, X.; Yang, Y. Carbon-Based Polyaniline Nanocomposites for Supercapacitors. In *Carbon-Based Polymer Nanocomposites for Environmental and Energy Applications*; Elsevier: Amsterdam, The Netherlands, 2018; pp. 489–535.
134. Zhang, S.; Shi, X.; Wróbel, R.; Chen, X.; Mijowska, E. Low-cost nitrogen-doped activated carbon prepared by polyethylenimine (PEI) with a convenient method for supercapacitor application. *Electrochim. Acta* **2019**, *294*, 183–191. [CrossRef]
135. Xu, G.; Wang, N.; Wei, J.; Lv, L.; Zhang, J.; Chen, Z.; Xu, Q. Preparation of Graphene Oxide/Polyaniline Nanocomposite with Assistance of Supercritical Carbon Dioxide for Supercapacitor Electrodes. *Ind. Eng. Chem. Res.* **2012**, *51*, 14390–14398. [CrossRef]
136. Biswas, S.; Drzal, L.T. Multilayered Nanoarchitecture of Graphene Nanosheets and Polypyrrole Nanowires for High Performance Supercapacitor Electrodes. *Chem. Mater.* **2010**, *22*, 5667–5671. [CrossRef]
137. Park, J.H.; Ko, J.; Park, O.O.; Kim, D.-W. Capacitance properties of graphite/polypyrrole composite electrode prepared by chemical polymerization of pyrrole on graphite fiber. *J. Power Sources* **2002**, *105*, 20–25. [CrossRef]
138. Chee, W.; Lim, H.N.; Harrison, I.; Chong, K.F.; Zainal, Z.; Ng, C.; Huang, N. Performance of Flexible and Binderless Polypyrrole/Graphene Oxide/Zinc Oxide Supercapacitor Electrode in a Symmetrical Two-Electrode Configuration. *Electrochim. Acta* **2015**, *157*, 88–94. [CrossRef]
139. Lim, Y.; Tan, Y.; Lim, H.; Huang, N.M.; Tan, W.; Yarmo, M.; Yin, C.-Y. Potentiostatically deposited polypyrrole/graphene decorated nano-manganese oxide ternary film for supercapacitors. *Ceram. Int.* **2014**, *40*, 3855–3864. [CrossRef]
140. Xiong, P.; Huang, H.; Wang, X. Design and synthesis of ternary cobalt ferrite/graphene/polyaniline hierarchical nanocomposites for high-performance supercapacitors. *J. Power Sources* **2014**, *245*, 937–946. [CrossRef]
141. Han, C.; Shi, R.; Zhou, N.; Li, H.; Xu, L.; Zhang, T.; Li, J.; Kang, F.; Wang, G.; Li, B. High-Energy and High-Power Nonaqueous Lithium-Ion Capacitors Based on Polypyrrole/Carbon Nanotube Composites as Pseudocapacitive Cathodes. *ACS Appl. Mater. Interfaces* **2019**, *11*, 15646–15655. [CrossRef] [PubMed]
142. Chen, T.; Dai, L. Flexible and wearable wire-shaped microsupercapacitors based on highly aligned titania and carbon nanotubes. *Energy Storage Mater.* **2016**, *2*, 21–26. [CrossRef]
143. Ma, W.; Chen, S.; Yang, S.; Chen, W.; Cheng, Y.; Guo, Y.; Peng, S.; Ramakrishna, S.; Zhu, M.-F. Hierarchical MnO_2 nanowire/graphene hybrid fibers with excellent electrochemical performance for flexible solid-state supercapacitors. *J. Power Sources* **2016**, *306*, 481–488. [CrossRef]
144. Jiang, S.; Shi, T.; Zhan, X.; Long, H.; Xi, S.; Hu, H.; Tang, Z. High-performance all-solid-state flexible supercapacitors based on two-step activated carbon cloth. *J. Power Sources* **2014**, *272*, 16–23. [CrossRef]
145. Zhao, D.; Chen, C.; Zhang, Q.; Chen, W.; Liu, S.; Wang, Q.; Liu, Y.; Li, J.; Yu, H. High Performance, Flexible, Solid-State Supercapacitors Based on a Renewable and Biodegradable Mesoporous Cellulose Membrane. *Adv. Energy Mater.* **2017**, *7*. [CrossRef]
146. Wang, K.; Zhang, X.; Li, C.; Zhang, H.; Sun, X.; Xu, N.; Ma, Y. Flexible solid-state supercapacitors based on a conducting polymer hydrogel with enhanced electrochemical performance. *J. Mater. Chem. A* **2014**, *2*, 19726–19732. [CrossRef]
147. Yu, J.; Wu, J.; Wang, H.; Zhou, A.; Huang, C.; Bai, H.; Li, L. Metallic Fabrics as the Current Collector for High-Performance Graphene-Based Flexible Solid-State Supercapacitor. *ACS Appl. Mater. Interfaces* **2016**, *8*, 4724–4729. [CrossRef]

148. Wu, P.; Cheng, S.; Yao, M.; Yang, L.; Zhu, Y.; Liu, P.; Xing, O.; Zhou, J.; Wang, M.; Luo, H.; et al. A Low-Cost, Self-Standing $NiCo_2O_4$@CNT/CNT Multilayer Electrode for Flexible Asymmetric Solid-State Supercapacitors. *Adv. Funct. Mater.* **2017**, *27*. [CrossRef]
149. Ye, K.-H.; Liu, Z.-Q.; Xu, C.-W.; Li, N.; Chen, Y.-B.; Su, Y.-Z. MnO_2/reduced graphene oxide composite as high-performance electrode for flexible supercapacitors. *Inorg. Chem. Commun.* **2013**, *30*, 1–4. [CrossRef]
150. Liu, C.; Zhao, S.; Lu, Y.; Chang, Y.; Xu, D.; Wang, Q.; Dai, Z.; Bao, J.; Han, M. 3D Porous Nanoarchitectures Derived from SnS/S-Doped Graphene Hybrid Nanosheets for Flexible All-Solid-State Supercapacitors. *Small* **2017**, *13*. [CrossRef] [PubMed]
151. Liao, Q.; Li, N.; Jin, S.; Yang, G.; Wang, C. All-Solid-State Symmetric Supercapacitor Based on Co_3O_4 Nanoparticles on Vertically Aligned Graphene. *ACS Nano* **2015**, *9*, 5310–5317. [CrossRef] [PubMed]
152. Zhu, G.; He, Z.; Chen, J.; Zhao, J.; Feng, X.; Ma, Y.; Sun, P.; Wang, L.; Huang, W. Highly conductive three-dimensional MnO2–carbon nanotube–graphene–Ni hybrid foam as a binder-free supercapacitor electrode. *Nanoscale* **2014**, *6*, 1079–1085. [CrossRef] [PubMed]
153. Qin, T.; Peng, S.; Hao, J.; Wen, Y.; Wang, Z.; Wang, X.; He, D.; Zhang, J.; Hou, J.; Cao, G. Flexible and Wearable All-Solid-State Supercapacitors with Ultrahigh Energy Density Based on a Carbon Fiber Fabric Electrode. *Adv. Energy Mater.* **2017**, *7*. [CrossRef]
154. Ko, W.-Y.; Chen, Y.-F.; Lu, K.-M.; Lin, K.-J. Porous honeycomb structures formed from interconnected MnO_2 sheets on CNT-coated substrates for flexible all-solid-state supercapacitors. *Sci. Rep.* **2016**, *6*. [CrossRef] [PubMed]
155. Parvez, K.; Wu, Z.; Li, R.; Liu, X.; Graf, R.; Feng, X.; Müllen, K. Exfoliation of Graphite into Graphene in Aqueous Solutions of Inorganic Salts. *J. Am. Chem. Soc.* **2014**, *136*, 6083–6091. [CrossRef]
156. Wang, X.; Sumboja, A.; Foo, W.L.; Yan, C.Y.; Tsukagoshi, K.; Lee, P.S. Rational design of a high performance all solid state flexible micro-supercapacitor on paper. *RSC Adv.* **2013**, *3*, 15827–15833. [CrossRef]
157. Yao, B.; Yuan, L.; Xiao, X.; Zhang, J.; Qi, Y.; Zhou, J.; Zhou, J.; Hu, B.; Chen, W. Paper-based solid-state supercapacitors with pencil-drawing graphite/polyaniline networks hybrid electrodes. *Nano Energy* **2013**, *2*, 1071–1078. [CrossRef]
158. Dai, S.; Xu, W.; Xi, Y.; Wang, M.; Gu, X.; Guo, N.; Hu, C. Charge storage in KCu7S4 as redox active material for a flexible all-solid-state supercapacitor. *Nano Energy* **2016**, *19*, 363–372. [CrossRef]
159. Lee, H.; Kwon, J.; Suh, Y.D.; Moon, H.; Hong, S.; Yeo, J.; Ko, S.H. All-solid-state flexible supercapacitors by fast laser annealing of printed metal nanoparticle layers. *J. Mater. Chem. A* **2015**, *3*, 8339–8345. [CrossRef]
160. Yun, J.; Kim, D.; Lee, G.; Ha, J.S. All-solid-state flexible micro-supercapacitor arrays with patterned graphene/MWNT electrodes. *Carbon* **2014**, *79*, 156–164. [CrossRef]
161. Kang, Y.J.; Chung, H.; Kim, M.-S.; Kim, W. Enhancement of CNT/PET film adhesion by nano-scale modification for flexible all-solid-state supercapacitors. *Appl. Surf. Sci.* **2015**, *355*, 160–165. [CrossRef]
162. Hu, H.; Zhang, K.; Li, S.; Ji, S.; Ye, C. Flexible, in-plane, and all-solid-state micro-supercapacitors based on printed interdigital Au/polyaniline network hybrid electrodes on a chip. *J. Mater. Chem. A* **2014**, *2*, 20916–20922. [CrossRef]
163. Xu, X.; Shi, W.; Li, P.; Ye, S.; Ye, C.; Ye, H.; Lu, T.; Zheng, A.; Zhu, J.; Xu, L.; et al. Facile Fabrication of Three-Dimensional Graphene and Metal–Organic Framework Composites and Their Derivatives for Flexible All-Solid-State Supercapacitors. *Chem. Mater.* **2017**, *29*, 6058–6065. [CrossRef]
164. Liu, X.; Qian, T.; Xu, N.; Zhou, J.; Guo, J.; Yan, C. Preparation of on chip, flexible supercapacitor with high performance based on electrophoretic deposition of reduced graphene oxide/polypyrrole composites. *Carbon* **2015**, *92*, 348–353. [CrossRef]
165. Liu, S.; Xie, J.; Li, H.; Wang, Y.; Yang, H.Y.; Zhu, T.; Zhang, S.; Cao, G.; Zhao, X. Nitrogen-doped reduced graphene oxide for high-performance flexible all-solid-state micro-supercapacitors. *J. Mater. Chem. A* **2014**, *2*, 18125–18131. [CrossRef]
166. Cao, X.; Zheng, B.; Shi, W.; Yang, J.; Fan, Z.; Luo, Z.; Rui, X.; Chen, B.; Yan, Q.; Zhang, H. Reduced Graphene Oxide-Wrapped MoO3Composites Prepared by Using Metal-Organic Frameworks as Precursor for All-Solid-State Flexible Supercapacitors. *Adv. Mater.* **2015**, *27*, 4695–4701. [CrossRef]
167. Sun, G.; An, J.; Chua, C.K.; Pang, H.; Zhang, J.; Chen, P. Layer-by-layer printing of laminated graphene-based interdigitated microelectrodes for flexible planar micro-supercapacitors. *Electrochem. Commun.* **2015**, *51*, 33–36. [CrossRef]

168. Xie, J.; Sun, X.; Zhang, N.; Xu, K.; Zhou, M.; Xie, Y. Layer-by-layer β-Ni(OH)2/graphene nanohybrids for ultraflexible all-solid-state thin-film supercapacitors with high electrochemical performance. *Nano Energy* **2013**, *2*, 65–74. [CrossRef]
169. Fei, H.; Yang, C.; Bao, H.; Wang, G. Flexible all-solid-state supercapacitors based on graphene/carbon black nanoparticle film electrodes and cross-linked poly(vinyl alcohol)–H_2SO_4 porous gel electrolytes. *J. Power Sources* **2014**, *266*, 488–495. [CrossRef]
170. Li, X.; Zhao, T.; Chen, Q.; Li, P.; Wang, K.; Zhong, M.; Wei, J.; Wu, D.; Wei, B.; Zhu, H. Flexible all solid-state supercapacitors based on chemical vapor deposition derived graphene fibers. *Phys. Chem. Chem. Phys.* **2013**, *15*, 17752–17757. [CrossRef] [PubMed]
171. Shao, Y.; Wang, H.; Zhang, Q.; Li, Y. Fabrication of large-area and high-crystallinity photoreduced graphene oxide films via reconstructed two-dimensional multilayer structures. *NPG Asia Mater.* **2014**, *6*, e119. [CrossRef]
172. Chen, J.; Xu, J.; Zhou, S.; Zhao, N.; Wong, C.-P. Facile and scalable fabrication of three-dimensional Cu(OH)2nanoporous nanorods for solid-state supercapacitors. *J. Mater. Chem. A* **2015**, *3*, 17385–17391. [CrossRef]
173. Niu, H.; Yang, X.; Jiang, H.; Zhou, D.; Li, X.; Zhang, T.; Liu, J.; Wang, Q.; Qu, F. Hierarchical core–shell heterostructure of porous carbon nanofiber@$ZnCo_2O_4$ nanoneedle arrays: Advanced binder-free electrodes for all-solid-state supercapacitors. *J. Mater. Chem. A* **2015**, *3*, 24082–24094. [CrossRef]
174. Siwal, S.S.; Zhang, Q.; Sun, C.; Thakur, V.K. Graphitic Carbon Nitride Doped Copper-Manganese Alloy as High-Performance Electrode Material in Supercapacitor for Energy Storage. *Nanomaterials* **2019**, *10*, 2. [CrossRef] [PubMed]

Communication

Direct Pre-lithiation of Electropolymerized Carbon Nanotubes for Enhanced Cycling Performance of Flexible Li-Ion Micro-Batteries

Vinsensia Ade Sugiawati [1], Florence Vacandio [2], Neta Yitzhack [3], Yair Ein-Eli [3,4] and Thierry Djenizian [1,5,*]

1 Mines Saint-Etienne, Center of Microelectronics in Provence, Department of Flexible Electronics, F-13541 Gardanne, France; vinsensia.sugiawati@gmail.com

2 CNRS, Electrochemistry of Materials Research Group, Aix Marseille Université, MADIREL, UMR 7246, F-13397 Marseille CEDEX 20, France; florence.vacandio@univ-amu.fr

3 Department of Materials Science and Engineering, Technion-Israel Institute of Technology, Haifa 3200003, Israel; neta_y@campus.technion.ac.il (N.Y.); eineli@technion.ac.il (Y.E.-E.)

4 The Nancy & Stephan Grand Technion Energy Program, Technion-Israel Institute of Technology, Haifa 3200003, Israel

5 Al-Farabi Kazakh National University, Center of Physical-Chemical Methods of Research and Analysis, Tole bi str., 96A, Almaty 050040, Kazakhstan

* Correspondence: thierry.djenizian@emse.fr

Received: 9 January 2020; Accepted: 7 February 2020; Published: 11 February 2020

Abstract: Carbon nanotubes (CNT) are used as anodes for flexible Li-ion micro-batteries. However, one of the major challenges in the growth of flexible micro-batteries with CNT as the anode is their immense capacity loss and a very low initial coulombic efficiency. In this study, we report the use of a facile direct pre-lithiation to suppress high irreversible capacity of the CNT electrodes in the first cycles. Pre-lithiated polymer-coated CNT anodes displayed good rate capabilities, studied up to 30 C and delivered high capacities of 850 mAh g^{-1} (313 μAh cm^{-2}) at 1 C rate over 50 charge-discharge cycles.

Keywords: carbon nanotubes; polymer electrolyte; Li-ion micro-batteries; flexible anode; pre-lithiation

1. Introduction

Li-ion batteries (LIBs) have been successfully employed in a wide range of applications, such as electric vehicles, microelectronic devices, etc., due to their remarkable properties such as high energy density, lack of memory effect, long cycle life, low self-discharge and high thermal resistance [1–3]. A large variety of carbon-based materials for LIBs have been widely investigated, such as graphene, fullerene and carbon nanotubes (CNT) [4–10]. For instance, a high capacity of a sandwich-like and porous $NiCo_2O_4$@reduced graphene oxide (rGO) nanocomposite serving as anode material was reported by Huang's group [11]. Kumar et al. synthesized octahedral iron oxide nanocrystals on reduced graphene oxide nanosheets by a microwave-assisted process [12]. Zhang et al. demonstrated that electrodeposition can be used to prepare $NiCo_2O_4$/graphene [13], while Chen et al. prepared graphene hybrid nanosheet arrays via a one-pot procedure [14]; these synthesized graphene electrodes delivered good electrochemical performances. More recently, rapid research progress has been made in exploring flexible anode materials delivering high storage capacity and remarkable long-term cyclability [15–17]. As an allotrope of carbon, CNT electrodes offer several outstanding properties, such as excellent flexibility, fast charge transport, large surface-to-volume ratio, good chemical stability, high electrical conductivity and high reversible capacity [18–26]. Several methods to synthesize CNT have been demonstrated, e.g., chemical vapor deposition (CVD) [25,27], pyrolysis [28], arc discharge [29,30], laser ablation [31] and electrolysis [32]. These methods allowed the growth of CNT with various morphology, structure and properties.

As a potential flexible anode material, our previous report showed that the CNT reached a higher specific capacity (more than 700 mAh g^{-1}) compared to a traditional graphite anodes (372 mAh g^{-1}) [33]. Also, Yoon et al. [34] reported that heat-treated CNT can deliver a reversible capacity of 446 mAh g^{-1} at 0.5 C, with a low Initial Coulombic Efficiency (ICE) of 9.6%. Zhou et al. [35] reported that CNT showing bamboo-like structure delivered a reversible capacity of 135 mAh g^{-1} with ICE of 17.3%. Li et al. studied a high-concentration, nitrogen-doped CNT anode providing reversible capacity of 494 mAh g^{-1} [21], while Welna et al. [36] demonstrated that vertically aligned MWCNT-based anodes showed an excellent lithium storage capacity of 980 mAh g^{-1} in the initial cycle which stabilized after first few cycles, delivering a discharge capacity of 750 mAh g^{-1}. Despite CNT exhibiting remarkable features as flexible anode materials, major challenges, such as the huge surface area of the CNT promoting a high capacity loss in the first initial cycles, needs to be first tackled [37,38]. Consequently, CNT-based electrodes have a low ICE due to the solid electrolyte interphase (SEI) film formation upon the initial lithiation. This SEI film formation consumes a high amount of Li ions during the first discharge, which further limits the electrochemical performance of the cells, particularly when the CNT anode is coupled with the cathode material in full-cell configuration [39–41].

Several pre-lithiation methods have been investigated to compensate the severe capacity loss of the anode materials, for instance, Seong et al. reported that SiO anode was effectively pre-lithiated using stabilized lithium metal powder (SLMP) [42], Liu et al. studied pre-lithiated Si nanowire anode via a self-discharge mechanism [43] and Scott et al. also reported a complete diminishing of the initial capacity loss for carbon electrodes using n-butyllithium in hexane [44]. Wu et al. successfully fabricated a PPy/Li_2S/KB cathode by pre-lithiation and the lithiated cathode exhibited high capacity of 1000 mAh g^{-1} with a coulombic efficiency around 95% at 0.2 C [45]. Among these methods, the self-discharge or direct pre-lithiation presents several advantages, such as low cost, fast pre-lithiation process (less time consuming), easy operation and an excellent lithiation efficiency [39,43]. However, there have been only few attempts to reduce the high irreversible capacity of CNT as anodes [39] and no study on pre-lithiated CNT coating with a polymer film.

Recently, our groups demonstrated the crucial impact of the electropolymerization of p-sulfonated poly(allyl phenyl ether) (SPAPE) electrolyte on the electrochemical performances of CNT anodes [33]. A high reversible capacity of 750 mAh g^{-1} (276 μAh cm^{-2}) at 1 C rate with the ICE of 10.4% can be obtained and interestingly, the areal capacity of SPAPE-coated CNT is enhanced by 67% compared to the pristine ones [33]. However, due to their low ICE, in the current study we report a simple direct pre-lithiation method to alleviate the initial irreversible capacity of the electropolymerized CNT anodes for flexible micro-batteries. Indeed, the long-term and extensive cycling performance of the pre-lithiated CNT is achieved over 500 cycles at a 10 C rate and its excellent rate capability is also present even at very high current density (1 to 30 C rates).

2. Materials and Methods

2.1. Materials

CNT were provided by Tortech NanoFibers Ltd. (Ma'alot Tarshiha, Israel), having a density of 0.613 g cm^{-3}, a thickness of 30 μm and a porosity of approximately 70% [3]. The samples were initially washed in iso-propyl alcohol before being used as anode material, as previously reported [19,37]. A Cu target was purchased from Neyco (VANVES, France) purity: 99.9%). Lithium bis(trifluoromethane)sulfonimide (LiTFSI), 1 M $LiPF_6$ (EC: DEC, *v*/*v*) and dimethylsulfoxide (DMSO) were purchased from Sigma–Aldrich (St. Quentin Fallavier Cedex, France).

2.2. Synthesis of the Positive Electrodes

$LiNi_{0.5}Mn_{1.5}O_4$ (LNMO) serving as cathode was synthesized by a sol–gel method as described in [46]. To prepare the composite cathode, LNMO powder was mixed with carbon black (Super P) and polyvinylidene fluoride (PVDF) at the ratio of 80:10:10 and ground in a mortar for 20 min. N-Methyl-2-pyrrolidone (NMP) was added in the powder mixture to obtain a paste. The paste was subsequently spread on an aluminum disk as current collector with a diameter of 8 mm. The electrode was dried at 80 °C and was kept under vacuum at 110 °C for 10 h.

2.3. Elctropolymerization and Direct Pre-Lithiation of CNT Electrodes

The metallic Cu thin film was deposited onto the CNT surface utilizing a Cu target by radio frequency sputtering (MP300 model, PLASSYS, Marolles en Hurepoix, France), as described in our previous work [33]. The 300 nm-thin layer of sputter-deposited Cu film was utilized as the backside connection of the electropolymerization reaction. Electropolymerization of the SPAPE polymer electrolyte onto CNT was conducted by cyclic voltammetry (CV) in a three-electrode electrochemical cell using a VersaSTAT 3 potentiostat (Princeton Applied Research, Elancourt, France) with a Pt electrode as the counter and Ag/AgCl (3 M KCl) as the reference electrode [33,47]. Cyclic voltammetry was performed onto the CNT electrodes in order to polymerize the sulfonated aromatic precursor. An electrolyte solution containing 5.2×10^{-3} M of the synthesized monomer was mixed with 0.5 M LiTFSI as a supporting electrolyte and DMSO as a solvent. The CV experiments were carried out at room temperature for 10 cycles at the scan rate of 20 mV s^{-1} in the potential window of −0.9 to −1.8 V vs. Ag/AgCl (3 M KCl). After electropolymerization, a simple pre-lithiation treatment was achieved by pressing a Li foil and the CNT soaked with 2 drops of electrolyte composed of 1 M $LiPF_6$ (EC: DEC, *v*/*v*). Various pre-lithiation durations were investigated i.e., 1, 3, 15 and 30 min, respectively. The pre-lithiated CNT were then tested as anode in half-cell and full-cell configuration. It is important to note that the used Li foil was cleaned and reutilized.

2.4. Characterization and Measurements

The surface morphology of the CNT anode was examined using a field-emission scanning electron microscope (SEM, Ultra-55, Carl Zeiss, Oberkochen, Germany) and by transmission electron microscopy (TEM) (Tecnai G2, Thermofisher Scientific, Waltham, MA, US). The purity of CNT was examined by x-ray diffraction (XRD) using a Diffractometer D5000 (Siemens, Munich, Germany) with $CuK_{\alpha1}$ (λ= 1.5406 Å) radiation, then analyzed by comparing with the JCDS-ICDD database (Joint Committee on Powder Diffraction Standards - International Center for Diffraction Data) to check the purity of the samples. Raman spectra were recorded with XploRA Raman spectrometer (Horiba Scientific, Kyoto, Japan) equipped with a 532 nm laser. For the electrochemical performance tests, Pristine CNT and SPAPE-coated CNT having a surface area of 0.44 cm^2 were used as electrodes without the use of any binders and conductive additives, assembling using standard two-electrode Swagelok cells. All cells assembly were conducted in a glove box, filled with high purity argon (Ar) in which the moisture and oxygen contents were less than 2 ppm. Cyclic voltammetry tests were carried out using a VMP3 (Bio Logic, Seyssinet-Pariset, France) in the potential window of 0.01–2 V vs. Li/Li^+ with a scan rate of 0.2 mV s^{-1}. Galvanostatic charge–discharge cycles were performed with a VMP3 (Bio Logic) in the potential window between 0.01 and 2 V vs. Li/Li^+ and the current density for the CNT electrodes (pristine and pre-lithiated) were 0.12, 0.24, 0.60, 1.20, 1.8, and 3.6 mA cm^{-2}, respectively. The pristine CNT–LNMO and pre-lithiated CNT-LNMO full-cells were cycled at 1C and 2C rate in the potential window of 2.5–4.4 V.

3. Results and Discussion

3.1. Structural and Morphological Characterization

As shown in Figure 1a, the Raman spectrum of the CNT exhibits the two main bands characteristic of carbon materials in general, and to CNT in particular [48,49]. The G band (1573 cm^{-1}) is correlated with the stretching of the C–C bond (sp^2), and the D band (1341 cm^{-1}) is attributed to the presence of disorders in the sp^2 structure. The ratio between these two peaks (I_D/I_G) is often used to assess the relative content of defects in the CNT. The low D band intensity of the CNT used in this work ($I_D/I_G = 0.21 \pm 0.02$) indicates its relative purity. The crystallinity of the CNT was verified using X-ray diffraction and the patterns are given in Figure 1b. Two distinguishable diffraction peaks are clearly seen: a strong C (002) peak at approximately 26° represents the characteristic of graphite peak and the peak at approximately 43° is attributed to the (100) planes of the nanotube structure.

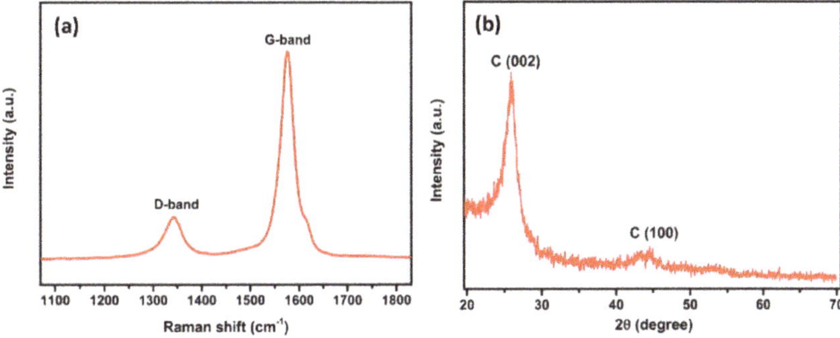

Figure 1. Raman spectra (**a**) and XRD pattern of the pristine carbon nanotubes (CNT) (**b**).

According to the SEM examinations, the pristine CNT are present as highly disoriented shaped nanotubes with diameters ranging from approximately 5 to 30 nm as shown in Figure 2a,b. After electropolymerization, densified and bundled CNT are obtained due to Van-der-Waals interactions among the neighboring tubes (Figure 3c,d). Moreover, we also observed that the electropolymerization process allows the densified and bundled CNT formation due to an interaction existing between the neighboring tubes [33]. The presence of nanoparticles attached on the pristine CNT surfaces is detected from the SEM and TEM images which is assumed to be a catalyst residue of metal impurities resulting from the synthesis process. The elemental energy dispersive x-ray spectroscopy (EDS) analysis spectra shows the presence of Fe particles which are attributed to the catalyst impurities and the Cu signal came from the Cu grid.

3.2. Cyclic Voltammetry

In the present study, we implemented a direct pre-lithiation method in order to suppress the irreversible capacity loss during the formation of SEI film in the first cycle [39]. A pre-lithiation process is schematically depicted in Figure 3a. The CNT were directly contacted with a Li metal film in a small amount of electrolyte. Hence, a self-Li-ion discharging process (lithiation) occurs due to the short being made; the different potential existing between CNT materials and Li metal becoming reduced to zero potential. In order to follow the effect of the pre-lithiation process at various contact times (1, 3, 15, and 30 min), we first investigated the electrochemical behavior of the pristine CNT and pre-lithiated CNT electrodes by comparing the cyclic voltammetry (CV) curves, as displayed in Figure 3b–f). It is clearly observed that prior to pre-lithiation, the CV curve exhibits an open circuit voltage (OCV) of ~3.1 V vs. Li/Li^+ and this would correspond to a purely delithiated state. During the first cycle, an obvious cathodic peak around 0.6 V attributed to the electrolyte decomposition is easily observed and corresponds to the formation and deposition of a SEI layer [50,51]. The peak disappearance in the subsequent cycles indicates a stable as-formed SEI is obtained. Another cathodic peak close to 0 V is attributed to Li^+ intercalation into CNT and oxidation peak located at 0.25 V corresponds to the Li^+ extraction process [21]. The broad oxidation peak ca. 1.25 V is also visible, corresponding to the extraction of Li ions from the cavities existing in the CNT structure and the small oxidation peak ca. 1.8 V might be ascribed to the reaction of lithium with hydrogen functional groups on the CNT surfaces [33].

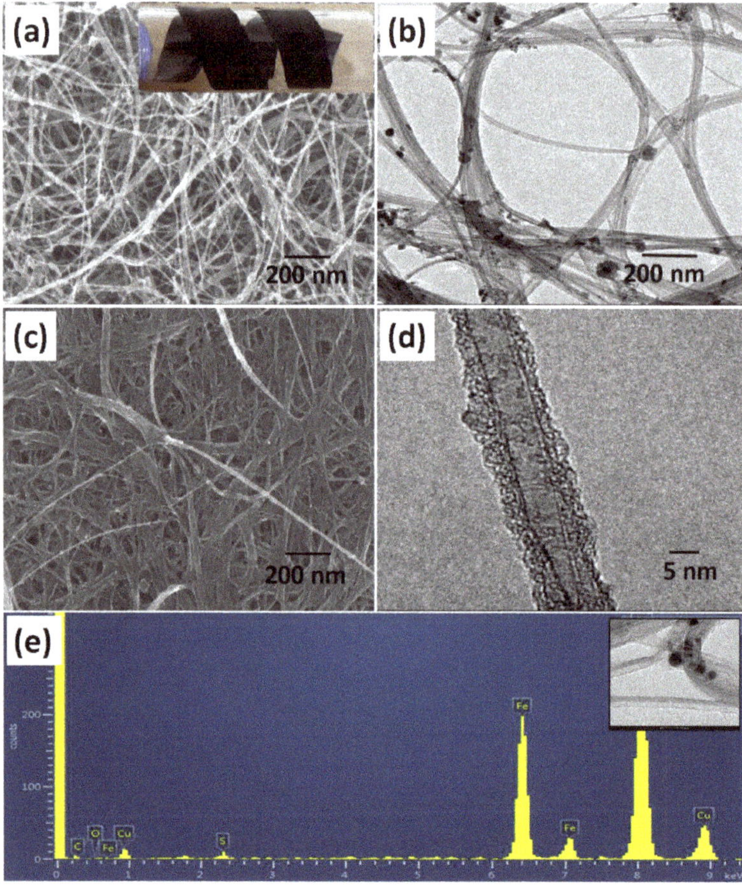

Figure 2. SEM image of pristine CNT (photograph of a flexible CNT in inset) (**a**); TEM image of pristine CNT (**b**); SEM image of p-sulfonated poly(allyl phenyl ether) (SPAPE)-coated CNT and (**c**) TEM image of SPAPE-coated CNT (**d**); and EDS spectra of pristine CNT (the inset is the corresponding TEM image) (**e**).

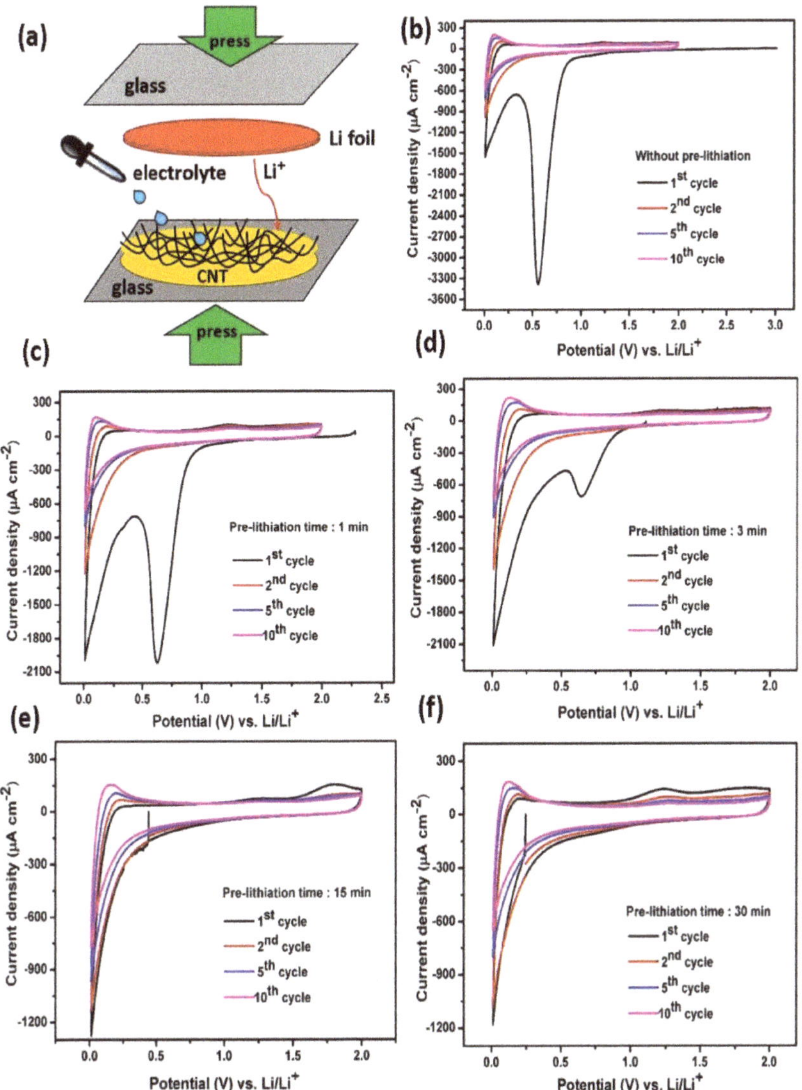

Figure 3. A schematic representation of a direct pre-lithiation method (**a**), cyclic voltammogram curves of the pristine CNT (**b**), pre-lithiated CNT for 1 min (**c**), pre-lithiated CNT for 3 min (**d**), pre-lithiated CNT for 15 min (**e**) and pre-lithiated CNT for 30 min (**f**) recorded at a scan rate of 0.2 mV s^{-1} in a potential range of 0.01–2 V vs. Li/Li^+.

After a pre-lithiation of 1 min, the OCV drops to 2.4 V vs. Li/Li^+, suggesting that a partial insertion of lithium ions into the CNT electrodes already occurred. We also observed the noticeable cathodic peak at ~0.6 V vs. Li/Li^+ being still present (with a lower absolute current density compared to pristine CNT), suggesting an incomplete SEI film formation on the CNT surfaces. Nonetheless, the direct pre-lithiation on the CNT surface provides beneficial effects, as the cathodic peak intensity of the SEI film formation has been successfully diminished. The observation is continued by examination of the CV curve after pre-lithiation for a period of 3 min and indeed, the OCV value decreased from 2.4 to

1.1 V vs. Li/Li$^+$. At this stage, the electrode still exhibits similarity to previous samples; namely, the cathodic peak at ~0.6 V vs. Li/Li$^+$ is visible but its intensity obviously weakens. We assume that the contact time of 3 min is insufficient to fully form a stable SEI film. Therefore, it was suggested that OCV should be below the cathodic peak potential of the SEI film (approximately 0.6 V vs. Li/Li$^+$). Thus, the contact time during pre-lithiation process should be longer than 3 min. To further investigate the different stages of OCV value, the CNT electrode was directly contacted with Li metal for 15 min. As seen, the OCV shifted to much lower voltage of 0.45 V vs. Li/Li$^+$ and the cathodic peak at ~0.6 V disappeared. After 30 min of pre-lithiation, CV curves show a slight decrease in the OCV, down to ~0.25 V vs. Li/Li$^+$ without any large cathodic peak. These results reveal that the SEI film has been successfully pre-formed on the CNT surfaces for a pre-lithiation time longer than 15 min. Compared to electrochemical pre-lithiation in which the lithiation reaction requires a slow rate (usually in 10 or 20 h), the direct pre-lithiation approach is much faster [20].

3.3. Galvanostatic Charge–Discharge Profiles

In order to gain a deeper insight into the effect of the pre-lithiation process at different contact times, the galvanostatic charge–discharge tests were performed in a half-cell battery configuration. The charge–discharge curves for five CNT samples recorded at a current density of 3.6 mA cm^{-2} (2 C) are illustrated in Figure 4a–e. As can be seen, the various pre-lithiation times result in different electrochemical performances when the CNT are used as flexible anode materials. The plotted data shows the first two cycles corresponding to the discharge and charge processes of the pristine CNT and pre-lithiated CNT electrodes (1, 3, 15, and 30 min). During the first discharge of the pristine CNT, the voltage drops rapidly from OCV to 0.8 V vs. Li/Li$^+$ with a large discharge plateau corresponding to the SEI film formation consisting of a mixture of organic and inorganic lithium compounds [52,53]. This pristine CNT provides a charge capacity of 8401 mAh g^{-1} (3091 μAh cm^{-2}) and a discharge capacity of 972 mAh g^{-1} (358 μAh cm^{-2}) at 2 C rate, holding ICE value of 11.57%. The very low ICE and a vast majority of irreversible capacity loss is related to the SEI formation [54]. In agreement with the CV results, the OCV of the pre-lithiated CNT electrodes decreases along with increment contact time.

After a short contact time of 1 min (Figure 4a), the OCV drops to ~2.4 V, which is lower than that of a pristine CNT electrode. The comparison of the first discharge capacity of 6802 mAh g^{-1} (2503 μAh cm^{-2}) and the first charge capacity of 843 mAh g^{-1} (310 μAh cm^{-2}) leads to an ICE of 12.38%. In good agreement with the previous CV results, the OCV of the cells utilizing a contact time of 3 min is 1.1 V vs. Li/Li$^+$ with a first discharge capacity of 5518 mAh g^{-1} (2030 μAh cm^{-2}) and a charge capacity of 1166 mAh g^{-1} (429 μAh cm^{-2}), yielding an improved ICE of 21.13%. Then, the ICE increases significantly up to 68.30% after a 15-min pre-lithiation treatment, corresponding to the first discharge and charge capacity of 1714 mAh g^{-1} (631 μAh cm^{-2}) and 1170 mAh g^{-1} (431 μAh cm^{-2}), respectively. Eventually, after pre-lithiation for 30 min, the capacity of the first discharge and charge cycles are 956 mAh g^{-1} (352 μAh cm^{-2}) and 1179 mAh g^{-1} (434 μAh cm^{-2}), respectively, resulting in a preloaded capacity of 223 mAh g^{-1} (82 μAh cm^{-2}) instead of capacity loss. Indeed, the charge–discharge potential profiles of the pre-lithiated CNT electrodes are consistent with the CV curves. These results suggest that longer pre-lithiation periods (i.e., 15 min) result in fully lithiated CNT electrodes and one should also note the remarkable enhancement in the ICE of the CNT electrodes. Herein, we highlight that the degree of pre-lithiation needs to be carefully controlled. From the CV and galvanostatic curves, 15-min pre-lithiation showed high capacity and high ICE value. Pre-lithiation for 30 min is assumed lead to over-lithiation which would result in lithium plating, short circuits, and also increase the side reactions during cycling due to the excessive Li-ions on the anode surface [55]. Additionally, pre-lithiation for 1 and 3 min are insufficient due to the observable SEI formation features. As a result, this SEI depletes the cyclable lithium from the cathode material in full-cell configuration.

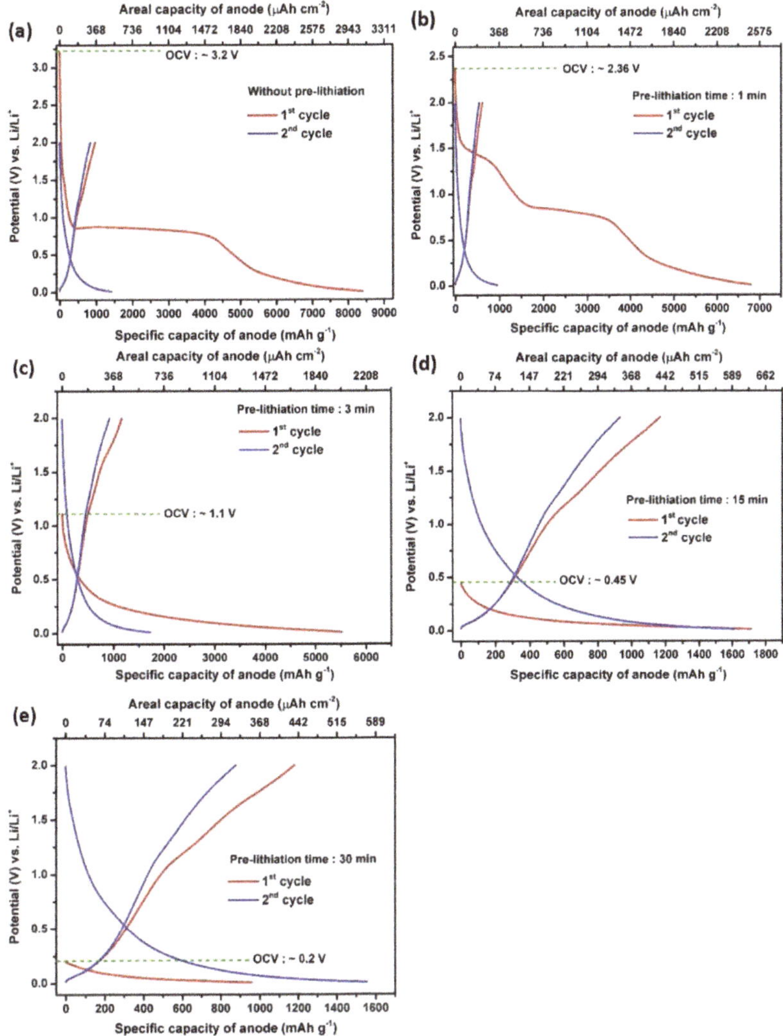

Figure 4. Initial charge–discharge potential profiles of the pristine CNT (**a**), pre-lithiated CNT for 1 min (**b**), pre-lithiated CNT for 3 min (**c**), pre-lithiated CNT for 15 min (**d**) and pre-lithiated CNT for 30 min (**e**) recorded at 2 C rate.

3.4. Morphology of CNT Electrode after Direct Pre-Lithiation

SEM images show the influence of direct pre-lithiation on the CNT surface morphology. The significant change of the morphology occurred, as displayed in Figure 5. After a very short contact time (1 min pre-lithiation), the bundled and densified morphology of CNT is preserved with only a few modifications of the CNT surfaces (Figure 5a,b). Surface examination is continued for 15 min pre-lithiation (Figure 5c,d), where the CV curves and galvanostatic cycling results support a stable SEI film formation, being pre-formed on the CNT surface upon self-discharge. Interestingly, and as expected, a polymer-like film covering the CNT surface can be assigned to the growth of the SEI thin film. After a 15 min pre-lithiation process, $(CH_2OCO_2Li)_2$, polyethylene oxide, Li_2CO_3, LiF, Li_2O, etc.

products are presumably accumulated over the CNT surface due to the insertion and absorption of the Li ions into the CNT structures, in addition to the electrolyte reduction [50].

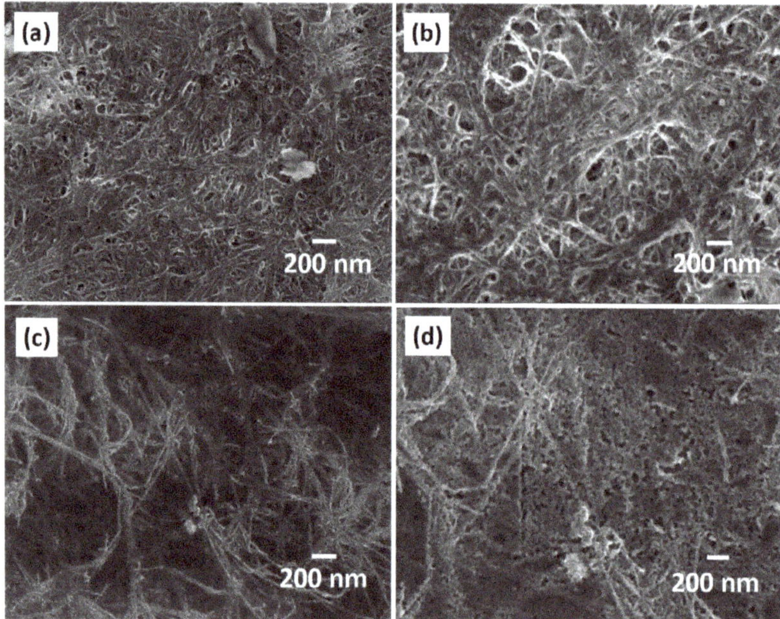

Figure 5. SEM images of a pre-lithiated CNT after 1 min (**a**,**b**) and pre-lithiated CNT after 15 min (**c**,**d**).

3.5. Electrochemical Performance in Half-Cells and Full-Cells

To verify the significant role of pre-lithiation prior to cell assembling and their functionalization with the polymer coating, four CNT electrodes were charged and discharged in the range of 0.01–2 V vs. Li/Li^+ to evaluate their cycle life. Figure 6a shows the comparison of the pristine and pre-lithiated CNT. As seen, for all samples, the CNT anodes show good cycling stability, even up to 500 cycles. For the pristine CNT samples, SPAPE-coated CNT attains higher reversible capacity of 463 mAh g^{-1} (170 μAh cm^{-2}) compared to pristine CNT (242 mAh g^{-1}, 89 μAh cm^{-2}), while for the pre-lithiated samples, the pristine CNT and SPAPE-coated CNT yielded a reversible capacity of 356 mAh g^{-1} (131 μAh cm^{-2}) and 508 mAh g^{-1} (187 μAh cm^{-2}), respectively over 500 cycles at 10 C rate (Figure 4b). Thus, it is clear that coating CNT with the polymer electrolyte via electropolymerization reaction has a beneficial impact, resulting in the improvement of the cell performance [33,46,56,57]. The enhancement can be attributed to the combination of two effects: the larger electrode/electrolyte interface resulting in improved charge transport and a better penetration of the polymer electrolyte onto the carbon nanotube surfaces [33]. The fact that SPAPE can improve the kinetics of charge/discharge has been evidenced by electrochemical impedance spectroscopy in one of our previous works [47].

By comparing the 1st and 2nd discharge capacity of both pristine and pre-lithiated samples, the importance of direct pre-lithiation prior to cell assembling can be demonstrated (Figure 6a,b). For the non-pristine CNT, the 1st and 2nd discharge capacity of the pristine CNT are 4298 mAh g^{-1} (1581 μAh cm^{-2}) and 681 mAh g^{-1} (251 μAh cm^{-2}), while SPAPE-coated CNT provides 1st discharge capacity of 6699 mAh g^{-1} (2465 μAh cm^{-2}) and 2nd discharge capacity of 955 mAh g^{-1} (351 μAh cm^{-2}). Both pristine samples exhibited very high initial capacity loss. This irreversible capacity is considered as a critical issue, notably when assembling a full-cell configuration, since lithium is irreversibly consumed after the first lithiation. In agreement with previous findings [45,54,58], the high capacity loss can be significantly reduced after pre-lithiation. For the pre-lithiated CNT, the 1st and 2nd discharge capacity of the pristine CNT are 286 mAh g^{-1} (97 μAh cm^{-2}) and 861 mAh g^{-1} (317 μAh cm^{-2}), while SPAPE-coated CNT gives 1st discharge capacity of 981 mAh g^{-1} (361 μAh cm^{-2}) and 2nd discharge capacity of 1188 mAh g^{-1} (437 μAh cm^{-2}). Figure 6c–f show the charge and discharge profiles of the CNT anodes under four fabrication conditions: pristine CNT, SPAPE-coated CNT, pre-lithiated pristine CNT and pre-lithiated SPAPE-coated CNT, respectively. The charge and discharge profiles for all samples show pronounced sloping curves, with a cell voltage of ~0.5 V vs. Li/Li^+. Moreover, the overlapping of the galvanostatic curves suggests a good electrochemical reversibility of the CNT anode, particularly after direct pre-lithiation.

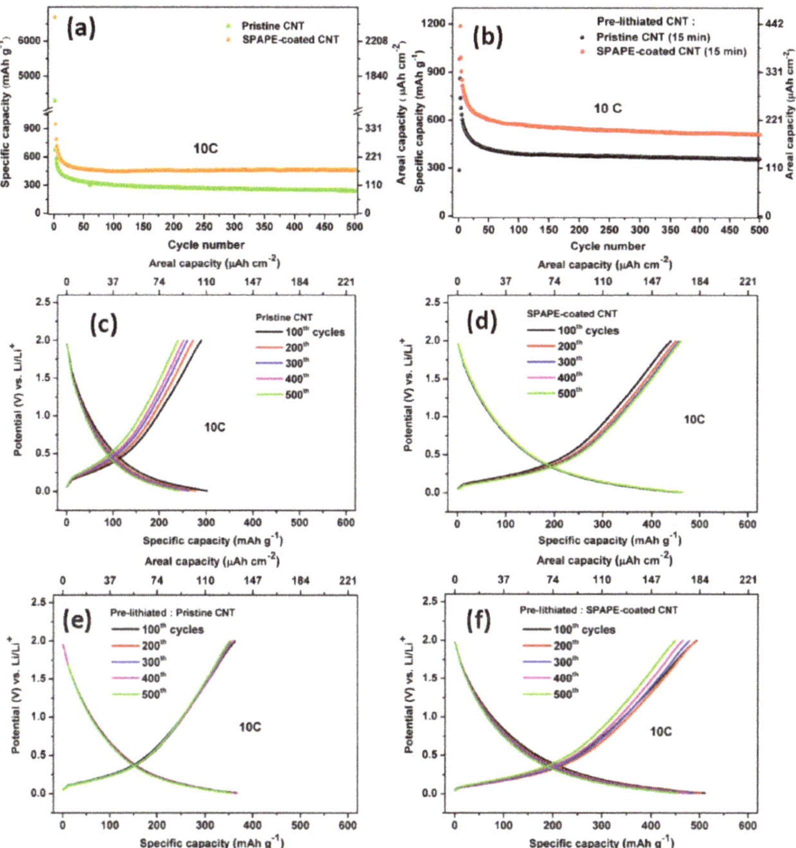

Figure 6. Long-term cycling performance of the (**a**) pristine and (**b**) pre-lithiated CNT anodes; galvanostatic charge–discharge curves of the (**c**) pristine CNT, (**d**) SPAPE-coated CNT, (**e**) pre-lithiated pristine CNT and (**f**) pre-lithiated SPAPE-coated CNT in the potential window of 0.01–2 V at a constant current density of 1.2 mA cm^{-2}.

To further examine the rate capability performance of the pre-lithiated CNT electrodes, galvanostatic charge–discharge cycling was carried out at progressively increased current density, ranging from 0.12 to 3.6 mA cm^{-2}. Figure 7a,b shows the galvanostatic charge–discharge profiles of the pre-lithiated CNT with polymer coating at a potential range of 0.01–2 V at 1 C for 50 cycles. After 50th cycle the coulombic efficiency reaches approximately 93%, corresponding to a charge capacity of 792 mAh g^{-1} (291 μAh cm^{-2}) and the discharge capacity is 850 mAh g^{-1} (313 μAh cm^{-2}) at 1 C.

Their reversible capacity is >2-fold higher compared to the storage capacity of graphite (372 mAh g^{-1}). We note that capacity fading occurs in the few initial cycles which could possibly due to the defects on CNT, impurities and irreversible lithium loss due to the side reactions upon cycling. However, the capacity is stabilized after 10 cycles and compared to pristine CNT which showed a severe capacity decay (~10 times) [33], the pre-lithiated CNTs show better cycling performance.

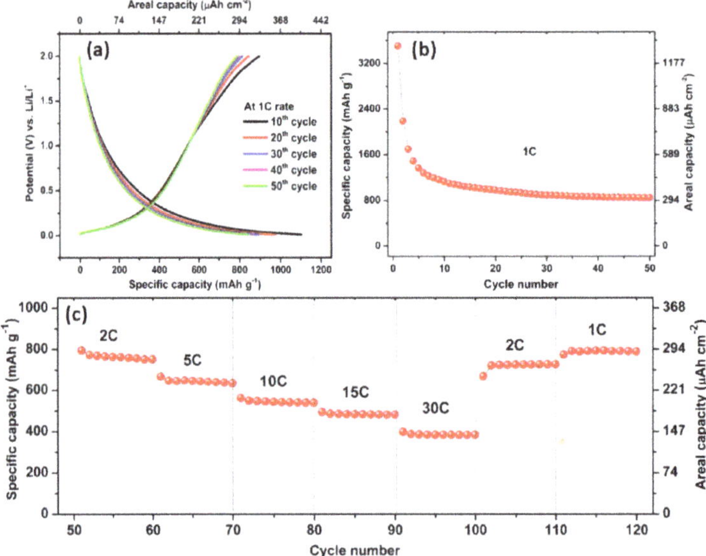

Figure 7. Typical galvanostatic charge–discharge potential profiles vs. Li/Li^+ for the 10th, 20th, 30th, 40th, and 50th cycle against the capacity of the pre-lithiated CNT electrode for 15 min at 1 C rate (**a**), rate capabilities of the pre-lithiated CNT electrode for 15 min at 1 C rate (**b**) and multiple C-rates (**c**).

Remarkably, even at stepwise accelerated rate, the pre-lithiated CNT could still provide excellent capacities of 764 mAh g^{-1} (281 μAh cm^{-2}) at 2 C, 647 mAh g^{-1} (238 μAh cm^{-2}) at 5 C, 545 mAh g^{-1} (200 μAh cm^{-2}) at 10 C, 484 mAh g^{-1} (178 μAh cm^{-2}) at 15 C and 385 mAh g^{-1} (142 μAh cm^{-2}) at 30 C (Figure 7c). Indeed, when the current density turned back from high to low current densities, the capacity can be greatly recovered over 120 cycles. Herein, we demonstrated that a simple direct pre-lithiation can improve the electrochemical performance of the flexible CNT anode by diminishing their extensive and large irreversible capacity, as well as providing a lithium supply to compensate the lithium loss during cycling. As a result, the CNT electrode shows good cycling stability over 500 cycles and excellent rate capabilities, even at high C-rates.

The promising performances of the CNT suggested their use as potential flexible anode material for application in full lithium-ion batteries using high-voltage cathode material, i.e., LNMO, in order to achieve high energy density. The galvanostatic charge–discharge profiles of the full-cell battery using both pristine CNT and pre-lithiated CNT as anodes are presented in the Figure 8. The battery was cycled at 1 C in the potential window of 2.5–4.4 V. Since the mass limitation is controlled by CNT, the capacity is reported versus the anode material and the mass was calculated considering a porosity of 70%. In the first cycle, the pristine CNT cell exhibits an initial charge capacity of 1856 μAh cm^{-2} and a discharge capacity of 76 μAh cm^{-2} with a relatively low coulombic efficiency of ca. 4.1% (Figure 8a), ascribed to the abovementioned SEI formation at the initial charging. This SEI film reduced the amount of active lithium ion upon cycling.

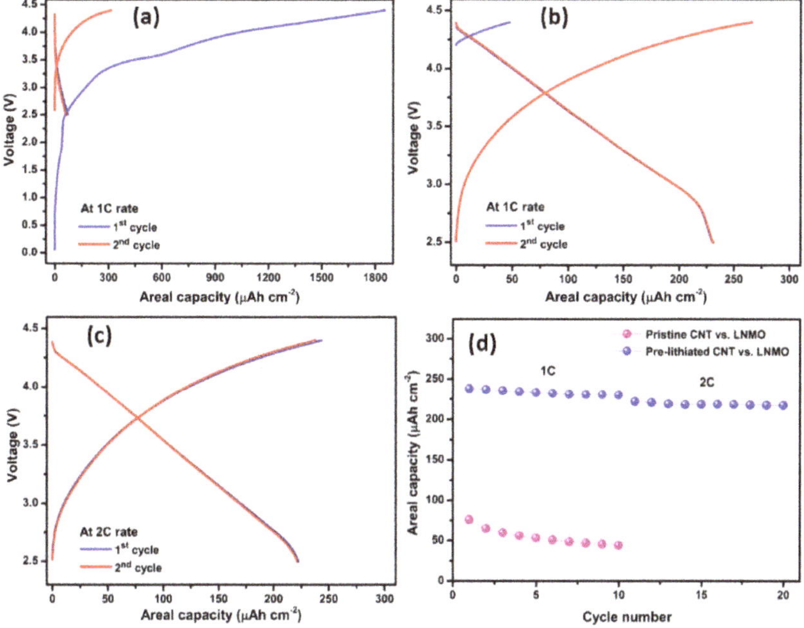

Figure 8. Typical galvanostatic charge–discharge potential profiles for the 1st and 2nd cycle at 1 C rate for (**a**) pristine CNT–LNMO, (**b**) pre-lithiated CNT-LNMO, (**c**) galvanostatic charge–discharge profiles of pre-lithiated CNT–LNMO at 2 C rate, (**d**) cycling performance of the pristine and pre-lithiated CNT anodes in the full-cell configuration.

In contrast, as seen in Figure 8b, the pre-lithiated CNT cell gives a superior electrochemical performance, delivering a discharge capacity of 238 μAh cm^{-2} with a cell voltage of about 3.75 V at 1 C. Another crucial point is that the OCV for the pristine CNT starts from ca. 0 V, while the OCV of the pre-lithiated CNT is ca. 4.2 V which means the SEI layer has been pre-formed on the CNT surface. For the pristine CNT, the capacity decreases 42% after 10th cycle at 1 C. The reason for this is likely ascribed to the relatively low coulombic efficiency of the pristine CNT in the half-cell (11.57%). At 2 C (Figure 8c), the pre-lithiated CNT reaches capacity of 223 μAh cm^{-2}. The important differences between pristine CNT and pre-lithiated CNT became obvious after 10 cycles charge–discharge cycling as seen in Figure 8d. In the full-cell configuration, the capacity of pre-lithiated CNT is 4 times higher compared to pristine CNT.

4. Conclusions

In summary, the electropolymerization of SPAPE polymer electrolyte into carbon nanotubes has been conducted by cyclic voltammetry. The enhanced electrochemical performance of SPAPE-coated CNT compared to pristine ones, due to the high electrode/electrolyte interface area leads to the improved charge transfer. In addition, this study also clearly showed the positive effect of direct pre-lithiation to suppress the initial irreversible capacity of CNT. Light-weight free-standing and flexible CNT without any binder and conductive additives have been successfully utilized as anode materials in Li-ion batteries, demonstrating both a stable and a high reversible capacity of 508 mAh g^{-1} (187 μAh cm^{-2}) at a 10 C rate over 500 cycles. Pre-lithiation of CNT via a self-discharge mechanism improves the first cycle coulombic efficiency from 11.57% to 68.30% after 15 min pre-lithiation period and reaches >100% subsequent to 30 min of pre-lithiation period. Moreover, by coupling the CNT anode with a high voltage LNMO spinel cathode, a 3.75 V full-cell presented a high capacity of 238 μAh cm^{-2} at a 1 C rate with the coulombic efficiency of ca. 90% in the initial cycle. This simple, fast and inexpensive method enables the achievement of high capacity flexible anodes for micro-batteries.

Author Contributions: V.A.S. Writing—original draft preparation. V.A.S. and N.Y. investigation. T.D., Y.E.-E. and F.V. conceptualization, supervision and writing—review & editing. All authors have read and agreed to the published version of the manuscript.

Funding: This research was funded by the Planning & Budgeting Committee of the Council of High Education and the Prime Minister's Office of Israel, in the framework of the INREP project and the support and funding the Grand Energy Technion Program (GTEP).

Acknowledgments: We acknowledge the Région SUD for financial support. We are grateful to the University of Roma Torgata for providing the SPAPE monomer and Tortech NanoFibers Ltd. (Israel) for producing and supplying the CNT.

Conflicts of Interest: The authors declare no conflict of interest.

References

1. Li, H.; Wang, Z.; Chen, L.; Huang, X. Research on Advanced Materials for Li-ion Batteries. *Adv. Mater.* **2009**, *21*, 4593–4607. [CrossRef]
2. Sugiawati, V.A.; Vacandio, F.; Perrin-Pellegrino, C.; Galeyeva, A.; Kurbatov, A.P.; Djenizian, T. Sputtered Porous Li-Fe-P-O Film Cathodes Prepared by Radio Frequency Sputtering for Li-ion Microbatteries. *Sci. Rep.* **2019**, *9*, 11172. [CrossRef] [PubMed]
3. Nitta, N.; Wu, F.; Lee, J.T.; Yushin, G. Li-ion battery materials: Present and future. *Mater. Today* **2015**, *18*, 252–264. [CrossRef]
4. Kumar, R.; Matsuo, R.; Kishida, K.; Abdel-Galeil, M.M.; Suda, Y.; Matsuda, A. Homogeneous reduced graphene oxide supported NiO-MnO_2 ternary hybrids for electrode material with improved capacitive performance. *Electrochim. Acta* **2019**, *303*, 246–256. [CrossRef]
5. Liu, X.M.; dong Huang, Z.; woon Oh, S.; Zhang, B.; Ma, P.C.; Yuen, M.M.F.; Kim, J.K. Carbon nanotube (CNT)-based composites as electrode material for rechargeable Li-ion batteries: A review. *Compos. Sci. Technol.* **2012**, *72*, 121–144. [CrossRef]
6. Li, L.; Yang, H.; Zhou, D.; Zhou, Y. Progress in Application of CNTs in Lithium-Ion Batteries. *J. Nanomater.* **2014**. [CrossRef]
7. Teprovich, J.A.; Weeks, J.A.; Ward, P.A.; Tinkey, S.C.; Huang, C.; Zhou, J.; Zidan, R.; Jena, P. Hydrogenated C60 as High-Capacity Stable Anode Materials for Li Ion Batteries. *ACS Appl. Energy Mater.* **2019**, *2*, 6453–6460. [CrossRef]
8. Qiao, L.; Sun, X.; Yang, Z.; Wang, X.; Wang, Q.; He, D. Network structures of fullerene-like carbon core/nano-crystalline silicon shell nanofibers as anode material for lithium-ion batteries. *Carbon* **2013**, *54*, 29–35. [CrossRef]
9. Kumar, R.; Sahoo, S.; Joanni, E.; Singh, R.K.; Tan, W.K.; Kar, K.K.; Matsuda, A. Recent progress in the synthesis of graphene and derived materials for next generation electrodes of high performance lithium ion batteries. *Prog. Energy Combust. Sci.* **2019**, *75*, 100786. [CrossRef]
10. Yadav, S.K.; Kumar, R.; Sundramoorthy, A.K.; Singh, R.K.; Koo, C.M. Simultaneous reduction and covalent grafting of polythiophene on graphene oxide sheets for excellent capacitance retention. *RSC Adv.* **2016**, *6*, 52945–52949. [CrossRef]
11. Wang, W.; Song, X.; Gu, C.; Liu, D.; Liu, J.; Huang, J. A high-capacity $NiCo_2O_4$@reduced graphene oxide nanocomposite Li-ion battery anode. *J. Alloys Compd.* **2018**, *741*, 223–230. [CrossRef]
12. Kumar, R.; Singh, R.K.; Alaferdov, A.V.; Moshkalev, S.A. Rapid and controllable synthesis of Fe_3O_4 octahedral nanocrystals embedded-reduced graphene oxide using microwave irradiation for high performance lithium-ion batteries. *Electrochim. Acta* **2018**, *281*, 78–87. [CrossRef]
13. Zhang, C.; Yu, J.S. Morphology-Tuned Synthesis of $NiCo_2O_4$-Coated 3D Graphene Architectures Used as Binder-Free Electrodes for Lithium-Ion Batteries. *Chem. Eur. J.* **2016**, *22*, 4422–4430. [CrossRef] [PubMed]
14. Chen, Y.; Zhu, J.; Qu, B.; Lu, B.; Xu, Z. Graphene improving lithium-ion battery performance by construction of $NiCo_2O_4$/graphene hybrid nanosheet arrays. *Nano Energy* **2014**, *3*, 88–94. [CrossRef]
15. Wang, C.; Wang, X.; Lin, C.; Zhao, X.S. Lithium Titanate Cuboid Arrays Grown on Carbon Fiber Cloth for High-Rate Flexible Lithium-Ion Batteries. *Small* **2019**, *15*, 1902183. [CrossRef] [PubMed]

16. Zhang, Z.; Zhang, M.; Lu, P.; Chen, Q.; Wang, H.; Liu, Q. CuO nanorods growth on folded Cu foil as integrated electrodes with high areal capacity for flexible Li-ion batteries. *J. Alloys Compd.* **2019**, *809*, 151823. [CrossRef]
17. Nasreldin, M.; Delattre, R.; Ramuz, M.; Lahuec, C.; Djenizian, T.; de Bougrenet de la Tocnaye, J.L. Flexible Micro-Battery for Powering Smart Contact Lens. *Sensors* **2019**, *19*, 2062. [CrossRef]
18. Yitzhack, N.; Auinat, M.; Sezin, N.; Ein-Eli, Y. Carbon nanotube tissue as anode current collector for flexible Li-ion batteries—Understanding the controlling parameters influencing the electrochemical performance. *APL Mater.* **2018**, *6*, 111102. [CrossRef]
19. Yehezkel, S.; Auinat, M.; Sezin, N.; Starosvetsky, D.; Ein-Eli, Y. Distinct Copper Electrodeposited Carbon Nanotubes (CNT) Tissues as Anode Current Collectors in Li-ion Battery. *Electrochim. Acta* **2017**, *229*, 404–414. [CrossRef]
20. Cai, M.; Sun, X.; Chen, W.; Qiu, Z.; Chen, L.; Li, X.; Wang, J.; Liu, Z.; Nie, Y. Performance of lithium-ion capacitors using pre-lithiated multiwalled carbon nanotubes/graphite composite as negative electrode. *J. Mater. Sci.* **2018**, *53*, 749–758. [CrossRef]
21. Li, X.; Liu, J.; Zhang, Y.; Li, Y.; Liu, H.; Meng, X.; Yang, J.; Geng, D.; Wang, D.; Li, R.; et al. High concentration nitrogen doped carbon nanotube anodes with superior Li+ storage performance for lithium rechargeable battery application. *J. Power Sources* **2012**, *197*, 238–245. [CrossRef]
22. Aguiló-Aguayo, N.; Amade, R.; Hussain, S.; Bertran, E.; Bechtold, T. New Three-Dimensional Porous Electrode Concept: Vertically-Aligned Carbon Nanotubes Directly Grown on Embroidered Copper Structures. *Nanomaterials* **2017**, *7*, 438. [CrossRef] [PubMed]
23. Peng, X.X.; Qiao, X.; Luo, S.; Yao, J.A.; Zhang, Y.F.; Du, F.P. Modulating Carrier Type for Enhanced Thermoelectric Performance of Single-Walled Carbon Nanotubes/Polyethyleneimine Composites. *Polymers* **2019**, *11*, 1295. [CrossRef] [PubMed]
24. Wang, F.; Feng, L.; Lu, M. Mechanical Properties of Multi-Walled Carbon Nanotube/Waterborne Polyurethane Conductive Coatings Prepared by Electrostatic Spraying. *Polymers* **2019**, *11*, 714. [CrossRef] [PubMed]
25. Kumar, R.; Singh, R.K.; Tiwari, V.S.; Yadav, A.; Savu, R.; Vaz, A.R.; Moshkalev, S.A. Enhanced magnetic performance of iron oxide nanoparticles anchored pristine/N-doped multi-walled carbon nanotubes by microwave-assisted approach. *J. Alloys Compd.* **2017**, *695*, 1793–1801. [CrossRef]
26. Kumar, R.; Yadav, R.M.; Awasthi, K.; Tiwari, R.S.; Srivastava, O.N. Effect of nitrogen variation on the synthesis of vertically aligned bamboo-shaped c–n nanotubes using sunflower oil. *Int. J. Nanosci.* **2011**, *10*, 809–813. [CrossRef]
27. Gao, H.; Hou, F.; Zheng, X.; Liu, J.; Guo, A.; Yang, D.; Gong, Y. Electrochemical property studies of carbon nanotube films fabricated by CVD method as anode materials for lithium-ion battery applications. *Vacuum* **2015**, *112*, 1–4. [CrossRef]
28. Hou, G.; Chauhan, D.; Ng, V.; Xu, C.; Yin, Z.; Paine, M.; Su, R.; Shanov, V.; Mast, D.; Schulz, M.; et al. Gas phase pyrolysis synthesis of carbon nanotubes at high temperature. *Mater. Des.* **2017**, *132*, 112–118. [CrossRef]
29. Arora, N.; Sharma, N.N. Arc discharge synthesis of carbon nanotubes: Comprehensive review. *Diam. Relat. Mater.* **2014**, *50*, 135–150. [CrossRef]
30. Kumar, R.; Singh, R.K.; Dubey, P.K.; Yadav, R.M.; Singh, D.P.; Tiwari, R.S.; Srivastava, O.N. Highly zone-dependent synthesis of different carbon nanostructures using plasma-enhanced arc discharge technique. *J. Nanopart. Res.* **2015**, *17*, 24. [CrossRef]
31. Chrzanowska, J.; Hoffman, J.; Małolepszy, A.; Mazurkiewicz, M.; Kowalewski, T.A.; Szymanski, Z.; Stobinski, L. Synthesis of carbon nanotubes by the laser ablation method: Effect of laser wavelength. *Phys. Status Solidi B* **2015**, *252*, 1860–1867. [CrossRef]
32. Schwandt, C.; Dimitrov, A.T.; Fray, D.J. High-yield synthesis of multi-walled carbon nanotubes from graphite by molten salt electrolysis. *Carbon* **2012**, *50*, 1311–1315. [CrossRef]
33. Sugiawati, V.A.; Vacandio, F.; Ein-Eli, Y.; Djenizian, T. Electrodeposition of polymer electrolyte into carbon nanotube tissues for high performance flexible Li-ion microbatteries. *APL Mater.* **2019**, *7*, 031506. [CrossRef]
34. Yoon, S.; Lee, S.; Kim, S.; Park, K.W.; Cho, D.; Jeong, Y. Carbon nanotube film anodes for flexible lithium ion batteries. *J. Power Sources* **2015**, *279*, 495–501. [CrossRef]

35. Zou, L.; Lv, R.; Kang, F.; Gan, L.; Shen, W. Preparation and application of bamboo-like carbon nanotubes in lithium ion batteries. *J. Power Sources* **2008**, *184*, 566–569. [CrossRef]
36. Welna, D.T.; Qu, L.; Taylor, B.E.; Dai, L.; Durstock, M.F. Vertically aligned carbon nanotube electrodes for lithium-ion batteries. *J. Power Sources* **2011**, *196*, 1455–1460. [CrossRef]
37. Yehezkel, S.; Auinat, M.; Sezin, N.; Starosvetsky, D.; Ein-Eli, Y. Bundled and densified carbon nanotubes (CNT) fabrics as flexible ultra-light weight Li-ion battery anode current collectors. *J. Power Sources* **2016**, *312*, 109–115. [CrossRef]
38. De las Casas, C.; Li, W. A review of application of carbon nanotubes for lithium ion battery anode material. *J. Power Sources* **2012**, *208*, 74–85. [CrossRef]
39. Lee, S.; Song, H.; Hwang, J.Y.; Jeong, Y. Directly-prelithiated carbon nanotube film for high-performance flexible lithium-ion battery electrodes. *Fibers Polym.* **2017**, *18*, 2334–2341. [CrossRef]
40. Holtstiege, F.; Bärmann, P.; Nölle, R.; Winter, M.; Placke, T. Pre-Lithiation Strategies for Rechargeable Energy Storage Technologies: Concepts, Promises and Challenges. *Batteries* **2018**, *4*, 4. [CrossRef]
41. Di Lecce, D.; Andreotti, P.; Boni, M.; Gasparro, G.; Rizzati, G.; Hwang, J.Y.; Sun, Y.K.; Hassoun, J. Multiwalled Carbon Nanotubes Anode in Lithium-Ion Battery with $LiCoO_2$, $Li[Ni_{1/3}Co_{1/3}Mn_{1/3}]O_2$, and $LiFe_{1/4}Mn_{1/2}Co_{1/4}PO_4$ Cathodes. *ACS Sustain. Chem. Eng.* **2018**, *6*, 3225–3232. [CrossRef]
42. Seong, I.W.; Kim, K.T.; Yoon, W.Y. Electrochemical behavior of a lithium-pre-doped carbon-coated silicon monoxide anode cell. *J. Power Sources* **2009**, *189*, 511–514. [CrossRef]
43. Liu, N.; Hu, L.; McDowell, M.T.; Jackson, A.; Cui, Y. Prelithiated silicon nanowires as an anode for lithium ion batteries. *ACS Nano* **2011**, *5*, 6487–6493. [CrossRef]
44. Scott, M.G.; Whitehead, A.H.; Owen, J.R. Chemical Formation of a Solid Electrolyte Interface on the Carbon Electrode of a Li-Ion Cell. *J. Electrochem. Soc.* **1998**, *145*, 1506–1510. [CrossRef]
45. Wu, Y.; Yokoshima, T.; Nara, H.; Momma, T.; Osaka, T. A pre-lithiation method for sulfur cathode used for future lithium metal free full battery. *J. Power Sources* **2017**, *342*, 537–545. [CrossRef]
46. Plylahan, N.; Letiche, M.; Barr, M.K.S.; Djenizian, T. All-solid-state lithium-ion batteries based on self-supported titania nanotubes. *Electrochem. Commun.* **2014**, *43*, 121–124. [CrossRef]
47. Ferrari, I.V.; Braglia, M.; Djenizian, T.; Knauth, P.; Di Vona, M.L. Electrochemically engineered single Li-ion conducting solid polymer electrolyte on titania nanotubes for microbatteries. *J. Power Sources* **2017**, *353*, 95–103. [CrossRef]
48. Dresselhaus, M.S.; Dresselhaus, G.; Saito, R.; Jorio, A. Raman spectroscopy of carbon nanotubes. *Phys. Rep.* **2005**, *409*, 47–99. [CrossRef]
49. Dresselhaus, M.S.; Jorio, A.; Hofmann, M.; Dresselhaus, G.; Saito, R. Perspectives on Carbon Nanotubes and Graphene Raman Spectroscopy. *Nano Lett.* **2010**, *10*, 751–758. [CrossRef]
50. Aravindan, V.; Lee, Y.S.; Madhavi, S. Best Practices for Mitigating Irreversible Capacity Loss of Negative Electrodes in Li-Ion Batteries. *Adv. Energy Mater.* **2017**, *7*, 1602607. [CrossRef]
51. Zhao, J.; Lu, Z.; Liu, N.; Lee, H.W.; McDowell, M.T.; Cui, Y. Dry-air-stable lithium silicide–lithium oxide core–shell nanoparticles as high-capacity prelithiation reagents. *Nat. Commun.* **2014**, *5*, 5088. [CrossRef] [PubMed]
52. Wan, Y.; Wang, L.; Chen, Y.; Xu, X.; Wang, Y.; Teng, C.; Zhou, D.; Chen, Z. A high-performance tin dioxide@carbon anode with a super high initial coulombic efficiency via a primary cell prelithiation process. *J. Alloys Compd.* **2018**, *740*, 830–835. [CrossRef]
53. Sun, H.; Xin, G.; Hu, T.; Yu, M.; Shao, D.; Sun, X.; Lian, J. High-rate lithiation-induced reactivation of mesoporous hollow spheres for long-lived lithium-ion batteries. *Nat. Commun.* **2014**, *5*, 1–8. [CrossRef]
54. Sun, Y.; Lee, H.W.; Seh, Z.W.; Liu, N.; Sun, J.; Li, Y.; Cui, Y. High-capacity battery cathode prelithiation to offset initial lithium loss. *Nat. Energy* **2016**, *1*, 15008. [CrossRef]
55. Abe, Y.; Saito, T.; Kumagai, S. Effect of Prelithiation Process for Hard Carbon Negative Electrode on the Rate and Cycling Behaviors of Lithium-Ion Batteries. *Batteries* **2018**, *4*, 71. [CrossRef]
56. Kyeremateng, N.A.; Dumur, F.; Knauth, P.; Pecquenard, B.; Djenizian, T. Electropolymerization of copolymer electrolyte into titania nanotube electrodes for high-performance 3D microbatteries. *Electrochem. Commun.* **2011**, *13*, 894–897. [CrossRef]

57. Sugiawati, V.A.; Vacandio, F.; Djenizian, T.; Galeyeva, A.; Kurbatov, A.P. Superior Electrochemical Performance of Electropolymerized Self-Organized TiO_2 Nanotubes Fabricated by Anodization of Ti Grid. *Front. Phys.* **2019**, *7*. [CrossRef]
58. Vargas, Ó.; Caballero, Á.; Morales, J.; Rodríguez-Castellón, E. Contribution to the Understanding of Capacity Fading in Graphene Nanosheets Acting as an Anode in Full Li-Ion Batteries. *ACS Appl. Mater. Interfaces* **2014**, *6*, 3290–3298. [CrossRef]

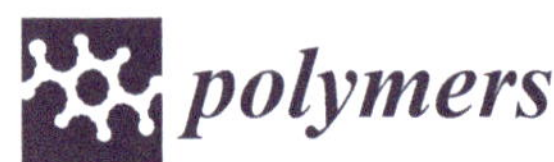

Article

Formation Features of Hybrid Nanocomposites Based on Polydiphenylamine-2-Carboxylic Acid and Single-Walled Carbon Nanotubes

Sveta Zhiraslanovna Ozkan *, Galina Petrovna Karpacheva, Aleksandr Ivanovich Kostev and Galina Nikolaevna Bondarenko

A.V. Topchiev Institute of Petrochemical Synthesis, Russian Academy of Sciences, 29 Leninsky prospect, 119991 Moscow, Russia

* Correspondence: ozkan@ips.ac.ru

Received: 31 May 2019; Accepted: 10 July 2019; Published: 13 July 2019

Abstract: Hybrid nanocomposites based on electroactive polydiphenylamine-2-carboxylic acid (PDPAC) and single-walled carbon nanotubes (SWCNTs) were obtained for the first time. Polymer-carbon nanomaterials were synthesized via in situ oxidative polymerization of diphenylamine-2-carboxylic acid (DPAC) in the presence of SWCNTs by two different ways. Hybrid SWCNT/PDPAC nanocomposites were prepared both in an acidic medium and in the heterophase system in an alkaline medium. In the heterophase system, the monomer and the SWCNTs are in the organic phase (chloroform) and the oxidant (ammonium persulfate) is in an aqueous solution of ammonium hydroxide. The chemical structure, as well as the electrical and thermal properties of the developed SWCNT/PDPAC nanocomposite materials were investigated.

Keywords: polydiphenylamine-2-carboxylic acid; single-walled carbon nanotubes; conjugated polymers; in situ oxidative polymerization; hybrid nanocomposites

1. Introduction

One of the most promising areas of development in the nanotechnology industry is the creation of nanomaterials that offer properties required by modern technologies. Researchers are particularly interested in hybrid nanocomposites that include carbon nanotubes (CNTs) [1–5] dispersed in a conjugated polymer matrix [6–9]. Conjugated polymers are a special class of polymer materials, the characteristic feature of which is the delocalization of π-electrons along the conjugation chain [10–12]. The specific electronic structure of polymers with a system of conjugated double bonds determines the electrophysical, electrochemical, optical, and other properties of these materials [13,14]. Due to the electronic interaction of polymer and carbon constituents in such systems, fundamentally new or enhanced properties, as compared to initial components, can be expected to appear (thermal stability, mechanical strength, electrical and thermal conductivity, etc.). This increases the potential range of their practical application and opens opportunities for solving numerous technological problems in electronics, microsystem technology, energy industry, engineering, medicine, etc. Due to the complementary properties, nanocomposite materials based on CNTs and conjugated polymers are promising for use in organic electronics, health care, for creating optoelectronic devices, thin film transistors, memory modules, electrochemical power sources, supercapacitors, sensors, displays, etc.

Nowadays, many methods for the production of nanocomposites based on CNTs and conjugated polymers have been developed. Polyaniline (PANI), polypyrrole, and polythiophene, as well as their derivatives are used as polymer components [6–9]. A serious problem in obtaining these nanocomposites is the tendency of CNTs to aggregate. The CNTs' aggregation prevents their homogeneous distribution in a polymer matrix and therefore does not allow the required properties to be obtained. The cause for this lies in the insolubility of CNTs in organic solvents and the lack of compatibility of the polymer with CNTs. The use of electrodeposition of polymers onto the CNT surface [15–20], electrochemical polycondensation of monomers on the surface of oriented CNTs [21], and dissolving polymers in a CNT suspension in an organic solvent [22,23] to produce hybrid materials does not allow a high dispersion of the carbon component in the polymer matrix.

In situ oxidative polymerization in the presence of CNTs is an effective method for the prevention of CNT aggregation [24–30]. The use of an ultrasound ensures a uniform CNT distribution in the reaction medium and prevents the nanotubes' coagulation during polymerization [31–33]. Studies of oxidative polymerization of aniline in the presence of CNTs showed that PANI forms a uniform polymer coating on the surface of carbon nanomaterials [34–38]. The formation of polymer chains on the CNT surface causes a strong π–π^* interaction [23,36,39,40] between the nanomaterial components. The π–π^* interaction occurs due to the charge transfer from the quinoid units of PANI to the aromatic structures of CNTs. This enhances the electron transport in the nanocomposite material compared to the pure polymer, thus improving its electrical properties [25,31,32]. The electrical conductivity of MWCNT/PANI increased with MWCNT loading, especially at low loadings, typically below 5 wt %. It was attributed to the induced crystallinity of PANI at the MWCNT surface [25].

Also, high dispersion of the CNT distribution can be achieved by CNT surface modification due to the covalent bonding of functional groups [41–45]. In situ polymerization in the presence of covalently functionalized CNTs not only provides the formation of a homogeneous nanostructure, but also prevents phase micro-separation. The PANI chains deposited on the surface of carboxylic acid functionalized multi-walled carbon nanotubes (MWCNTs) have longer conjugation lengths than the pure PANI [45]. This could be attributed to the site-selective interaction between the conjugated structure of PANI via the quinoid ring and the π-bonded surface of functionalized MWCNTs [36].

A comparative study of MWCNT/PANI with SWCNT/PANI nanocomposite films showed that the electrical conductivity was higher with the use of SWCNTs. This is due to the formation of the core-shell SWCNT-PANI structure and the absence of the agglomerated polymerized PANI islands, which are less conductive. The composite films reached their percolation threshold at the 0.1 wt % MWCNT loading and at the 0.05 wt % SWCNT loading [46].

Earlier, we obtained a hybrid nanomaterial based on poly-3-amine-7-methylamine-2-methylphenazine (PAMMP) and single-walled carbon nanotubes (SWCNTs) [47,48]. The nanocomposite synthesis was carried out in an aqueous solution of acetonitrile via oxidative polymerization of 3-amine-7-dimethylamine-2-methylphenazine hydrochloride in the presence of SWCNTs. A shift of the skeletal oscillation frequencies of PAMMP indicated the π–π^* interaction between the phenazine units of the polymer and SWCNTs.

In this research paper, nanocomposite materials based on polydiphenylamine-2-carboxylic acid (PDPAC) and SWCNTs were prepared for the first time. Hybrid SWCNT/PDPAC nanomaterials were synthesized via in situ oxidative polymerization of diphenylamine-2-carboxylic acid (DPAC) in the presence of SWCNTs in an alkaline and an acidic media. Single-walled carbon nanotubes were chosen because they are characterized by a smaller diameter and better dispersibility in the polymer matrix. Due to this, a greater amount of polymer formed on the CNT surface, forming the core-shell structure [46]. PDPAC is a novel conjugated polyacid synthesized by the authors [49–51], which is the N-substituted polyaniline containing an aromatic substituent with a carboxyl group in the *orto* position. The presence of the functional group can be expected to determine the electronic interaction between CNTs and the polymer not only through the main polymer chain but also through side substitutes. The influence of the reaction medium pH and the SWCNT content on the structure, morphology, thermal stability, and electrical properties of nanocomposites was established. The use of PDPAC in the nanocomposites expands the range of conjugated polymers involved in the creation of novel materials for modern technologies.

2. Experimental

2.1. Materials

Ammonium persulfate (analytical grade) was purified by recrystallization from distilled water by a known procedure [52]. Diphenylamine-2-carboxylic acid (N-phenylanthranilic acid) ($C_{13}H_{11}O_2N$) (analytical grade), aqueous ammonia (reagent grade), sulfuric acid (reagent grade), chloroform (reagent grade), and DMF (Acros Organics) were used as received without any additional purification. The aqueous solutions of reagents were prepared with the use of bidistilled water. SWCNTs from Carbon Chg, Ltd. (Moscow, Russia), with values of d = 1.4 to 1.6 nm, and l = 0.5 to 1.5 μm, were produced by the electric arc discharge technique with Ni/Y catalyst.

2.2. Preparation of SWCNT/PDPAC Nanocomposites

The SWCNT/PDPAC nanocomposites were prepared by two ways. The synthesis of the nanocomposite in the heterophase system in an alkaline medium (SWCNT/PDPAC-1) was carried out as follows. First, the required amount of monomer (DPAC) (0.1 mol/L, 0.64 g) was dissolved in a mixture of an organic solvent—chloroform (15 mL) and alkali (NH_4OH) (0.5 mol/L, 2.3 mL). The SWCNTs were added to the resulting solution. The content of carbon nanotubes was C_{SWCNT} = 1 to 3, 10 wt % relative to the monomer weight. The process was carried out at room temperature with constant intensive stirring for 1 h. Then, for the in situ oxidative polymerization of DPAC in the presence of SWCNTs, an aqueous solution (15 mL) of an oxidizing agent (ammonium persulfate) (0.2 mol/L, 1.368 g) was added to the SWCNT/DPAC suspension in a mixture of chloroform and NH_4OH. Ammonium persulfate was added in one go, without gradual dosing of reagents. The SWCNT/DPAC suspension was pre-cooled to 0 °C by using the LOIP FT-311-25 cryothermostat (Saint-Petersburg, Russia). The volume ratio of organic and aqueous phases was 1:1 (V_{total} = 30 mL). The synthesis was carried out for 3 h at 0 °C under intensive stirring using an electronic stirrer with a RW 16 basic upper drive (Ika Werke, Germany). A narrow cylinder-shaped round-bottom two-neck flask was used for increasing the efficiency of stirring. When the reaction was completed, the mixture was precipitated in a 10-fold excess of a 2% solution of H_2SO_4. The resulting product was filtered off, washed repeatedly with distilled water to remove residual amounts of reagents, and vacuum-dried over KOH to constant weight. The yield of the SWCNT/PDPAC-1 nanocomposite was 0.44 g (67.56%) with C_{SWCNT} = 2 wt %.

The synthesis of the nanocomposite in an acidic medium (SWCNT/PDPAC-2) was carried out as follows. First, the SWCNTs were added to the monomer solution (0.1 mol/L, 0.64 g) in 5 M H_2SO_4. The SWCNT/DPAC suspension was stirred in an ultrasonic bath (UZV-2414, Vologda, Russia) at room temperature for 0.5 h. The mixture was heated to 25 °C. The content of carbon nanotubes was C_{SWCNT} = 1 to 3, 10 wt % relative to the monomer weight. Then, for the oxidative polymerization of DPAC in the presence of SWCNTs, aqueous solution of ammonium persulfate (0.2 mol/L, 1.368 g) in the same solvent (1/4 of the total volume, V_{total} = 30 mL) was added to the SWCNT/DPAC suspension. The oxidizer was added drop-wise under intensive stirring. The SWCNT/DPAC suspension was pre-cooled to 0 °C. The synthesis continued for 3 h with intense stirring at 0 °C. When the synthesis was completed, the reaction mixture was precipitated in 200 mL of distilled water. The resulting product was filtered off, and washed repeatedly with 1% solution of H_2SO_4 to remove residual reagents. The product was vacuum-dried over $CaCl_2$ to constant weight. The yield of the SWCNT/PDPAC-2 nanocomposite was 0.57 g (86.47%) at C_{SWCNT} = 3 wt %.

2.3. Synthesis of PDPAC

For comparison with nanocomposites, polydiphenylamine-2-carboxylic acid (PDPAC) was prepared under the same conditions. The PDPAC-1 was synthesized via oxidative polymerization in the heterophase system in an alkaline medium ($C_{monomer}$ = 0.1 mol/L, $C_{oxidizer}$ = 0.2 mol/L, C_{alkali} = 0.5 mol/L) [49]. The PDPAC-2 was obtained in the homogeneous acidic medium ($C_{monomer}$ = 0.1 mol/L, $C_{oxidizer}$ = 0.2 mol/L, C_{acid} = 5 mol/L) [50,51].

2.4. Characterization

Attenuated total reflection (ATR) FTIR spectra of the samples in the attenuated total reflectance mode were measured by using a HYPERION-2000 IR microscope (Bruker, Karlsruhe, Germany). The microscope was coupled with the Bruker IFS 66v FTIR spectrometer (Karlsruhe, Germany). The optical range was 600 to 4000 cm^{-1} (150 scans, ZnSe crystal, resolution of 2 cm^{-1}).

Electronic absorption spectra of samples in DMF were registered by using an UV-1700 spectrophotometer (Shimadzu, Kyoto, Japan) in the range of 190 to 1100 nm.

The ^{13}C solid-state CP/MAS NMR spectra were recorded on an Infinity INOVA 500 NMR spectrometer (Varian Inc., Palo Alto, California, USA). CP/MAS NMR spectra for ^{13}C nuclei were registered by using the direct polarization (90° pulse at ^{13}C was 2.5 μs, time lapse between scans was 30 s) [53]. Crystalline adamantane was used as the secondary external standard of the chemical shift scale [54,55].

An electron microscopic study was performed by using a Zeiss Supra 25 FE-SEM field emission scanning electron microscope (Carl Zeiss AG, Jena, Germany).

An X-ray diffraction study was performed in ambient atmosphere by using a Difray-401 X-ray diffractometer (Scientific Instruments Joint Stock Company, Saint-Petersburg, Russia) with Bragg-Brentano focusing on CrK_{α} radiation with λ = 0.229 nm.

Thermogravimetric analysis (TGA) was carried out with a Mettler Toledo TGA/DSC1 (Columbus, Ohio, USA) in the dynamic mode. The test temperature ranged from 30 to 1000 °C (100 mg sample, 10 °C/min heating rate, air and argon atmosphere, 10 mL/min argon flow velocity). The samples were analyzed in an Al_2O_3 crucible.

Differential scanning calorimetry (DSC) was performed on a Mettler Toledo DSC823^e calorimeter (Columbus, Ohio, USA). The investigation was carried out from room temperature to 350 °C at a heating rate of 10 °C/min under the nitrogen flow of 70 mL/min.

A 6367A precision LCR meter (Microtest Co., Ltd., New Taipei City, Taiwan) was used to measure the *ac* conductivity in the range of frequency of 0.1 Hz to 1.15 MHz.

3. Results and Discussion

3.1. Synthesis and Characterization of Materials

Two methods of obtaining polymer-carbon hybrid nanocomposites based on thermostable polydiphenylamine-2-carboxylic acid (PDPAC) [49–51] and single-walled carbon nanotubes (SWCNTs) were proposed. Hybrid nanomaterials were synthesized via in situ oxidative polymerization of diphenylamine-2-carboxylic acid (DPAC) in the presence of SWCNTs in the heterophase system in an alkaline medium (pH 11.4) (SWCNT/PDPAC-1) and in an acidic medium (pH 0.3) (SWCNT/PDPAC-2). Meanwhile, in the heterophase system, in an alkaline medium, the monomer and the SWCNTs were in the organic phase (chloroform) and the oxidant (ammonium persulfate) was in an aqueous solution of ammonium hydroxide. For comparison, polymers of diphenylamine-2-carboxylic acid were synthesized under the same conditions. The PDPAC-1 was obtained in NH_4OH solution in the presence of chloroform. The PDPAC-2 was prepared in 5 M H_2SO_4.

Earlier, we showed that during DPAC polymerization in a sulfuric acid solution, the growth of a polymer chain occurs via the C–C bonding into the *para* position of phenyl rings relative to nitrogen [50,51]. The ATR FTIR spectrum of the polymer prepared in 5 M H_2SO_4 shows absorption bands at 892 and 803 cm^{-1} related to out-of-plane bending vibrations of δ_{C-H} bonds of 1,4-disubstituted and 1,2,4-trisubstituted benzene rings. The chemical structure of PDPAC-2 obtained in 5 M H_2SO_4 (pH 0.3) has the following form:

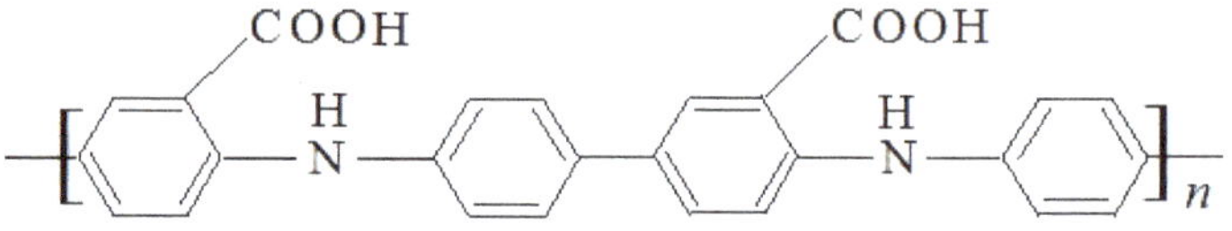

Polymerization of DPAC in the heterophase system in an alkaline medium results in another type of bonding. The ATR FTIR spectrum of the polymer obtained in ammonium hydroxide solution in the presence of chloroform shows the broadening and shift to 753 cm^{-1} of the absorption band at 746 cm^{-1}. Also, the ATR FTIR spectrum shows the absence of a band at 892 cm^{-1}, and the presence of a wide band at 828 cm^{-1}. It all indicates the presence of 1,2-disubstituted and 1,2,4-trisubstituted aromatic rings in the PDPAC-1 structure. This suggests that, during the polymerization of DPAC in an alkaline medium, the growth of a polymer chain proceeds via the C–C bonding in the 2- and 4-positions of phenyl rings relative to nitrogen [49]. The chemical structure of PDPAC-1 prepared in NH_4OH solution in the presence of chloroform (pH 11.4) has the following form:

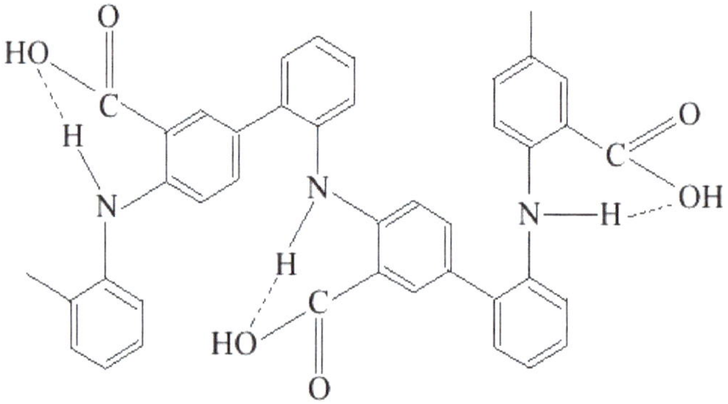

A comparison of ATR FTIR spectra of polymers with nanocomposites prepared under the same conditions was made. It was shown that all the main bands characterizing the chemical structure of PDPAC remain in the SWCNT/PDPAC nanocomposite FTIR spectra.

Figure 1 shows ATR FTIR spectra of nanocomposites obtained in NH_4OH solution in the presence of chloroform (SWCNT/PDPAC-1) and in 5 M H_2SO_4 (SWCNT/PDPAC-2), depending on the SWCNT concentration. Table 1 shows the assignment of main characteristic absorption bands in the ATR FTIR spectra of PDPAC and SWCNT/PDPAC nanocomposites according to the synthesis method.

Based on the data shown in Figure 1 and Table 1, it can be concluded that the chemical structure of the polymer matrix has a strong dependency on the pH of the reaction medium for the nanocomposite synthesis. It was established that during DPAC polymerization in an acidic medium (pH 0.3), in the presence of SWCNTs, the polymer chain grows via the C–C bonding into the *para* position of phenyl rings relative to nitrogen ($\delta_{C\text{–}H}$ = 862 and 802 cm^{-1}). The process is the same as during the DPAC polymerization under these conditions. In the SWCNT/PDPAC-2 nanocomposite, the absorption bands at 862 and 802 cm^{-1} are due to out-of-plane bending vibrations of $\delta_{C\text{–}H}$ bonds of 1,4-disubstituted and 1,2,4-trisubstituted benzene rings (Figure 1b).

During the PDPAC/SWCNT-1 nanocomposite synthesis in the heterophase system in an alkaline medium (pH 11.4) in the presence of SWCNTs, the polymer chain grows via the C–C bonding into the 2- and 4-positions of phenyl rings relative to nitrogen ($\delta_{C\text{–}H}$ = 828 and 748 cm^{-1}). The absorption bands at 828 and 748 cm^{-1} in the SWCNT/PDPAC-1 nanocomposite correspond to out-of-plane bending vibrations of the $\delta_{C\text{–}H}$ bonds of 1,2,4- and 1,2-substituted benzene rings (Figure 1a). The absorption bands at 1680 and 1215 cm^{-1} characterize the stretching vibrations of $\nu_{C=O}$ in COOH groups. At the same time, COOH groups ($\nu_{C=O}$ = 1680 and 1215 cm^{-1}) are associated with the N–H group

($\nu_{N\text{-}H}$ = 3175 cm^{-1}) of the main chain, the same as during the polymerization of DPAC in an alkaline medium [49]. Carboxyl groups along the entire polymer chain form intramolecular hydrogen bonds with amino groups. This is confirmed by the presence in the ATR FTIR spectra of the PDPAC/SWCNT-1 nanocomposite the absorption band at 3264 cm^{-1}. This band characterizes the associated COOH—N-H carboxyl groups with the hydrogen bond (Table 1).

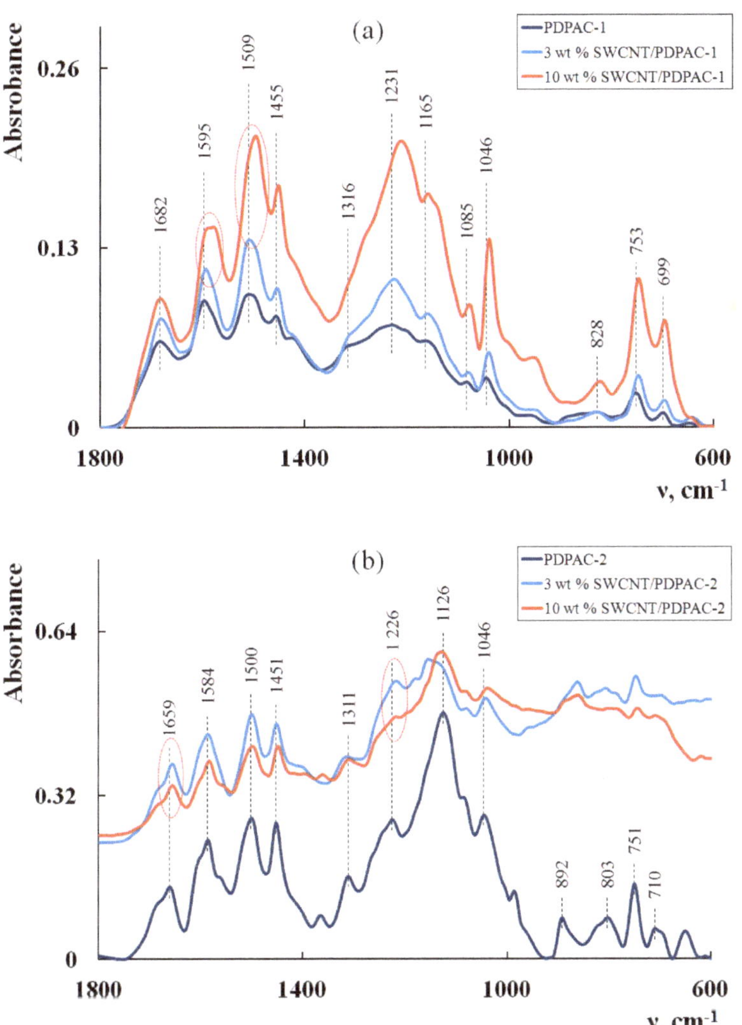

Figure 1. Attenuated total reflection (ATR) FTIR spectra of the PDPAC-1 (**a**) and PDPAC-2 (**b**), and SWCNT/PDPAC-1 (**a**) and SWCNT/PDPAC-2 nanocomposites (**b**).

Table 1. Assignment of main characteristic absorption bands in the FTIR spectra of materials.

Assignment of Absorption Bands	Frequency ν, cm^{-1}				
	PDPAC-1	**SWCNT/ PDPAC-1**	**PDPAC-2**	**SWCNT/ PDPAC-2**	**DPAC (monomer)**
Stretching vibrations of $\nu_{N–H}$	3236	3175	3303	3308	3337
H—N-H with hydrogen bond	3293	3264	-	-	-
Stretching vibrations of $\nu_{C–H}$ in an aromatic ring	3081	3062	3020	3033	3034
Stretching vibrations of $\nu_{C=O}$ in COOH	1682 1231	1680 1215	1659 1226	1655 1218	1658 1259
Stretching vibrations of $\nu_{C–C}$ in an aromatic ring	1595 1509	1585 1495	1584 1500	1582 1498	1575 1509
Stretching vibrations of $\nu_{C–N}$	1316	1313	1311	1310	1323
Out-of-plane bending vibrations of $\delta_{C–H}$ in an 1,2-substituted aromatic ring	753	748	751	746	746
Out-of-plane bending vibrations of $\delta_{C–H}$ in an 1,2,4-trisubstituted aromatic ring	828	828	803	802	-
Out-of-plane bending vibrations of $\delta_{C–H}$ in an 1,4-substituted aromatic ring	-	-	892	862	-
Out-of-plane bending vibrations of $\delta_{C–H}$ in a mono-substituted aromatic ring	699	696	710	709	697

The results of the electronic absorption spectra study also confirm the formation of hydrogen bonds in the chain of the polymer matrix in the PDPAC/SWCNT-1 nanocomposite. Figure 2 shows the electronic absorption spectra of SWCNT/PDPAC-1 and SWCNT/PDPAC-2 nanocomposites, depending on the SWCNT concentration.

As seen in Figure 2a, the maximum at λ_{max} = 550 nm is present in the electronic spectra of the SWCNT/PDPAC-1 nanocomposite. The peak at λ_{max} = 550 nm characterizes the electronic transitions of associated COOH—N-H carboxyl groups with the hydrogen bond. This maximum is absent in the electronic absorption spectra of the SWCNT/PDPAC-2 nanocomposite, where carboxyl groups are not associated with amino groups in the structure. At the same time, FTIR spectroscopy data show that carboxyl groups interact with SWCNTs in the acidic medium. ATR FTIR spectra of the SWCNT/PDPAC-2 nanocomposite (Figure 1b), if compared with the polymer PDPAC-2 spectrum, demonstrate a shift of the absorption bands at 1659 and 1226 cm^{-1} to 1655 and 1218 cm^{-1} related to stretching vibrations of $\nu_{C=O}$ bonds in the COOH groups (Table 1). This shift of the absorption bands by 4 to 8 cm^{-1} indicates the interaction of PDPAC-2 carboxyl groups with the SWCNT surface. This could be caused by the charge transfer through site-selective interaction between the PDPAC-2 carboxyl groups and the SWCNT aromatic structures [36,56].

A characteristic change in the ATR FTIR spectra of the SWCNT/PDPAC-1 nanocomposite compared with the polymer PDPAC-1 spectrum is that the increase in the CNT content results in a hypsochromic shift of the skeletal oscillation frequencies of PDPAC-1 by 10 to 14 cm^{-1} (Figure 1a). The FTIR spectra of the SWCNT/PDPAC-1 nanomaterial demonstrate a shift of the absorption bands at 1595 and 1509 cm^{-1} to 1585 and 1495 cm^{-1}, corresponding to the stretching vibrations of $\nu_{C–C}$ bonds in the aromatic rings (Table 1). This shift indicates the π–π* interaction of PDPAC-1 phenyl rings with the SWCNT aromatic structures. For such materials, the existence of the π–π* interaction between the surface of CNTs and the

PANI quinoid units (stacking effect) has been established [23,31,32,36,44]. The formation of polymer on the surface of CNTs provides a charge transfer from the polymer chain to the CNTs [23,36,44]. This is revealed in the shift of the skeletal oscillation frequencies of the polymer.

CP/MAS ^{13}C NMR data confirm the growth pattern of the polymer chain proposed above. Figure 3 shows the CP/MAS ^{13}C NMR spectra of the PDPAC-1 and PDPAC-2 as well as SWCNT/PDPAC-1 and SWCNT/PDPAC-2 nanocomposites, prepared at C_{SWCNT} = 10 wt %.

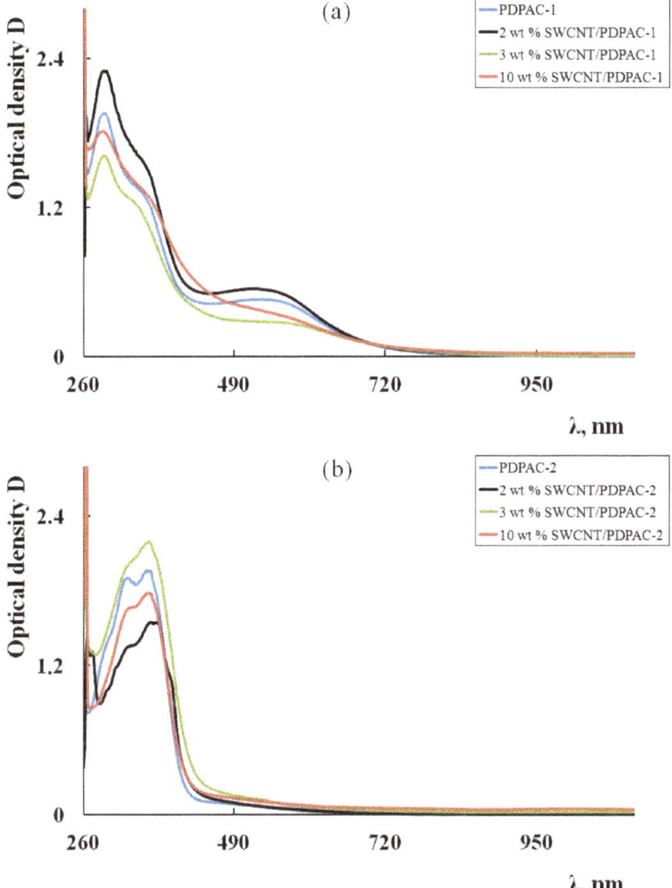

Figure 2. Electronic absorption spectra of the PDPAC-1 (**a**) and PDPAC-2 (**b**), and SWCNT/PDPAC-1 (**a**) and SWCNT/PDPAC-2 nanocomposites (**b**).

The solid-state MAS ^{13}C NMR spectrum of the SWCNT/PDPAC-1 nanocomposite retains all signals characterizing the polymer PDPAC-1 (Figure 3a). Both spectra show broad signals from 105 to 155 ppm, with the maximum at δ_C = 129 ppm characterizing carbon centers in benzene rings. The signal in the region of δ_C = 149 ppm corresponds to carbon atoms of the C–NH groups. The significant broadening of all spectrum signals, especially those around δ_C = 129 ppm, is a characteristic change in the CP/MAS ^{13}C NMR spectrum of the SWCNT/PDPAC-1 nanocomposite, compared with the PDPAC-1 spectrum. This indicates the interaction of carbon centers in benzene rings with carbon nanotubes, which leads to a decrease in the relaxation time, T_1, of these centers.

The CP/MAS ^{13}C NMR spectrum of the SWCNT/PDPAC-2 nanocomposite, compared with the polymer PDPAC-2 spectrum, shows the increase in signal intensity at 126 and 132 ppm (Figure 3b). This is related to the fact that the paramagnetic centers of CNT reduce the relaxation times, T_1, of carbon atoms in the polymer. At the same time, previously unobservable quaternary carbon atoms begin to manifest themselves in the NMR spectrum with the increase in the CNT content. The signal at δ_C = 146 ppm corresponds to the carbon atoms of the C–NH groups. The wide signal at δ_C = 170 ppm characterizes the carboxyl groups.

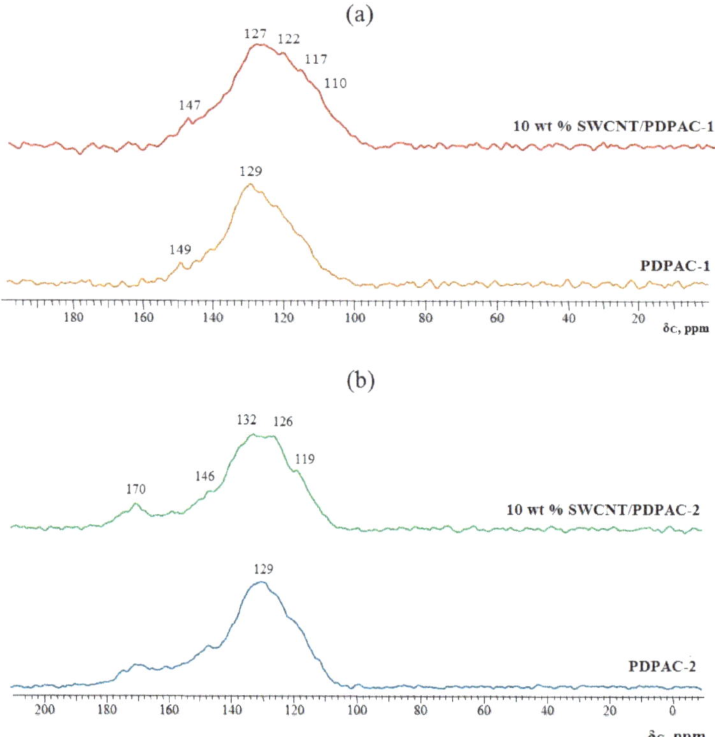

Figure 3. CP/MAS ^{13}C NMR spectra of the PDPAC-1 (**a**) and PDPAC-2 (**b**), and SWCNT/PDPAC-1 (**a**) and SWCNT/PDPAC-2 nanocomposites (**b**).

3.2. Morphology of Nanocomposites

The morphology and structure of the obtained hybrid nanomaterials were studied by means of FE-SEM and XRD. Figure 4 shows electron microscopic images of the SWCNT/PDPAC nanocomposites. According to the FE-SEM data, carbon nanotubes are distributed in an amorphous polymer matrix (Figure 4).

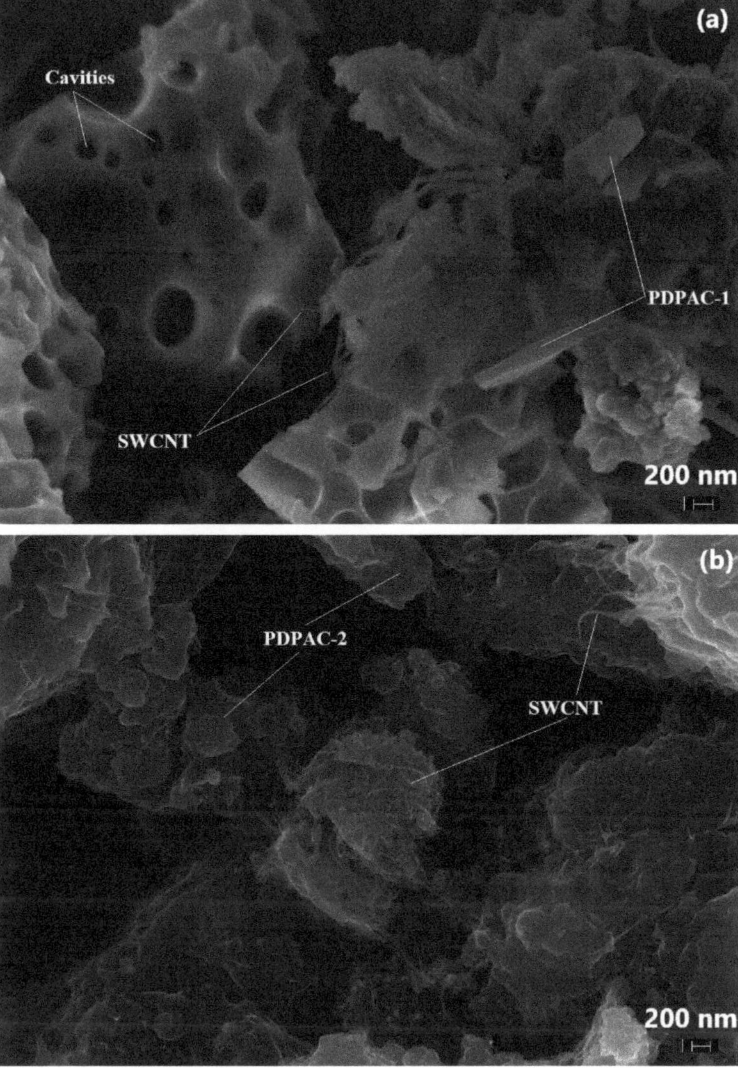

Figure 4. Field emission (FE)-SEM images of the SWCNT/PDPAC-1 (**a**) and SWCNT/PDPAC-2 nanocomposites (**b**).

Figure 5 shows the diffraction patterns of the SWCNT/PDPAC nanocomposites compared with PDPAC. According to the XRD data, the SWCNT/PDPAC nanocomposites, as well as PDPAC polymers, are amorphous irrespective of the preparing method. The XRD patterns of nanocomposites show an amorphous halo at scattering angles $2\theta = 20$ to 47°, and a second diffuse halo at $2\theta = 60$ to 70° characterizing the polymer matrix. The absence of the carbon phase reflection peak in the diffractograms of the nanocomposites is explained by the impossibility of obtaining a diffraction pattern from a single SWCNT plane (Figure 5).

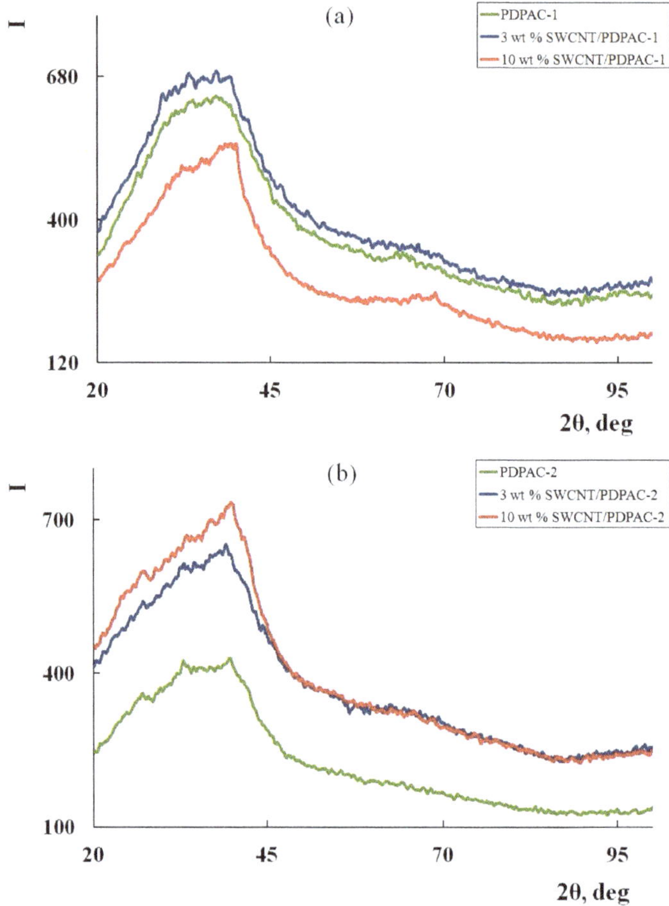

Figure 5. X-ray diffractograms of the PDPAC-1 (**a**) and PDPAC-2 (**b**), and SWCNT/PDPAC-1 (**a**) and SWCNT/PDPAC-2 nanocomposites (**b**).

As seen in Figure 4, according to the FE-SEM data, the morphology of the SWCNT/PDPAC nanocomposites depends on the pH of the synthesis reaction medium. As well as of the original PDPAC-1 polymer [49], when the SWCNT/PDPAC-1 nanocomposite is synthesized in the heterophase system, the presence of an organic solvent (chloroform) in an alkaline medium leads to the formation of the polymer matrix morphology with pronounced cavities. These cavities are formed in the places of chloroform drops as the monomer transits from the organic phase to the aqueous one (shown in Figure 4a).

3.3. Thermal Properties of Materials

The thermal stability of the hybrid SWCNT/PDPAC nanocomposites depending on the synthesis method and the SWCNT concentration was studied by the TGA and DSC methods. Figure 6 shows TGA thermograms of the SWCNT/PDPAC nanocomposites compared with PDPAC up to 1000 °C in air and argon flow. The carbon nanotube content in nanocomposites was C_{SWCNT} = 3 and 10 wt % relative to the weight of monomer. Table 2 gives the main thermal characteristics of the materials.

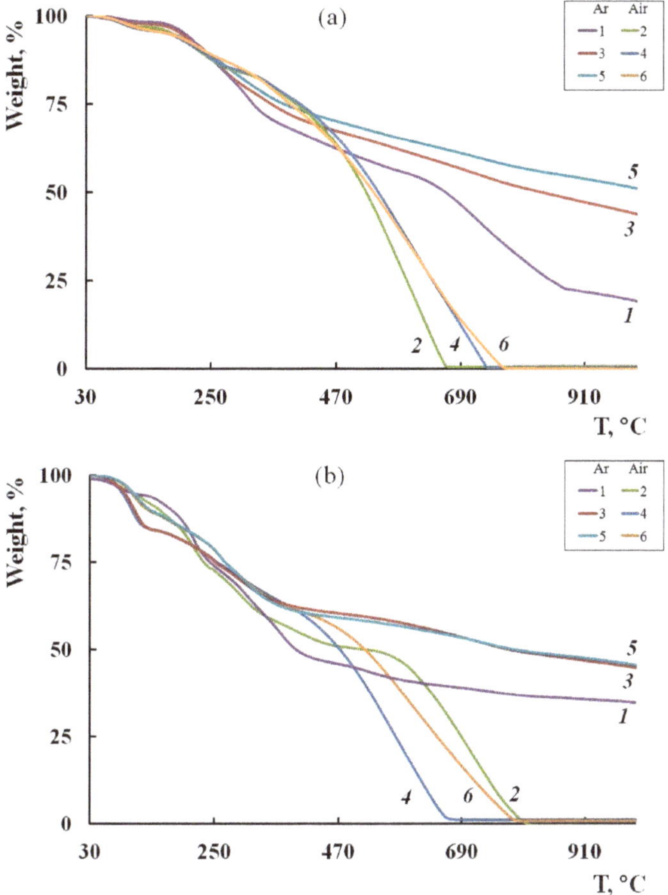

Figure 6. TGA thermograms of the PDPAC-1 (1,2a) and PDPAC-2 (1,2b), and SWCNT/PDPAC-1 (**a**) and SWCNT/PDPAC-2 nanocomposites (**b**), prepared at C_{SWCNT} = 3 (3,4) and 10 wt % (5,6) at heating of up to 1000°C.

Table 2. Thermal characteristics of materials.

Property	PDPAC-1	SWCNT/PDPAC-1		PDPAC-2	SWCNT/PDPAC-2	
		3 wt %	**10 wt %**		**3 wt %**	**10 wt %**
* $T_{5\%}$, °C	185/205	174/197	173/178	104/102	88/92	103/107
** $T_{50\%}$, °C	523/663	544/834	536/>1000	517/396	473/774	518/789
*** Residue, %	20	44	51	35	45	46

* $T_{5\%}$, ** $T_{50\%}$—5 and 50% weight losses (air/Ar), *** residue at 1000 °C in the Ar flow.

As can be seen in Figure 6, the weight loss curves of the obtained materials have a stepwise pattern. In this case, the weight loss at low temperatures is associated with the removal of moisture, which is also confirmed by the DSC data. Figure 7 presents DSC thermograms of the SWCNT/PDPAC nanocomposites. An endothermic peak at ~90 to 97 °C is related to the residual moisture removal. The removal of moisture is confirmed by the absence of this endothermic peak on the DSC thermograms of nanocomposites registered after re-heating in an inert atmosphere.

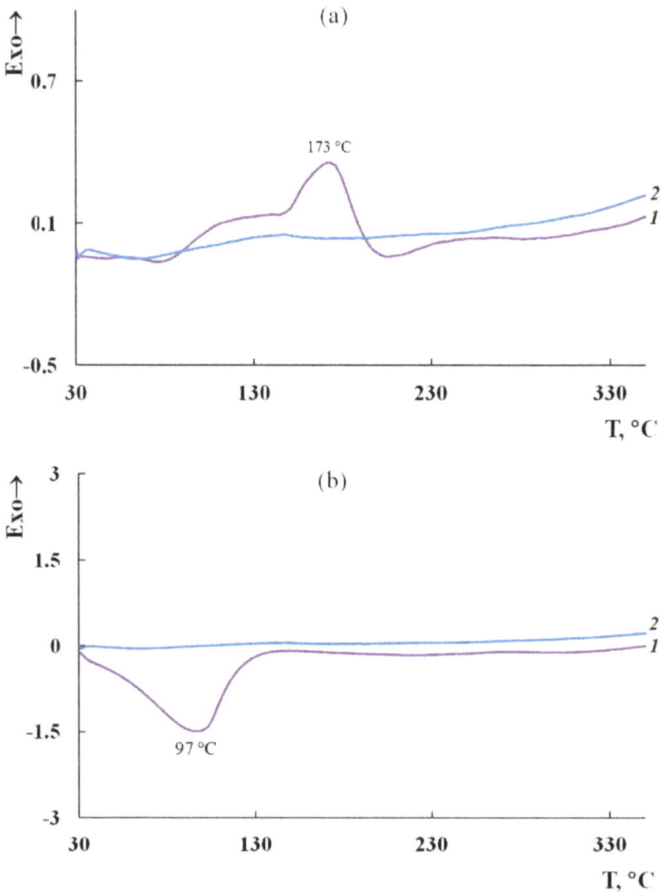

Figure 7. DSC thermograms of the SWCNT/PDPAC-1 (**a**) and SWCNT/PDPAC-2 nanocomposites (**b**) upon heating in the nitrogen flow to 350 °C (1—first heating, 2—second heating).

As seen in Figure 6a, both in the PDPAC-1 polymer and in the SWCNT/PDPAC-1 nanocomposites, the weight loss at ~170 °C is connected with the removal of COOH groups [49], despite the fact that they are in the associated COOH—N-H state. In this temperature range, the DSC thermogram has an exothermic peak associated with decomposition. The removal of COOH groups is confirmed by the absence of an exothermic peak at 173 °C on the DSC thermogram plotted after re-heating to 350 °C (Figure 7a). This behavior is also typical of PDPAC-2 obtained in a solution of sulfuric acid [50].

The removal of COOH groups in the SWCNT/PDPAC-1 nanocomposites is also confirmed by FTIR spectroscopy data. Figure 8a presents the ATR FTIR spectra of the 3 wt % SWCNT/PDPAC-1 nanocomposite before and after heating in air at 200 and 300 °C. A comparative analysis of the ATR FTIR spectra of the original nanocomposite and nanocomposites heated to 200 and 300 °C was made. It was shown that the intensity of the bands at 1680 and 1228 cm^{-1}, characterizing the COOH groups, decreases as the temperature rises. Additionally, at 300 °C, the band at 1680 cm^{-1} disappears completely. At the same time, the removal of COOH groups begins at temperatures above 150 °C.

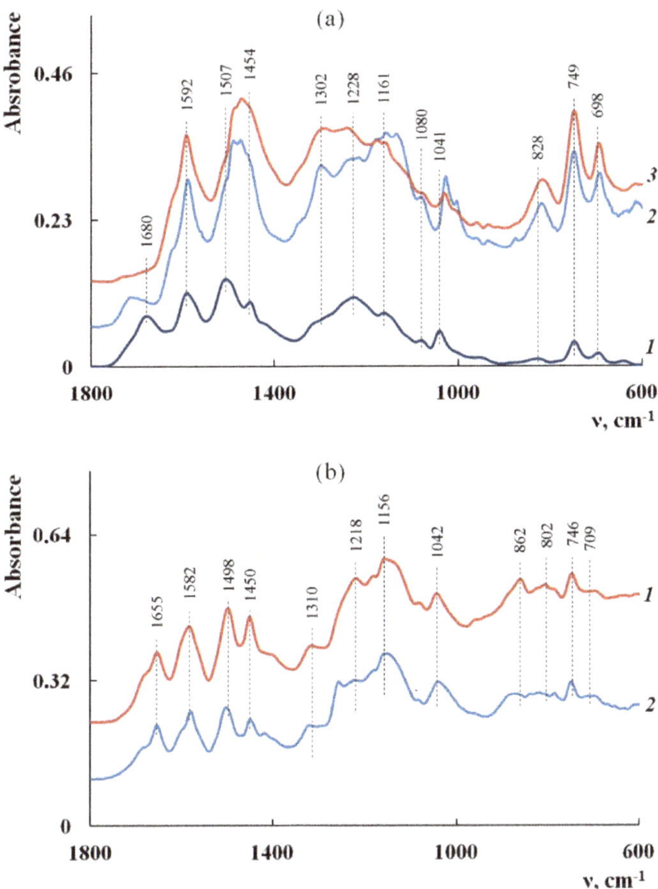

Figure 8. ATR FTIR spectra of the SWCNT/PDPAC-1 (**a**) and SWCNT/PDPAC-2 nanocomposites (**b**) before (1) and after heating in air at 200 (2) and 300 °C (3).

As can be seen in Figure 6a, the nanocomposite loses half of the original weight in an inert atmosphere at 834 °C for 3 wt % SWCNT/PDPAC-1, whereas for the PDPAC-1 polymer, this is 663 °C. At 1000 °C, the residue is 44% for 3 wt % SWCNT/PDPAC-1 and 51% for 10 wt % SWCNT/PDPAC-1. The processes of thermo-oxidative degradation of the SWCNT/PDPAC-1 nanocomposites do not depend on the SWCNT content and begin at 350 °C, as well as of the original PDPAC-1 polymer. In air, a 50% weight loss in the SWCNT/PDPAC-1 nanocomposites is observed at 536 to 544 °C. The neat PDPAC-1 polymer loses half of the initial weight in air at 523 °C (Table 2).

As seen in Figure 6b, the absence of weight loss at ~170 °C in the SWCNT/PDPAC-2 nanocomposites is due to the interaction of PDPAC-2 carboxyl groups with the aromatic structures of SWCNTs in the acidic medium. The DSC thermogram presented in Figure 7b does not show thermal effects in this temperature range. The FTIR spectroscopy data confirm the absence of weight loss in the range of 150 to 200 °C. The ATR FTIR spectrum of the 3 wt % SWCNT/PDPAC-2 nanocomposite heated up to 200 °C retains the absorption bands at 1655 and 1218 cm^{-1}, which characterize the COOH groups (Figure 8b).

A comparative TGA analysis of the PDPAC-2 polymer and SWCNT/PDPAC-2 nanocomposites obtained at C_{SWCNT} = 3 and 10 wt % was made. It revealed that the main processes of thermal oxidative degradation of nanocomposites begin at 395 and 435 °C, respectively. That is significantly lower than the starting temperature of PDPAC-2 polymer degradation at 570 °C (Table 2). This is due to the fact that in PDPAC-2, after the removal of COOH groups at ~170 °C with the temperature growth, there is the process of additional polymerization, which is induced by atmospheric oxygen [50]. It should be noted that in the argon atmosphere in this temperature range, there is a weight loss in PDPAC-2. In air, a 50% weight loss in the polymer and nanocomposites is observed at 517 °C for PDPAC-2 and 473 °C for 3 wt % SWCNT/PDPAC-2 and 518 °C for 10 wt % SWCNT/PDPAC-2. In an inert atmosphere, PDPAC-2 loses half of its original weight at 396 °C, whereas the SWCNT/PDPAC-2 nanocomposites lose it at 774 to 789 °C. The residue of the SWCNT/PDPAC-2 nanocomposites at 1000 °C is 45% to 46%, which is higher than the same value for PDPAC-2 (35%).

3.4. Electrical Characterization of Materials

The frequency dependences of the *ac* conductivity (σ_{ac}) for the PDPAC and the SWCNT/PDPAC nanocomposites were studied. Figure 9 shows the dependence of the conductivity for SWCNT/PDPAC-1 and SWCNT/PDPAC-2 nanocomposites prepared at C_{SWCNT} = 3 and 10 wt % on the *ac* frequency compared with PDPAC-1 and PDPAC-2. Table 3 gives the *ac* conductivity (σ_{ac}) of materials.

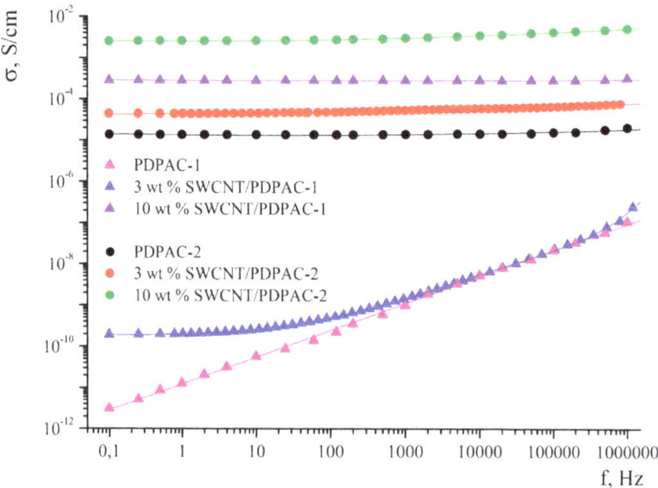

Figure 9. Frequency dependence of the conductivity for the PDPAC and SWCNT/PDPAC nanocomposites.

Table 3. The *ac* conductivity (σ_{ac}) of materials.

Property	PDPAC-1	SWCNT/PDPAC-1		PDPAC-2	SWCNT/PDPAC-2	
		3 wt %	10 wt %		3 wt %	10 wt %
* σ_1, S/cm	3.10×10^{-12}	1.93×10^{-10}	2.85×10^{-4}	1.35×10^{-5}	4.48×10^{-5}	2.52×10^{-3}
** σ_2, S/cm	1.12×10^{-7}	1.87×10^{-6}	3.32×10^{-4}	2.17×10^{-5}	8.07×10^{-5}	5.15×10^{-3}

* σ_1, ** σ_2—The *ac* conductivity at 0.1 Hz and 1.15 MHz.

As seen in Figure 9, the PDPAC-1 polymer shows the linear dependence of the conductivity on frequency typical of non-conductive materials [57]. The 3 wt % SWCNT/PDPAC-1 nanocomposite material shows a gradual increase in electrical conductivity over the entire researched frequency range (0.1–10^6 Hz). As the frequency grows, the electrical conductivity of the 3 wt % SWCNT/PDPAC-1 material increases by four orders of magnitude from 1.93×10^{-10} to 1.87×10^{-6} S/cm. This nature of frequency dependence of conductivity indicates a hopping mechanism of charge transfer [58,59]. With an increase in the content of SWCNTs from 3 to 10 wt %, the *ac* conductivity increases by six orders of magnitude from 1.93×10^{-10} to 2.85×10^{-4} Sm/cm (Table 3). In this case, there is a weak frequency dependence of the conductivity at C_{SWCNT} = 10 wt %. This is due to the fact that the 10 wt % SWCNT/PDPAC-1 nanocomposite has passed its percolation threshold.

As for the 10 wt % SWCNT/PDPAC-1, the SWCNT/PDPAC-2 nanocomposites show very weak dependence of the conductivity, σ_{ac}, on the frequency (Figure 9). As the *ac* frequency grows, the conductivity of the 3 wt % SWCNT/PDPAC-2 nanocomposite increases only from 4.48×10^{-10} to 8.07×10^{-4} S/cm. However, it should be noted that in the low-frequency range, the conductivity of the 3 wt % SWCNT/PDPAC-2 nanocomposite is significantly higher (by five orders of magnitude) than the conductivity of the 3 wt % SWCNT/PDPAC-1 material. This is due to the fact that during the nanocomposite synthesis in an acidic medium, doping of the polymer component occurs, which makes the main contribution to the 3 wt % SWCNT/PDPAC-2 nanocomposite conductivity. Also, as seen in Figure 9 and Table 3, the *ac* conductivity of the doped PDPAC-2 is much higher than the neutral PDPAC-1 conductivity. The weak frequency dependence of the SWCNT/PDPAC-2 electrical conductivity can be associated with a small value of the imaginary part of the complex dielectric capacitivity, ε", characteristic of the conductive materials. Therefore, its contribution to conductivity is manifested only at high frequencies [57,60].

Therefore, the results indicate that prepared hybrid thermally stable and electrically conductive nanomaterials have potential applications in the manufacture of supercapacitors [61,62], rechargeable batteries [63,64], sensors [41,45,65–68], sorbents [69], field-emission devices [70], anti-corrosion coatings [71], dye-sensitized solar cells [72–79], etc.

4. Conclusions

Polymer-carbon nanocomposites based on polydiphenylamine-2-carboxylic acid (PDPAC) and single-walled carbon nanotubes (SWCNTs) were synthesized for the first time in the heterophase system in an alkaline medium and in an acidic medium. The dependence of the chemical structure and morphology of the polymer matrix on the pH of the reaction medium of the nanocomposites' synthesis was shown. It was found that during the polymerization in 5 M H_2SO_4 (pH 0.3), the polymer chains grow via the C–C bonding into the *para* position of the phenyl rings relative to nitrogen ($\delta_{C\text{–}H}$ = 892 and 803 cm^{-1}). In the heterophase system in an alkaline medium (pH 11.4), the growth of the polymer chain occurs via the C–C bonding into the 2- and 4-positions of the phenyl rings relative to nitrogen ($\delta_{C\text{–}H}$ = 828 and 753 cm^{-1}). The presence of an organic solvent in the reaction medium leads to a change in the polymer matrix morphology. As a result, cavities formed in the place of chloroform drops. The resulting SWCNT/PDPAC nanocomposites are electrically conductive and thermally stable. The nanocomposites' *ac* conductivity (σ_{ac}) relates to the polymer component nature in the hybrid nanomaterial and the CNTs' presence. In the low-frequency range, because of doping of the polymer component in an acidic medium, the conductivity of the SWCNT/PDPAC-2 nanocomposites is significantly higher than the SWCNT/PDPAC-1 material's conductivity. It was shown that the SWCNT/PDPAC nanocomposites show weak dependence of the conductivity, σ_{ac}, on the frequency. In an inert atmosphere, the residue of the SWCNT/PDPAC nanocomposites at 1000 °C was 44% to 51% for SWCNT/PDPAC-1 and 45% to 46% for SWCNT/PDPAC-2.

Author Contributions: S.Z.O. proposed synthesis methods for novel hybrid nanocomposites, conducted the experiments, analyzed the data, interpreted the results, and prepared the manuscript paper. G.P.K. revised the manuscript. A.I.K. conducted the experiments and the investigation of nanocomposites by electron spectroscopy, and analyzed the data. G.N.B. conducted the investigation of nanocomposites by FTIR spectroscopy.

Funding: This research received no external funding.

Acknowledgments: This work was carried out within the State Program of TIPS RAS. The equipment from the collective exploitation center "New petrochemical processes, polymer composites and adhesives" was used.

Conflicts of Interest: The authors declare no conflicts of interest.

References

1. Iijima, S. Helical microtubules of graphitic carbon. *Nature* **1991**, *354*, 56–58. [CrossRef]
2. Hamon, M.A.; Chen, J.; Hu, H.; Chen, Y.; Itkis, M.E.; Rao, A.M.; Eklund, P.C.; Haddon, R. Dissolution of single-walled carbon nanotubes. *Adv. Mater.* **1999**, *11*, 834–840. [CrossRef]
3. Riggs, J.E.; Guo, Z.; Carroll, D.L.; Sun, Y.-P. Strong luminescence of solubilized carbon nanotubes. *J. Am. Chem. Soc.* **2000**, *122*, 5879–5880. [CrossRef]
4. Baughman, R.H.; Zakhidov, A.A.; de Heer, W.A. Carbon nanotubes—The route toward applications. *Science* **2002**, *297*, 787–792. [CrossRef] [PubMed]
5. Dai, L.; Mau, A.V.H. Controlled synthesis and modification of carbon nanotubes and C_{60}: Carbon nanostructures for advanced polymeric composite materials. *Adv. Mater.* **2001**, *13*, 899–913. [CrossRef]
6. Lota, K.; Khomenko, V.; Frackowiak, E. Capacitance properties of poly(3,4-ethylene- dioxythiophene)/carbon nanotubes composites. *J. Phys. Chem. Solids* **2004**, *65*, 295–301. [CrossRef]
7. Wu, T.M.; Lin, Y.W.; Liao, C.S. Preparation and characterization of polyaniline/multiwalled carbon nanotube composites. *Carbon* **2005**, *43*, 734–740. [CrossRef]
8. Wu, T.M.; Lin, Y.W. Synthesis, characterization, and electrical properties of polypyrrole/multiwalled carbon nanotube composites. *J. Polym. Sci. Polym. Chem.* **2006**, *44*, 6449–6457. [CrossRef]
9. Li, C.; Bai, H.; Shi, G. Conducting polymer nanomaterials: Electrosynthesis and applications. *Chem. Soc. Rev.* **2009**, *38*, 2397–2409. [CrossRef]
10. Sapurina, I.; Stejskal, J. The mechanism of the oxidative polymerization of aniline and the formation of supramolecular polyaniline structures. *Polym. Int.* **2008**, *57*, 1295–1325. [CrossRef]
11. Sapurina, I.Y.; Stejskal, J. The effect of pH on the oxidative polymerization of aniline and the morphology and properties of products. *Russ. Chem. Rev.* **2010**, *79*, 1123–1143. [CrossRef]
12. Song, E.; Choi, J.W. Conducting polyaniline nanowire and its applications in chemiresistive sensing. *Nanomaterials* **2013**, *3*, 498–523. [CrossRef] [PubMed]
13. Mittal, G.; Dhand, V.; Rhee, K.Y.; Park, S.; Lee, W.R. A review on carbon nanotubes and graphene as fillers in reinforced polymer nanocomposites. *J. Ind. Eng. Chem.* **2015**, *21*, 11–25. [CrossRef]
14. Oueiny, C.; Berlioz, S.; Perrin, F.-X. Carbon nanotube–polyaniline composites. *Prog. Polym. Sci.* **2014**, *39*, 707–748. [CrossRef]
15. Luo, J.; Niu, H.-J.; Wu, W.-J.; Wang, C.; Bai, X.-D.; Wang, W. Enhancement of the efficiency of dye-sensitized solar cell with multi-wall carbon nanotubes/polythiophene composite counter electrodes prepared by electrodeposition. *Solid State Sci.* **2012**, *14*, 145–149. [CrossRef]
16. Gao, M.; Huang, S.M.; Dai, L.M.; Wallace, G.; Gao, R.P.; Wang, Z.L. Aligned coaxial nanowires of carbon nanotubes sheathed with conducting polymers. *Angew. Chem. Int. Ed.* **2000**, *39*, 3664–3667. [CrossRef]
17. Hughes, M.; Shaffer, M.S.P.; Renouf, A.C.; Singh, C.; Chen, G.Z.; Fray, J.; Windle, A.H. Electrochemical capacitance of nanocomposite films formed by coating aligned arrays of carbon nanotubes with polypyrrole. *Adv. Mater.* **2002**, *14*, 382–385. [CrossRef]
18. Chen, G.Z.; Shaffer, M.S.P.; Coleby, D.; Dixon, G.; Zhou, W.Z.; Fray, D.J.; Windle, A.H. Carbon nanotube and polypyrrole composites: Coating and doping. *Adv. Mater.* **2000**, *12*, 522–526. [CrossRef]
19. Wu, M.Q.; Snook, G.A.; Gupta, V.; Shaffer, M.; Fray, D.J.; Chen, G.Z. Electrochemical fabrication and capacitance of composite films of carbon nanotubes and polyaniline. *J. Mater. Chem.* **2005**, *15*, 2297–2303. [CrossRef]
20. Peng, C.; Snook, G.A.; Fray, D.J.; Shaffer, M.S.P.; Chen, G.Z. Carbon nanotube stabilised emulsions for electrochemical synthesis of porous nanocomposite coatings of poly[3,4-ethylene-dioxythiophene]. *Chem. Commun.* **2006**, *44*, 4629–4631. [CrossRef]
21. Okotrub, A.V.; Asanov, I.P.; Galkin, P.S.; Bulusheva, L.G.; Chekhova, G.N.; Kurenya, A.G.; Shubin, Y.V. Composites based on polyaniline and aligned carbon nanotubes. *Polym. Sci. B* **2010**, *52*, 101–108. [CrossRef]
22. Jin, L.; Bowe, C.; Zhou, O. Alignment of carbon nanotubes in a polymer matrix by mechanical stretching. *Appl. Phys. Lett.* **1998**, *73*, 1197–1199. [CrossRef]
23. Zengin, H.; Zhou, W.; Jin, J.; Czerw, R.; Smith, D.W., Jr.; Echegoyen, L.; Carroll, D.L.; Foulger, S.H.; Ballato, J. Carbon nanotube doped polyaniline. *Adv. Mater.* **2002**, *14*, 1480–1483. [CrossRef]

24. Gerasin, V.A.; Antipov, E.M.; Karbushev, V.V.; Kulichikhin, V.G.; Karpacheva, G.P.; Talroze, R.V.; Kudryavtsev, Y.V. New approaches to the development of hybrid nanocomposites: From structural materials to high-tech applications. *Russ. Chem. Rev.* **2013**, *82*, 303–332. [CrossRef]
25. Yazdi, M.K.; Motlagh, G.H.; Garakani, S.S.; Boroomand, A. Effects of multiwall carbon nanotubes on the polymerization model of aniline. *J. Polym. Res.* **2018**, *25*, 265. [CrossRef]
26. Ham, H.T.; Choi, Y.S.; Jeong, N.; Chung, I.J. Single-wall carbon nanotubes covered with polypyrrole nanoparticles by the miniemulsion polymerization. *Polymer* **2005**, *46*, 6308–6315. [CrossRef]
27. Park, Ch.; Ounaies, Z.; Watson, K.A.; Crooks, R.E.; Smith, J., Jr.; Lowther, S.E.; Connell, J.W.; Siochi, E.J.; Harrison, J.S.; Terry, L.S. Dispersion of single wall carbon nanotubes by in situ polymerization under sonication. Chem. *Phys. Lett.* **2002**, *364*, 303–308.
28. Jeevananda, T.; Siddaramaiah, D.; Lee, T.S.; Lee, J.H.; Samir, O.M.; Somashekar, R. Polyaniline-multiwalled carbon nanotube composites: Characterization by WAXS and TGA. *J. Appl. Polym. Sci.* **2008**, *109*, 200–210. [CrossRef]
29. Zelikman, E.; Narkis, M.; Siegmann, A.; Valentini, L.; Kenny, J.M. Polyaniline/multiwalled carbon nanotube systems: Dispersion of CNT and CNT/PANI interaction. *Polym. Eng. Sci.* **2008**, *48*, 1871–1877. [CrossRef]
30. Konyushenko, E.N.; Stejskal, J.; Trchova, M.; Hradil, J.; Kovarova, J.; Prokes, J.; Cieslar, M.; Hwang, J.-Y.; Chen, K.-H.; Sapurina, I. Multi-wall carbon nanotubes coated with polyaniline. *Polymer* **2006**, *47*, 5715–5723. [CrossRef]
31. Ginic-Markovic, M.; Matisons, J.G.; Cervini, R.; Simon, G.P.; Fredericks, P.M. Synthesis of new polyaniline/nanotube composites using ultrasonically initiated emulsion polymerization. *Chem. Mater.* **2006**, *18*, 6258–6265. [CrossRef]
32. Suckeveriene, R.Y.; Zelikman, E.; Mechrez, G.; Tzur, A.; Frisman, I.; Cohen, Y.; Narkis, M. Synthesis of hybrid polyaniline/carbon nanotube nanocomposites by dynamic interfacial inverse emulsion polymerization under sonication. *J. Appl. Polym. Sci.* **2011**, *120*, 676–682. [CrossRef]
33. Zelikman, E.; Suckeveriene, R.Y.; Mechrez, G.; Narkis, M. Fabrication of composite polyaniline/CNT nanofiber/s using an ultrasonically assisted dynamic inverse emulsion polymerization technique. *Polym. Adv. Technol.* **2010**, *21*, 150–152. [CrossRef]
34. Zhang, X.; Zhang, J.; Liu, Z. Tubular composite of doped polyaniline with multi-walled carbon nanotubes. *Appl. Phys. A Mater. Sci. Process.* **2004**, *80*, 1813–1817. [CrossRef]
35. Sainz, R.; Benito, A.M.; Martínez, M.T.; Galindo, J.F.; Sotres, J.; Baró, A.M.; Corraze, B.; Chauvet, O.; Maser, W.K. Soluble self-aligned carbon/polyaniline composites. *Adv. Mater.* **2005**, *17*, 278–281. [CrossRef]
36. Cochet, M.; Maser, W.K.; Benito, A.M.; Callejas, M.A.; Martínez, M.T.; Benoit, J.-M.; Schreiber, J.; Chauvet, O. Synthesis of a new polyaniline/nanotube composite: "*in-situ*" polymerization and charge transfer through site-selective interaction. *Chem. Commun.* **2001**, *37*, 1450–1451. [CrossRef]
37. Deng, M.; Yang, B.; Hu, Y. Polyaniline deposition to enhance the specific capacitance of carbon nanotubes for supercapacitors. *J. Mater. Sci.* **2005**, *40*, 5021–5023. [CrossRef]
38. Yu, Y.; Che, B.; Si, Zh.; Li, L.; Chen, W.; Xue, G. Carbon nanotube/polyaniline core-shell nanowires prepared by in situ inverse microemulsion. *Synth. Met.* **2005**, *150*, 271–277. [CrossRef]
39. Gull, N.; Khan, S.M.; Islam, A.; Zia, S.; Shafiq, M.; Sabir, A.; Munawar, M.A.; Butt, M.T.Z.; Jamil, T. Effect of different oxidants on polyaniline/single walled carbon nanotubes composites synthesized via ultrasonically initiated in-situ chemical polymerization. *Mater. Chem. Phys.* **2016**, *172*, 39–46. [CrossRef]
40. Dhand, C.; Solanki, P.R.; Datta, M.; Malhotra, B.D. Polyaniline/single-walled carbon nanotubes composite based triglyceride biosensor. *Electroanalysis* **2010**, *22*, 2683–2693. [CrossRef]
41. Qu, F.; Yang, M.; Jiang, J.; Shen, G.; Yu, R. Amperometric biosensor for choline based on layer-by-layer assembled functionalized carbon nanotube and polyaniline multilayer film. *Anal. Biochem.* **2005**, *344*, 108–114. [CrossRef] [PubMed]
42. Zhao, B.; Hu, H.; Yu, A.; Perea, D.; Haddon, R.C. Synthesis and characterization of water soluble single-walled carbon nanotube graft copolymers. *J. Am.Chem. Soc.* **2005**, *127*, 8197–8203. [CrossRef] [PubMed]
43. Philip, B.; Xie, J.; Abraham, J.K.; Varadan, V.K. Polyaniline/carbon nanotube composites: Starting with phenylamino functionalized carbon nanotubes. *Polym. Bull.* **2005**, *53*, 127–138. [CrossRef]
44. Xu, J.; Yao, P.; Liu, L.; Jiang, Zh.; He, F.; Li, M.; Zou, J. Synthesis and characterization of an organic soluble and conducting polyaniline-grafted multiwalled carbon nanotube core–shell nanocomposites by emulsion polymerization. *J. Appl. Polym. Sci.* **2010**, *118*, 2582–2591. [CrossRef]

45. Kar, P.; Choudhury, A. Carboxylic acid functionalized multi-walled carbon nanotube doped polyaniline for chloroform sensors. *Sens. Actuators B Chem.* **2013**, *183*, 25–33. [CrossRef]
46. Ibrahim, N.I.; Wasfi, A.S. A comparative study of polyaniline/MWCNT with polyaniline/SWCNT nanocomposite films synthesized by microwave plasma polymerization. *Synth. Met.* **2019**, *250*, 49–54. [CrossRef]
47. Ozkan, S.Z.; Karpacheva, G.P. Hybrid material based on poly-3-amine-7-methylamine-2-methylphenazine and single-walled carbon nanotubes and method of its production. *Polymers* **2018**, *10*, 544. [CrossRef] [PubMed]
48. Ozkan, S.Z.; Karpacheva, G.P.; Kolyagin, Yu.G. Hybrid nanocomposite based on poly-3-amine-7-methylamine-2-methylphenazine and single-walled carbon nanotubes. *Polym. Bull.* **2018**. [CrossRef]
49. Ozkan, S.Zh.; Eremeev, I.S.; Karpacheva, G.P.; Bondarenko, G.N. Oxidative polymerization of N-phenylanthranilic acid in the heterophase system. *Open, J. Polym. Chem.* **2013**, *3*, 63–69. [CrossRef]
50. Ozkan, S.Zh.; Bondarenko, G.N.; Karpacheva, G.P. Oxidative polymerization of diphenylamine-2-carboxylic acid: Synthesis, structure, and properties of polymers. *Polym. Sci. B* **2010**, *52*, 263–269. [CrossRef]
51. Ozkan, S.Zh.; Eremeev, I.S.; Karpacheva, G.P.; Prudskova, T.N.; Veselova, E.V.; Bondarenko, G.N.; Shandryuk, G.A. Polymers of dipheylamine-2-carboxylic acid: Synthesis, structure and properties. *Polym. Sci. B* **2013**, *55*, 107–115.
52. Karyakin, Yu.V.; Angelov, I.I. *Pure Chemical Reagents*; Khimiya: Moscow, Russia, 1974. (In Russian)
53. Thakur, R.S.; Kurur, N.D.; Madhu, P.K. Swept-frequency two-pulse phase modulation for heteronuclear dipolar decoupling in solid-state NMR. *Chem. Phys. Lett.* **2006**, *426*, 459–463. [CrossRef]
54. Earl, W.L.; Vanderhart, D.L. Measurement of ^{13}C chemical shifts in solids. *J. Magn. Reson.* **1982**, *48*, 35–54. [CrossRef]
55. Morcombe, C.R.; Zilm, K.W. Chemical shift referencing in MAS solid state NMR. *J. Magn. Reson.* **2003**, *162*, 479–486. [CrossRef]
56. Peng, H.; Alemany, L.B.; Margrave, J.L.; Khabashesku, V.N. Sidewall carboxylic acid functionalization of single-walled carbon nanotubes. *J. Am. Chem. Soc.* **2003**, *125*, 15174–15182. [CrossRef] [PubMed]
57. Eletskii, A.V.; Knizhnik, A.A.; Potapkin, B.V.; Kenny, J.M. Electrical characteristics of carbon nanotube doped composites. *Uspekhi Phyzicheskikh Nauk.* **2015**, *185*, 225–270. [CrossRef]
58. Rehwald, W.; Kiess, H.; Binggeli, B. Frequency dependent conductivity in polymers and other disordered materials. *Z. Phys. B Condens. Matter.* **1987**, *68*, 143–148. [CrossRef]
59. Dyre, J.C. The random free-energy barrier model for ac conduction in disordered solids. *J. Appl. Phys.* **1988**, *64*, 2456–2468. [CrossRef]
60. Coleman, J.N.; Curran, S.; Dalton, A.B.; Davey, A.P.; McCarthy, B.; Blau, W.; Barklie, R.C. Percolation-dominated conductivity in a conjugated-polymer-carbon-nanotube composite. *Phys. Rev. B* **1998**, *58*, R7492(R)–R7495(R). [CrossRef]
61. Potphode, D.D.; Sivaraman, P.; Mishra, S.P.; Patri, M. Polyaniline/partially exfoliated multi-walled carbon nanotubes based nanocomposites for supercapacitors. *Electrochim. Acta* **2015**, *155*, 402–410. [CrossRef]
62. Zhou, Z.; Wu, X.-F.; Hou, H. Electrospun carbon nanofibers surface-grown with carbon nanotubes and polyaniline for use as high-performance electrode materials of supercapacitors. *RSC Adv.* **2014**, *4*, 23622–23629. [CrossRef]
63. Ha, J.-S.; Lee, J.-M.; Lee, H.-R.; Huh, P.; Jo, N.-J. Polymer nanocomposite electrode consisting of polyaniline and modified multi-walled carbon nanotube for rechargeable battery. *J. Nanosci. Nanotechnol.* **2015**, *15*, 8977–8983. [CrossRef]
64. He, B.L.; Dong, B.; Wang, W.; Li, H.L. Performance of polyaniline/multi-walled carbon nanotubes composites as cathode for rechargeable lithium batteries. *Mater. Chem. Phys.* **2009**, *114*, 371–375. [CrossRef]
65. Kulkarni, M.V.; Kale, B.B. Studies of conducting polyaniline (PANI) wrapped-multiwalled carbon nanotubes (MWCNTs) nanocomposite and its application for optical pH sensing. *Sens. Actuators B Chem.* **2013**, *187*, 407–412. [CrossRef]
66. Roy, A.; Ray, A.; Sadhukhan, P.; Naskar, K.; Lal, G.; Bhar, R.; Sinha, C.; Das, S. Polyaniline-multiwalled carbon nanotube (PANI-MWCNT): Room temperature resistive carbon monoxide (CO) sensor. *Synth. Met.* **2018**, *245*, 182–189. [CrossRef]

67. Bachnav, S.G.; Patil, D.R. Study of polypyrrole-coated MWCNT nanocomposites for ammonia sensing at room temperature. *J. Mater. Sci. Chem. Eng.* **2015**, *3*, 30–44.
68. Xue, L.; Wang, W.; Guo, Y.; Wan, P. Flexible polyaniline/carbon nanotube nanocomposite film-based electronic gas sensors. *Sens. Actuators B Chem.* **2017**, *244*, 47–53. [CrossRef]
69. Dutra, F.V.A.; Pires, B.C.; Nascimento, T.A.; Borges, K.B. Functional polyaniline/multiwalled carbon nanotube composite as an efficient adsorbent material for removing pharmaceuticals from aqueous media. *J. Environ. Manag.* **2018**, *221*, 28–37. [CrossRef]
70. Jin, Y.W.; Jung, J.E.; Park, Y.J.; Choi, J.H.; Jung, D.S.; Lee, H.W.; Park, S.H.; Lee, N.S.; Kim, J.M.; Ko, T.Y.; et al. Triode-type field emission array using carbon nanotubes and a conducting polymer composite prepared by electrochemical polymerization. *J. Appl. Phys.* **2002**, *92*, 1065–1069. [CrossRef]
71. Deshpande, P.; Vathare, S.; Vagge, S.; Tomšík, E.; Stejskal, J. Conducting polyaniline/multi-wall carbon nanotubes composite paints on low carbon steel for corrosion protection: Electrochemical investigations. *Chem. Pap.* **2013**, *67*, 1072–1078. [CrossRef]
72. Xiao, Y.; Lin, J.-Y.; Wu, J.; Tai, S.-Y.; Yue, G.; Lin, T.-W. Dye-sensitized solar cells with high-performance polyaniline/multi-wall carbon nanotube counter electrodes electropolymerized by a pulse potentiostatic technique. *J. Power Sources* **2013**, *233*, 320–325. [CrossRef]
73. He, B.; Tang, Q.; Liang, T.; Li, Q. Efficient dye-sensitized solar cells from polyaniline-single wall carbon nanotube complex counter electrodes. *J. Mater. Chem. A* **2014**, *2*, 3119–3126. [CrossRef]
74. Saranya, K.; Rameez, M.; Subramania, A. Developments in conducting polymer based counter electrodes for dye-sensitized solar cells—An overview. *Eur. Polym. J.* **2015**, *66*, 207–227. [CrossRef]
75. Peng, S.; Wu, Y.; Zhu, P.; Thavasi, V.; Mhaisalkar, S.G.; Ramakrishna, S. Facile fabrication of polypyrrole/functionalized multiwalled carbon nanotubes composite as counter electrodes in low-cost dye-sensitized solar cells. *J. Photochem. Photobiol. A Chem.* **2011**, *223*, 97–102. [CrossRef]
76. Abdul Almohsin, S.; AL-Mutoki, S.M.; Li, Z. Electrochemical polymerization of PPy–MWCNT composite as a counter electrode for dye-sensitized solar cells. *J. Arkansas Acad. Sci.* **2012**, *66*, 31–35.
77. He, B.; Tang, Q.; Luo, J.; Li, Q.; Chen, X.; Cai, H. Rapid charge-transfer in polypyrrole–single wall carbon nanotube complex counter electrodes: Improved photovoltaic performances of dye-sensitized solar cells. *J. Power Sources* **2014**, *256*, 170–177. [CrossRef]
78. Wang, W.-Y.; Ting, P.-N.; Luo, S.-H.; Lin, J.-Y. Pulse-reversal electropolymerization of polypyrrole on functionalized carbon nanotubes as composite counter electrodes in dye-sensitized solar cells. *Electrochim. Acta* **2014**, *137*, 721–727. [CrossRef]
79. Niu, H.; Qin, S.; Mao, X.; Zhang, S.; Wang, R.; Wan, L.; Xua, J.; Miao, S. Axle-sleeve structured MWCNTs/polyaniline composite film as cost-effective counter-electrodes for high efficient dye-sensitized solar cells. *Electrochim. Acta* **2014**, *121*, 285–293. [CrossRef]

Article

Flexible Electrode Based on MWCNT Embedded in a Cross-Linked Acrylamide/Alginate Blend: Conductivity vs. Stretching

Jake Thibodeau and Anna Ignaszak *

Department of Chemistry, University of New Brunswick, Fredericton, NB E3B 5A3, Canada; n9tqn@unb.ca
* Correspondence: anna.ignaszak@unb.ca; Tel.: +1-506-261-9128

Received: 23 December 2019; Accepted: 6 January 2020; Published: 9 January 2020

Abstract: A polyacrylamide-alginate hydrogel electrolyte, blended with Multi-Walled Carbon Nanotubes (MWCNT) as an electronically conductive fraction, allows for the creation of a flexible, durable, and resilient electrode. The MWCNT content is correlated with mechanical characteristics such as stretch modulus, tensile resistance, and electrical conductivity. The mechanical analysis demonstrates tensile strength that is comparable to similar hydrogels reported in the literature, with increasing strength for MWCNT-embedded hydrogels. The impedance spectroscopy reveals that the total resistance of electrodes decreases with increasing MWCNT content upon elongation and that bending and twisting do not obstruct their conductivity. The MWCNT-inserted hydrogels show mixed ionic and electronic conductivities, both within a range of 1–4 × 10^{-2} S cm^{-1} in a steady state. In addition, the thermal stability of these materials increases with incrementing MWCNT content. This observation agrees with long-term charge-discharge cycling that shows enhanced electrochemical durability of the MWCNT-hydrogel hybrid when compared to pure hydrogel electrolyte. The hydrogel-carbon films demonstrate an increased interfacial double-layer current at a high MWCNT content (giving an area-specific capacitance of ~30 mF cm^{-2} at 2.79 wt.% of MWCNT), which makes them promising candidates as printable and flexible electrodes for lightweight energy storage applications. The maximum content of MWCNT within the polymer electrolyte was estimated at 2.79 wt.%, giving a very elastic polymer electrode with good electrical characteristics.

Keywords: flexible electrode; cross-linked acrylamide/alginate; carbon nanotubes; tensile strength; impedance spectroscopy

---

1. Introduction

Advances in hydrogel performance, when combined with interstitial conductive materials, have shown potential for incorporation with modern electronic applications [1,2]. Technological products, such as wearable electronics [3,4], personal sensors [5–7], and smart textiles [8,9], are examples where these novel materials can enhance their functionality. Successful implementation requires materials that are durable, flexible, and resilient, capable of meeting the conductivity and physical requirements. Modularity is a central asset, as many modern devices currently are rigid and bulky. Innovations may allow for new materials and devices that are able to conform to human mobility demands.

These electronics can be utilized either in an active or passive way. An active device responds to user input; for example, as two-way radio communications or live GPS data transmissions to emergency responders [10,11]. Passive devices do not require user input, but rather provide sensory data or function to the user, or to an external support point [12]. Examples include providing vital signs to medical staff (such as the Georgia Tech Wearable Motherboard, or Smart Shirt) [13,14], body armor cooling devices by police and military [15], or smart GPS tracking for the elderly suffering from dementia [16]. The applications can also be implemented into personal electronic devices—smartwatches and fitness trackers are increasingly common in everyday usage [17]. Additionally, new lightweight energy materials are highly sought after in applications such as exoskeletons for physical rehabilitation [18], biomimetic actuators [19], field-effect transistors [20], and chemical sensing for air pollution and other biohazards [21].

For this research, in order to determine a suitable basis for composition, we examined an acrylamide-alginate flexible material pioneered by Sun et al. [22]. Polymers, with the applications in mind, must endure various physical stresses, such as stretching, twisting, and bending. The double-network structure made of cross-linked polymers results in a greater fracture strength compared to single networks, with high retention of shape against temporary mechanical deformations [23]. This is attributed to the combined interactions of short-chain polymers providing mechanical strength, and long-chain polymers providing elasticity. Previous research in this area shows that short-chain polymers cross-linked covalently suffer permanent damage upon deformation of the material [22]. However, a known method of minimizing this unwanted effect is the replacement of the covalent cross-linking bonds with non-covalent (i.e., ionic) using multivalent ion coordinators [24]. This allows a formation of much weaker ionic linkage between the rigid polymer blocks, observed as reduced stiffness of the gel, and thus resulting in more stretchable material. Among multivalent cations that generated suitable ionic cross-linking were Ca^{2+}, Sr^{2+}, Ba^{2+}, Al^{3+}, and Fe^{3+} [25]. Calcium was chosen due to the relative minor environmental impact that it has upon degradation, along with a relatively strong ionic coordination strength. Initial formulations were based on similar work by Demianenko et al. [26], in part due to the utilization of UV irradiation in the curing process that shortens the synthesis time.

Multi-Walled Carbon Nanotubes (MWCNTs) were chosen as the electron-conducting component of the flexible electrode proposed in this work. MWCNT-alginate hydrogels have been previously explored by Joddar et al. [27], with a focus on mechanical properties—notably with overall declining performance above 1 mg/mL of MWCNT (to a tested max of 5 mg/mL). Hong et al. [28] integrated single-walled CNTs in poly (dimethyl siloxane) and analyzed their electrical characteristics in various circuits, as well as under differing mechanical strain. Sudha et al. [29] created an MWCNT-polyacrylamide hydrogel, exploring dehydration and rehydration properties, with a brief examination of their conductivities. Among other carbon allotropes, carbon nanotubes are particularly researched as they show several benefits when combined with various polymer electrolytes. CNTs not only facilitate good electronic conductivity, but also improve the overall mechanical strength and thermal stability of these hybrid materials [30]. In addition, they are excellent double-layer capacitors that cannot be outperformed by more affordable carbons. Thus, CNTs are of interest in energy harvesting and reposition applications [28,31].

In this study, we have investigated the conductivity and mechanical features of a polyacrylamide-alginate hydrogel electrolyte, combined with MWCNTs as the electronically conductive component. The effect of MWCNT content in the flexible electrode is verified by mechanical characteristics such as stretch modulus and tensile resistance. In addition, electrochemical characteristics such as conductivity, long-term electrochemical stability upon charging, potential stability, and thermal stability were examined. Mechanical impact and deformation on conductivity were analyzed by impedance spectroscopy upon elongation, bending, and twisting of the hydrogel electrolyte and MWCNT-embedded electrodes. On the macroscopic level, we determined the maximum loading of MWCNT within the hydrogel with an overall aim to create an electrode that has the flexibility to endure moderate strains and stress. Ultimately, this material can be used as a conducting platform in applications such as smart textiles, personal sensors, wearable electronics, and lightweight energy storage and conversion.

2. Materials and Methods

Materials: Calcium chloride dihydrate (99.0–105.0%) was purchased from Fisher Scientific (Ottawa, ON, Canada). Multi-walled carbon nanotube (thin and short, <5% metal oxide), potassium chloride (99.6%), *N*,*N*′-methylenebisacrylamide powder (≥99.5%), acrylamide powder (≥99%, HPLC grade), alginic acid sodium salt powder, *N*,*N*,*N*′,*N*′-tetramethylethylenediamine (≥99%, GC grade), and ammonium persulfate (≥98%) were obtained from Sigma Aldrich (Oakville, ON, Canada). Ultra-pure distilled water (18 MΩ) was used as the reaction solvent. Fluorine-doped tin oxide (FTO) glass with a surface resistivity of 7 Ω (Sigma Aldrich, Oakville, ON, Canada) was cut into 2.5 cm × 5.5 cm × 0.2 cm slides, acting as the current collector in electrochemical tests, and layered with copper foil at the electrode clip contact area.

Methods: Tensile tests were performed using an Instron 4465 Tensile test workstation equipped with 1 kN of load cell under air. The elongation rate was kept as 40 mm min^{-1}. Thermogravimetry analysis (TGA) was carried out on the TA Instruments analyzer model Q50 (TA Instruments, Mississauga, ON, Canada). The mass of samples ranged from 9.0 to 10.3 mg. N_2 gas was kept at the flow rate of 90 mL min^{-1}. A ramp temperature was 10 °C min^{-1} over the range from 25 to 600 °C. Fourier-transform infrared (FTIR) analysis was carried out using a Bruker spectrometer model ALPHAII with a single reflection ART diamond accessory (Bruker, Billerica, MA, USA). All hydrogels were dehydrated prior to TGA and FTIR tests by drying at 70 °C for 24 h in a laboratory oven.

Electrochemical tests were carried out using a CH Instruments electrochemical workstation (model CHI660E, CH Instruments Inc., Austin, TX, USA). All electrochemical characteristics were done using a two-electrode cell consisting of two FTO plates used as a current collector. The samples were square shape with a 12 mm edge, with a typical thickness of 4.5 mm (varying slightly based on the addition of MWCNT), and mounted between FTO plates. For the conductivity test upon elongation, one of the FTO plates was moved with a distance increment of 1 cm and the length and the thickness of the sample were measured using a caliper. The reported conductivity of materials is as-obtained (in siemens, S) and with the correction to the size (thickness, length, and width) of the stretched sample (in S cm^{-1}).

Cyclic voltammetry was acquired at a scan rate varying from 0.05 to 0.2 V s^{-1} and from −0.8 V to 0.8 V in the two-electrode cell (without a reference). Impedance spectroscopy was carried out in the frequency range of 1×10^5–1 Hz, with an amplitude of 0.005 V, and at the constant polarization of 0 V. ZView software (Scribner Associates Inc., Southern Pines, NC, USA) was used for the modeling of impedance spectra. The electrochemical stability was carried out by applying a 2500 scan of the charge-discharge at 0.5 mA current load, with a cathodic and anodic time of 30 s and (zero hold time giving a total time of the degradation test at 21 h). The charge-discharge was also carried out at 0.005, 0.01, 0.05, 0.1, 0.25, 0.5, 0.75, and 1.0 mA for all samples, with preliminary extended testing for pure/360 mg from 0.005 mA to 2.5 mA. However, changes at the interface material-current collector (assumed as beginning of degradation) were observed at 0.5 mA, which was chosen for the long-term durability test.

Synthesis of Hydrogel Electrolyte and Electrode: Ten milliliters of deionized water (that is a constant volume of water used for each hydrogel formulation) was purged with dry nitrogen for 30 min prior to use. Then, 1.4470 g of acrylamide, 0.7455 g of potassium chloride, and 0.0016 g of *N*,*N*′-methylenebisacrylamide were dissolved in 4 mL of deionized water. Separately, 0.1809 g of sodium alginate was dissolved in 3 mL of deionized water and combined with the first mixture under vigorous stirring. Furthermore, for the MWCNT-embedded electrodes, the designated amount of MWCNTs, was added (0.010–0.360 g) upon stirring. Afterward, 4.6 μL of *N*,*N*,*N*′,*N*′-tetramethylethylenediamine was injected into this mixture. In a separate flask, 0.0435 g of ammonium persulfate and 0.0927 g of calcium chloride dihydrate were dissolved in 3 mL of water, combined with the remaining reactants upon stirring, poured onto a petri dish (60 mm diameter), and placed in a UV cross-linker for 60 min of irradiation at a power of 300.000 μJ cm^{-2} (254 nm, model VWR® UV Crosslinker, VWR, Ville Mont-Royal, QC, Canada). All procedures were carried out under ambient conditions. The MWCNT content was varied resulting in the following formulations (names of hydrogels correspond to the mg of MWCNT): 40 mg refers to MWCNT concentration of 4 mg mL^{-1} (mL of water), that is 0.32 wt.% of MWCNT; 80 mg contains 8 mg mL^{-1} giving 0.63 wt.% of MWCNT; 120 mg has 12 mg mL^{-1} corresponding to 0.95 wt.% MWCNT; 240 mg has 24 mg mL^{-1}, that is 1.88 wt.% MWCNT; 360 mg has 36 mg mL^{-1}, that is 2.79 wt.% MWCNT; and 480 mg has 48 mg mL^{-1}, giving 3.69 wt.% MWCNT.

3. Results and Discussion

3.1. Structure and Thermal Stability: Effect of MWCNT Content on Interactions between Molecular Components and Morphology

The hydrogel matrix proposed in this work is formed via copolymerization of acrylamide and bis-acrylamide, with the optimal composition decided based on work reported by Sun et al. [22], and by a previous study carried out in our group [31]. Briefly, the reaction is vinyl addition polymerization via a free radical-generating system initiated by ammonium persulfate and TEMED. TEMED accelerates the rate of formation of radicals from persulfate that reacts with acrylamide monomers (resulting in radicals that react with neutral monomer and form a polymer). The extending polymer chains are randomly cross-linked by *N,N′*-methylenebisacrylamide (Figure 1; blue), resulting in stiff gels. In addition, in the presence of alginate and calcium chloride, two types of cross-linked polymer blocks are formed: ionically cross-linked alginate (Figure 1; red), and covalently cross-linked polyacrylamide. In an aqueous solution, guluronic acid units in alginate chains form ionic crosslinks through Ca^{2+} coordination. By contrast, in a polyacrylamide hydrogel, the polyacrylamide chains form a network by covalent crosslinks. There also exist alginate-polyacrylamide hybrid fractions formed when the two types of polymer networks are joined by covalent crosslinks (Figure 1; green).

The most problematic aspect of these combined materials is the homogenous incorporation of MWCNT within the polymer. This is because of the tendency of MWCNT to agglomerate under electrostatic forces. Magnetic stirring, sonication, and other homogenization methods are not effective. Using surfactants for MWCNT dispersion in the solution of monomers is more challenging, affects the rate of polymerization, and ultimately the chemical composition of hydrogels, their mechanical strength, and conductivity. As compared to our previous study [31], Nafion, acting as a surfactant in an aqueous suspension of MWCNT, has been replaced with the viscous solution of sodium alginate. The alginate can both participate in the cross-linking and stabilize the MWCNT in water. Based on the chemical structure of alginate, we can assume that improvement of MWCNT dispersion is related to the modification of the electrical surface charge of MWCNT, defined as zeta-potential (unlike Nafion, which is composed of both hydrophobic and hydrophilic fractions that are typical for the surfactant structure) [30]. This potential is often related to an electric surface potential (E_{Elec}) and contributes to the total surface energy, $E_{tot} = E_{VW} + E_{Elec}$. E_{tot} depends on the sum of attractive and repulsive forces between particles, which are both dependent on an inter-particle distance. The London-van der Waals contribution for two particles of the same material (E_{VW}) is always attractive, thus promoting the aggregation of suspended particles. On the contrary, the repulsive component, E_{Elec}, is related to the formation of an electric double-layer on the particle surface when immersed in a polar solvent. For example, in an aqueous suspension of functionalized MWCNT, negative charge is developed on the particle surface during ionization of the oxygen-rich functionalities (e.g., -COOH). Because of the surface charge, an electrostatic potential is created in the proximity of the nanoparticle, and a concentrated layer of counter ions, known as the Stern layer, is formed. As a result, the zeta potential at the MWCNT surface exceeds the total interaction potential barrier (E_{tot}), and at a given distance, the particle repulsion is stronger and more attractive than the van der Waals forces, resulting in the formation of homogeneous suspensions [32,33]. With this, the carbon nanotubes could be well-dispersed even at very high content without the addition of Nafion.

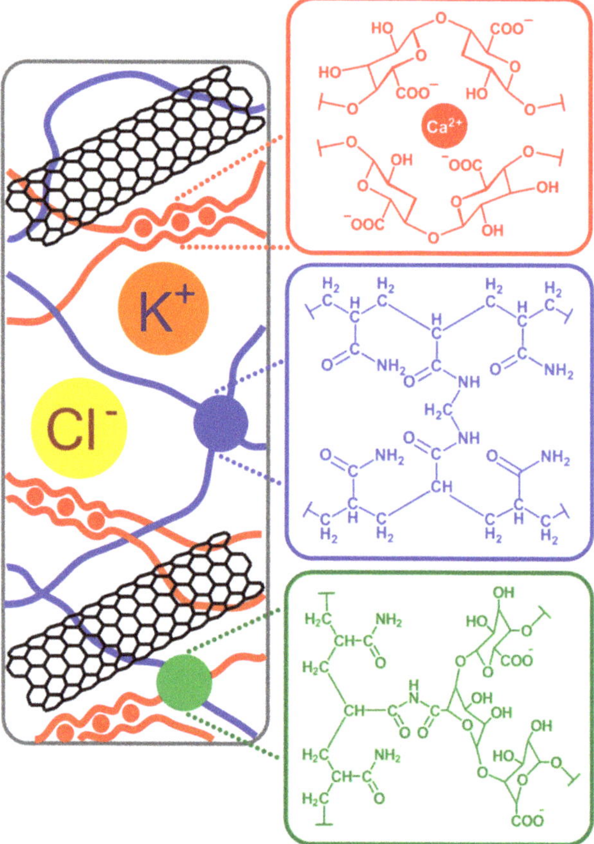

Figure 1. Structure of a Multi-Walled Carbon Nanotubes (MWCNT)-embedded hydrogel consisting of an alginate blocks formed via ionic cross-linking through Ca^{2+} (red); polyacrylamide blocks derived from covalent crosslinks through *N*,*N'*-methylenebisacrylamide (blue); alginate-polyacrylamide hybrid gel fraction: the two types of polymer network are joined by covalent crosslinks (green).

Figure 2 shows a thermal gravimetric scan of a pure hydrogel (A) and an MWCNT-reinforced hydrogel at two differing MWCNT contents (8 mg mL^{-1} and 36 mg mL^{-1}) plotted with the first derivative of weight loss (in %) as the function of applied temperature. The same tests were done for newly-prepared materials and the same films after a long-term electrochemical charge-discharge scanning (accelerated degradation). The latest was used to analyze possible chemical changes related to electrochemical degradation. All signals acquired below 200 °C are related to the removal of residual water. Regarding this, the hydrogel with the highest MWCNT content (Figure 2C) showed a negligible amount of water after drying. This is related to the fact that a significant weight fraction of this material is occupied by MWCNT, not by hydrophilic polymers. For the pure hydrogel (Figure 2A), an initial decomposition temperature (IDT) is detected at 221 °C and the final decomposition temperature (FDT) at 417 °C, resulting in 35% weight loss. The thermal degradation of polyacrylamide occurs in three pyrolysis phases. The first signal at IDT is attributed to the degradation of the pure and not cross-linked acrylamide monomer. Afterward, the decomposition of non-cross-linked polyacrylamide has a peak maximum at 264 °C. The main decomposition of the polyacrylamide cross-linked with *N*,*N'*-methylenebisacrylamide showed an onset temperature at 305 °C and endset at 440 °C. In the temperature range 221–310 °C, one ammonia molecule is liberated for every two amide groups, resulting in the formation of imide [22]. Subsequently, thermal degradation of imides and breaking of the polymer backbone occurs at higher temperatures. As discussed in previous work [22], pure alginate (not cross-linked with acrylamides) has two pyrolysis stages: the first thermal degradation process takes place in the temperature range 225–300 °C. The weight loss in the first stage is attributed to the degradation of the carboxyl groups (leading to CO_2 release). The second stage occurred at much higher temperatures (650–740 °C) and corresponds to depolymerization resulting in a carbonaceous residue. Since the first step of decomposition of alginate overlap with signals related to acrylamide monomers, chemical analysis from the degradation of hybrid gels is difficult. One can observe that the pure alginate-polyacrylamide hybrid gel (Figure 2A) clearly shows pyrolysis stages shifted from locations of single networks [22]. This qualitatively demonstrates the formation of new covalent bonds between alginate and polyacrylamide.

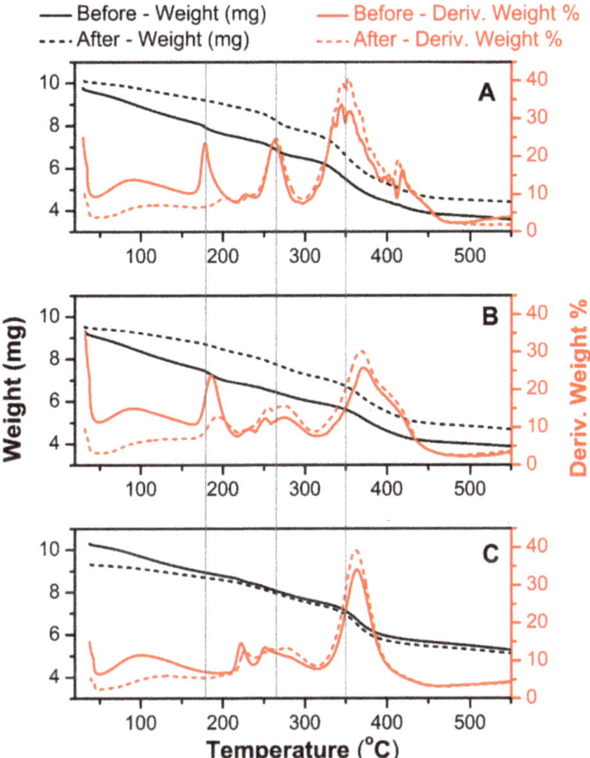

Figure 2. TGA (black) and first derivative of mass loss (red) of pure hydrogel electrolyte (**A**) and with 80 mg of MWCNT (**B**) and 360 mg (**C**). Solid lines refer to freshly prepared samples and dashed lines refer to samples after long-term electrochemical degradation using charge-discharge techniques.

MWCNT-embedded hydrogels (Figure 2B,C) exhibit improved thermal stability in comparison to pure hydrogels. This is manifested by the shift of decomposition temperature of the main polymer blend from 350 °C (Figure 2A—pure hydrogel) to 370 °C. The percent weight loss also decreases with increasing MWCNT content, especially for the highest carbon loading (Figure 2C). The higher resistance to heat in the presence of the MWCNT reinforced within the polyacrylamide/alginate network can be also related to weak chemical interactions between the MWCNT and the different reaction sites of both polymers [27,29]. For example, the latest study suggested that carboxyl groups present on the MWCNT surface can react with hydroxyl functionalities of alginate, resulting in hydrogen bridging [27]. These molecular interactions are believed to contribute to both better thermal stability of the MWCNT/polymer hybrid, and also to improved dispersion of the MWCNT in the solution of monomers and within the hydrogel itself. More probably, this chemical interaction is based on the esterification reaction and involves the mentioned carboxylic moieties from oxidized carbon and the hydroxyl groups of alginates. Nevertheless, this reactivity requires significant oxidization of the MWCNT surface (high content of -COOH), which was not found in the present study. The surface elemental analysis of MWCNT using an X-ray photoelectron did not show a significant difference in oxygen content (related to -COOH, -OH, or other oxygen-containing functional groups on the MWCNT surface; data not included) for as-obtained (unwashed) and the acid-washed MWCNT (3 M HNO_3, reflux 1 h). Thus, improved dispersion of carbon within the polymer observed in this work can be assumed as the combined effects of electrostatic interactions (predominant) and a minor contribution from the chemical bond formation between alginate and carboxylic groups present on the MWCNT surface.

Figure 3 demonstrates the FTIR spectrum of pure and MWCNT-containing hydrogels with distinctive signals assigned to acrylamide and alginate polymers. The FTIR spectra for all MWCNT-embedded hydrogels were similar, and corresponded to hydrogel signals, except an intensity of peaks decreased with increasing MWCNT content. FTIR showed only signals related to polymers, which were not influenced by the presence of MWCNT at any loading. Briefly, peaks at 1113 and 1350 cm^{-1} corresponded to the C-O stretch and at 1181 cm^{-1} were assigned to the C-N stretch, together with the N-H vibration at 2940 cm^{-1}. The vibrations at 1316, 1401, 1462, 2762, and 2855 cm^{-1} corresponded to the C-H bonding, arising from amines and amides present in the hydrogel structure. The band at 1606 cm^{-1} was related to the C=C vibration and the band at 1679 cm^{-1} was a signal of unreacted carbon double bonds from acrylamide and bis-acrylamide monomers. The amide stretch was also observed at 3442 cm^{-1} and the -OH stretch at 3178 cm^{-1}. In summary, FTIR did not demonstrate chemical interactions between MWCNT and organic fraction. This is presumably due to a very low carbon content per total hydrogel mass, calc. 2.79 wt.% of MWCNT for hydrogels containing 360 mg of carbon.

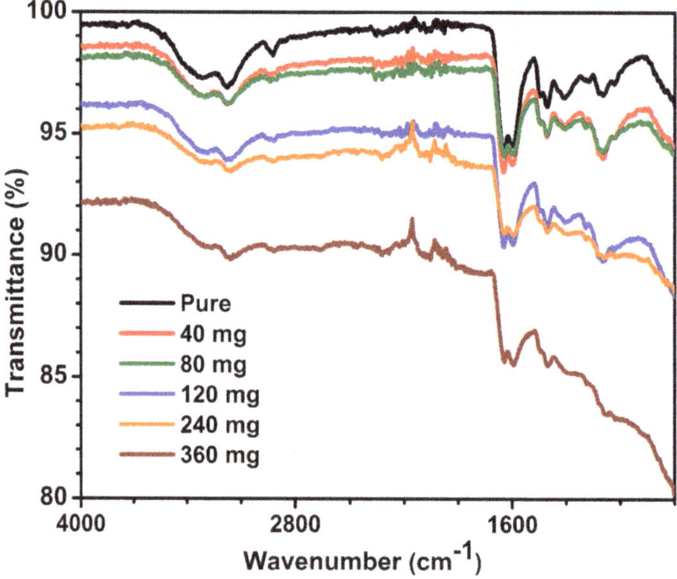

Figure 3. FTIR of pure and MWCNT-embedded hydrogels.

3.2. Resilience—Mechanical Analysis

Tensile characteristics of pure hydrogel and its combinations with MWCNT are shown in Figure 4. They were used to explore a maximum elongation length before mechanical failure (Figure 4A–D) and tensile strength acquired from the load sensor attached to one end of the assembly (Figure 4, bottom). With an adjustment for area difference (thickness of films was relatively similar for all samples), pure hydrogel tested at a maximum of 22.3 kPa (1.34 N), at a displacement of 271 mm. For the hydrogel with the highest MWCNT content (36 mg mL^{-1}), the applied sensor load was 28.8 kPa (1.76 N), at the displacement of 77 mm. Overall, the stretch modulus is sixteen times higher than that of the initial length for pure, and almost five times for MWCNT-embedded samples. In trend, an increase of MWCNT correlates to a general increase in tensile strength, with a more rapid decline in stretch modulus. A total force resistance of electrodes synthesized in this work was lower, yet resilience to deformation and stretch modulus held similar (17–21 factor) as compared to pure hydrogel with a similar composition as reported by others [22]. Although hydrogels fabricated in this work demonstrated lower force resistance to elongation, the comparison with the referred literature can be only qualitative since the results strongly depend on the thickness and geometry of the sample [22], the size of mounting clamps, as well as on the quality of the clamp (e.g., clamps with a sharp, uneven edge cause the material to be cut through faster). Joddar and colleagues reported the MWCNT embedded in alginate hydrogel (without acrylamide) at a very low load content [29]. The resilience ranges from 103.5 kPa for the pure hydrogel, 79.5 kPa (1 mg mL^{-1} MWCNT), 62 (3 mg mL^{-1} MWCNT), and 18 kPa (5 mg mL^{-1} MWCNT), with the decrease of total strength and stretch modulus with increasing MWCNT content observed. This conclusion contradicted their early central hypothesis that increasing the MWCNT content would increase the stiffness of the alginate gel [27]. This could be related to the weak van der Waals forces occurring between the MWCNT and alginate [27]. An opposite effect (increase in stiffness with increasing MWCNT content, Figure 4) is observed in this work. For hydrogels proposed in this work, the role of acrylamide blocks can improve resilience by additional alginate-acrylamide crosslinks (Figure 1), as well as the weak interactions (possibly hydrogen bringing) between alginate and acrylamide blocks with oxygen-containing functionalities present on the MWCNT surface. This suggests the participation of MWCNT strands within the polymer network.

As indicated in Figure 4E, there is a clear correlation between an increase in MWCNT content and an increase in tensile strength. Notably, the total MWCNT content of 36 mg mL^{-1} showed both the best mechanical and electrochemical characteristics. Further testing at 48 mg mL^{-1} indicated an oversaturation of the gel matrix, leading to unincorporated deposits of MWCNT on the surface after cross-linking was complete. The important consideration from this is the retention of physical properties of the hydrogel, with the addition of the MWCNT. Stretch was significantly reduced but retains enough capacity to serve practical applicability. An additional benefit is an increase in overall tensile force, indicating the contribution of MWCNT in the improvement of mechanical strength of hydrogels proposed in this work.

3.3. Impedance Analysis

For the polarizable electrode in the absence of redox reactions (non-faradaic measurements), the impedance signal will be dominated by the surface capacitance. This is true at least if the capacitance is low and the value of electrolyte solution is not too high. Given that electrodes in our study are symmetric, we can assume that the impedance between the electrodes can be modeled by electrolyte resistors and interfacial capacitance (Figure 5A). The impedance spectra of dry MWCNTs showed pure resistive behavior as represented by an absolute resistance that is independent of the frequency shown in the Bode diagram (red line in Figure 5B) and the phase shift remains at 0° in the whole frequency range (Figure 5C).

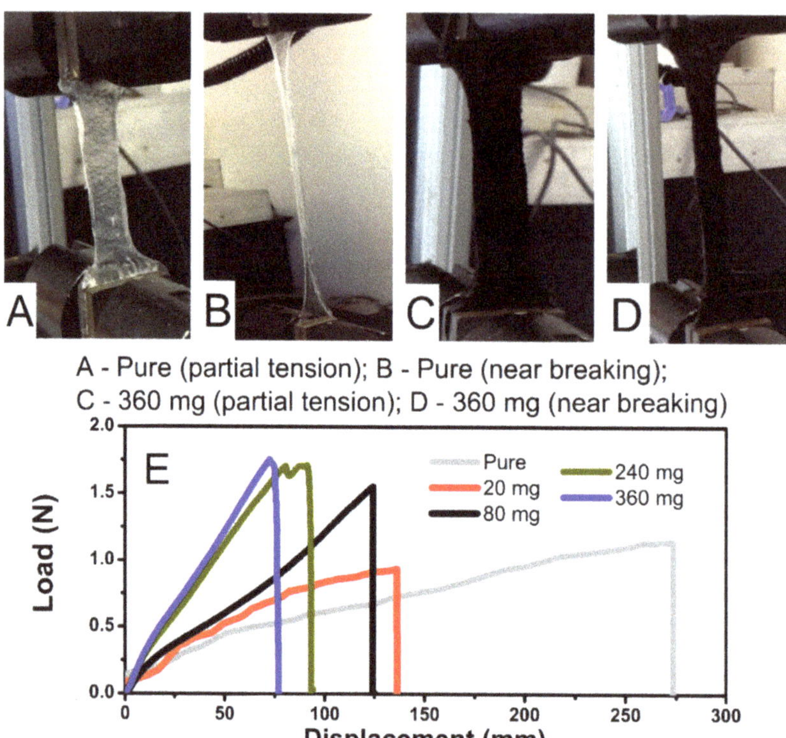

Figure 4. Mechanical tensile strength measurement set-up, showing pure sample under partial mechanical load (**A**) and near failure point (**B**). A 360 mg sample with partial mechanical load (**C**) and near failure point (**D**). Force loading and related displacement recorded during tensile measurements (**E**).

There are three distinct regions observed in the Bode plots for pure hydrogel (black spectrum). At the highest frequencies (between 10^5 and 10^4 Hz), resistive behavior is observed where the phase shift passes through 0° and the impedance magnitude is a horizontal line (because resistive impedance is not frequency-dependent). This represents the bulk resistance of the gel/electrolyte, which is used to calculate the hydrogel conductivities. In a broad range of frequency (800–1 Hz), capacitive behavior is observed as indicated by the negative |Z| slope (Figure 5B) and the phase angle approaching −90° (Figure 5C), which is due to double-layer capacitance formed at the interface of the electrolyte and FTO plates, and is typically observed in this frequency range. From the narrow range (400–800 Hz), diffusion behavior is observed as the phase angle approaches −45° (and the slope of impedance magnitude has changed). This represents the diffusion of ions through the finite thickness of the hydrogel. Considering all these components, the hydrogel EIS spectra can be fitted using an equivalent circuit shown in Figure 5A as an insert. Hydrogels containing 10, 20, and 40 mg of MWCNTs exhibited almost identical impedance magnitude over all frequencies when compared to pure hydrogel. Their slightly decreased conductivity is due to incorporated MWCNT, which has otherwise higher resistivity (80.7 Ohms, red spectrum) than pure electrolyte. When comparing pure electrolyte with the hydrogel containing 80 mg of MWCNTs, the most significant differences occurred in the mid and low frequencies, where interfacial and bulk ion diffusion is predominant.

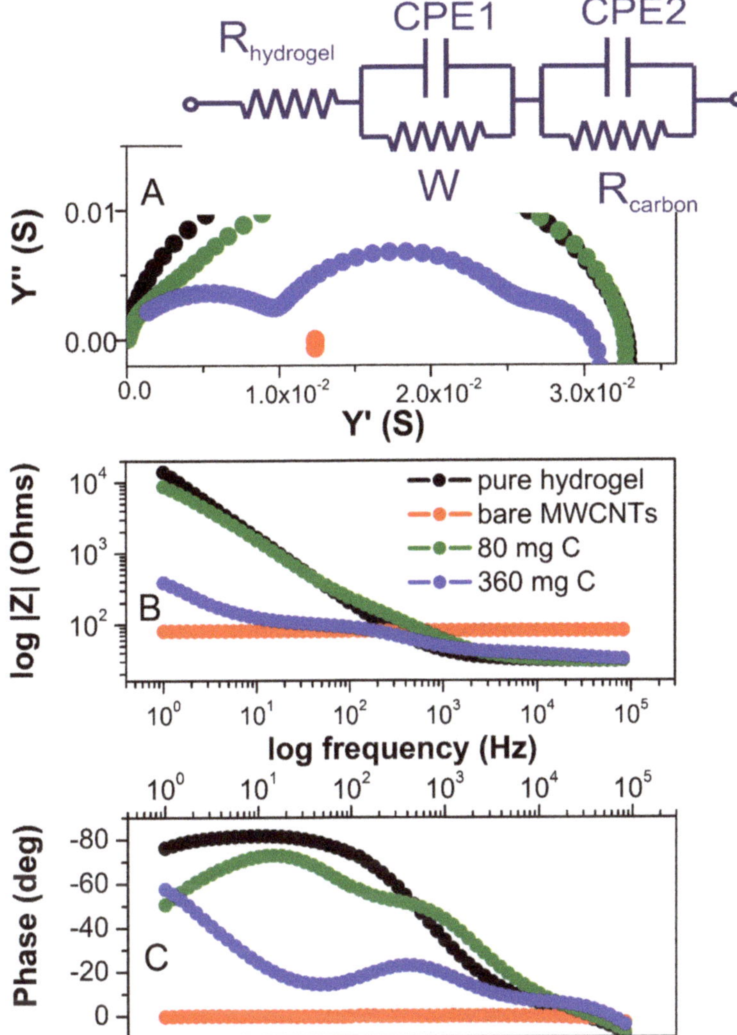

Figure 5. (**A**) Admittance spectra of MWCNT (red), pure hydrogel electrolyte (black), hydrogel with 80 mg MWCNT (green), and with 360 mg of MWCNT (blue). (**B**) Bode diagram of absolute impedance and (**C**) phase angle as the function of frequency. An electrical equivalent circuit representing the distribution of resistances and capacitances in the electrode was used for the fitting of the impedance spectra and is shown as an insert ($R_{hydrogel}$ represents a resistivity of pure hydrogel; CPE1 is a pseudo-capacitance of the hydrogel; *W* is a Warburg resistor related to diffusion of ions within hydrogel electrolyte; R_{carbon} and CPE2 are resistance and capacitance of the carbon fraction, respectively).

The weakly resolved semicircles in impedance and admittance spectra (blue and green spectra, Figure 5A) indicate multiple circuit elements. Clearly, at higher MWCNT content, the two conductivity components can be separated. The one at higher frequencies is related to the more conducive hydrogel electrolyte and the second semicircle (at the middle range of frequencies) represents an electronic conductivity of the MWCNT fraction. This component is more clearly manifested in samples with 360 mg of MWCNT, with the resistance of MWCNT fraction that is very close to that of dry MWCNT

(Figure 5, red). Yet, the total conductivity of the material is very similar to pure hydrogel over the broad range of MWCNT content. A significant change can be observed at the lowest frequency in both Bode diagrams. The magnitude of impedance decreases, and the phase angle changes from capacitive to diffusive (approaching −45°) for all hydrogels containing MWCNT. The Warburg is associated with a semi solid-state diffusion of ions in hydrogel. The Warburg coefficient representing transition time (τ, s) reflects the diffusion through the hydrogel is inversely related to the diffusion coefficient, according to Equation (1):

$$\tau = L^2/D, \tag{1}$$

where L (m) is an effective diffusion length and D ($m^2\ s^{-1}$) is a diffusion coefficient. The transition time (τ) is estimated based on fitting of an electrical equivalent circuit to the impedance spectra and obtaining Warburg resistor defined as:

$$Z = \frac{Rctnh(jT\omega)^P}{(jT\omega)^P}, \tag{2}$$

where Z is impedance, ω is frequency, R represents resistance, and P is a constant ($0 < P < 1$).

An effective diffusion length (L, m) is calculated using Equation (1) and the results are shown in Table 1. In general, the transition time τ, estimated from the fitting of the Warburg element, decreases with an increasing MWCNT content in the hydrogel. This increase for the electrode containing 360 mg of MWCNT is within two orders of magnitude of that of pure hydrogel. Consequently, an effective diffusion length for ion transport, L, decreases with increasing MWCNT amount. The trend in both, the time constant and the diffusion length, indicates that mass transport within the MWCNT-embedded hydrogels is improved as compared to the pure hydrogel. This can be rationalized as a positive effect of an increase of surface area between the electronic conductor (MWCNT) and the hydrogel electrolyte, resulting in the expansion of the double-layer interface within the material and the current collectors.

Table 1. Time constant (τ) representing the transition time of an ion transport within a hydrogel obtained from the fitting of electrical equivalent circuits (Figure 5A, insert) and the conductivity of the MWCNT fraction incorporated into the hydrogel electrolyte. A conductivity of the hydrogel fraction in a steady state (without elongation) is 5×10^{-2} S cm^{-1}, as demonstrated in Figure 6C.

Sample (mg of MWCNTs)	τ [1] (s)	L [1] (m)	Conductivity of MWCNT Fraction (S/cm)
0	5.72×10^{-5}	1.61×10^{-7}	-
20	1.53×10^{-5}	8.36×10^{-8}	-
40	2.17×10^{-7}	9.95×10^{-9}	0.014
80	1.82×10^{-7}	9.12×10^{-9}	0.010
120	1.14×10^{-7}	7.21×10^{-9}	0.014
240	4.18×10^{-7}	1.38×10^{-8}	0.015
360	3.36×10^{-7}	1.22×10^{-8}	0.023
Pure MWCNT	-	-	0.031

[1] $\tau = L^2/D$ (D = diffusion coefficient estimated for 0.1 M KCl in acrylamide hydrogel as 4.57×10^{-10} $m^2\ s^{-1}$ [34], L = an effective diffusion length for electrolyte species in acrylamide-based hydrogels.

When the concentration gradient vanishes at the hydrogel/current collector interface, the Warburg impedance transforms into a capacitive behavior for pure hydrogel (black spectrum, Figure 5A) and for all samples with an MWCNT content lower than 80 mg (green spectrum, Figure 5A). This is demonstrated as an almost constant phase angle reaching a steady value of 80° for pure hydrogel. The low-frequency region is noticeably affected by the amount of MWCNT embedded in hydrogel. This is manifested by a change from capacitive behavior observed for pure hydrogel electrolyte to a more diffusive behavior for the MWCNT-hydrogel films. An additional diffusion process may occur at the MWCNT-hydrogel interface together with the surface diffusion of the adsorbed ions at the material/current collector. With increasing MWCNT content, this effect is stronger. Consequently, the low-frequency region of the impedance spectra for the film containing 360 mg of MWCNT shows an infinitive diffusion due to a large MWCNT-hydrogel interface across the material.

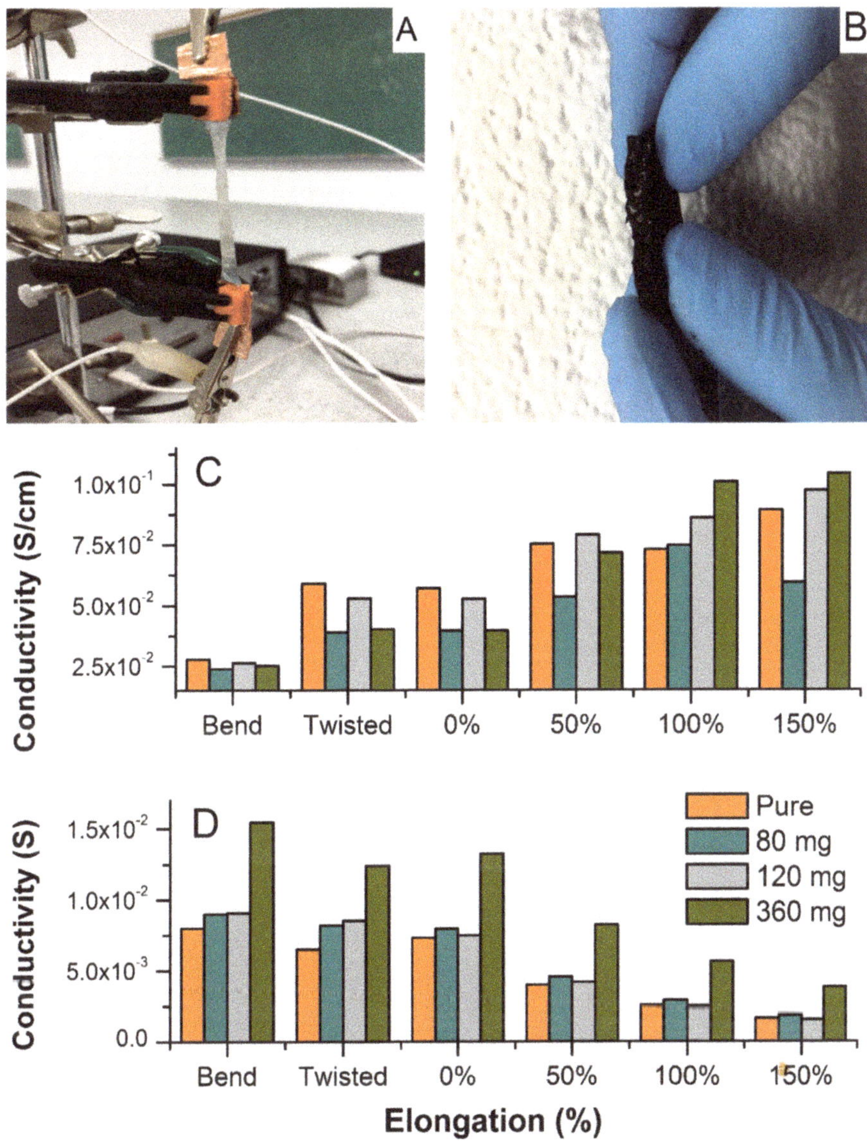

Figure 6. An electrochemical set-up for conductivity test upon elongation. (**A**) A cross-sectional photo of hydrogel-embedded MWCNTs (2.79 wt.% of MWCNT). (**B**) Total conductivity upon elongation, bending, and twisting normalized to the thickness, length, and width of the film upon elongation (**C**) and without normalization (**D**). The conductivity was estimated based on admittance analysis and example of spectra are shown in Figures 5A and 7.

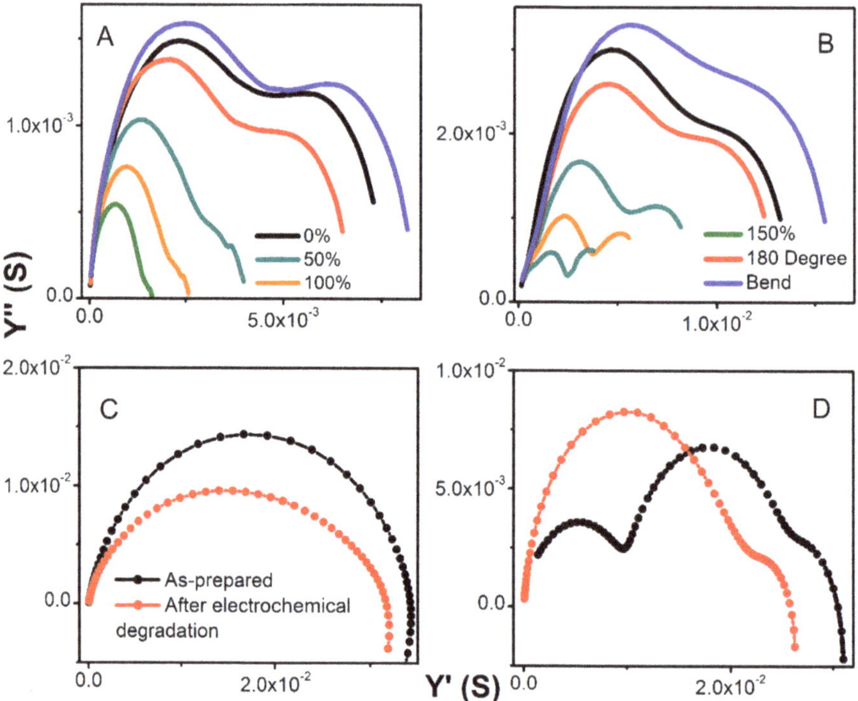

Figure 7. Admittance spectra upon elongation for pure hydrogel (**A**) and with 360 mg of MWCNT (**B**). Admittance spectra for as-prepared (black) and after a long-term charge-discharge (red) for pure hydrogel (**C**) and hydrogel with 360 mg of MWCNT (**D**).

Figure 6 shows the electrochemical set-up for testing conductivity upon elongation, bending, and twisting (A), and a cross-sectional photo of the MWCNT-hydrogel (2.79 wt.% of carbon), (B) together with a total conductivity of hydrogel electrodes and electrolytes normalized to the thickness, length, and width of the film upon deformation (C), and without normalization (D). Overall, the pure hydrogel electrolyte shows slightly higher specific conductivity as compared to the MWCNT-embedded films in steady state (Figure 6A). This decrease in total conductivity for the MWCNT-containing materials is because the resistance of pure MWCNTs is slightly higher than the hydrogel, as manifested in Figure 5A (red spectrum of bare MWCNT). However, the hydrogel with the highest MWCNT content has higher conductivity than the bare MWCNT. From the fitting of the admittance spectra, the conductivity of ionic fraction (hydrogel) and an electronic component (MWCNT) could be resolved and is included in Table 1. The ionic component (conductivity of pure hydrogel) is higher than the electronic one and remains constant regardless of the MWCNT content (5.2×10^{-2} S cm^{-1}). The electronic counterpart evolves around values obtained for the bare MWCNT. A minor increase with increasing MWCNT content results in mixed conductivity that is close to that of pure electrolyte for the film containing 360 mg of MWCNT. This demonstrates that the hydrogel does not suppress MWCNT conductivity. The hydrogel matrix shows conductivity similar to liquid electrolyte and the MWCNT response is similar to what is typically observed in the liquid electrolyte with a similar KCl concentration [31]. All films showed no influence of twisting at about 180 degrees on the specific conductivity (Figure 6C,D). Furthermore, impedance was recorded during bending, and demonstrated a weak decrease when normalized to new geometries (Figure 6C), or no difference without normalization (Figure 6D). Interestingly, the elongation from original shape (0% elongation corresponds to as-prepared film) causes an increase in specific conductivity for all hydrogels since their thickness decreased significantly (although they were stretched to more than 150% of their original length; Figure 7A,B shows an evolution of admittance spectra upon elongation for pure hydrogels and with 360 mg of MWCNT). All electrodes demonstrated excellent reproducibility of electrical characteristics upon multiple stretching and deformation during repetitions of the same test (unless breaking was anticipated).

Yet, hydrogels with higher MWCNT content (240 and 360 mg, Figure 6C) showed improved total conductivity upon elongation when compared to pure hydrogel. This indicates that MWCNT not only generates electronic conductivity, but also enhances the total conductivity of the hydrogel, especially when subjected to mechanical deformations. This can be rationalized as improved tensile strength and rigidness of the MWCNT-embedded films. With that, some indication is given that weak chemical interactions between the MWCNT surface (enriched with oxygen-containing functionalities, i.e., carboxylic, phenolic, hydroxyl groups) and the acrylamide are possible, and result in improved mechanical and electrical characteristics of the combined materials. These molecular interactions can be also confirmed by the shift of onset peaks of hydrogel in the presence of MWCNT in the TGA spectrum (Figure 2). In addition, optical observations revealed a uniform distribution of MWCNT within the hydrogel, even at high MWCNT content (Figure 6B). The electrode shows good homogeneity and carbon dispersion across the film. This could be rationalized as both electrostatic and chemical effects i.e., related to the negative charge of MWCNT surface and alginate, and/or weak interaction between carboxyl groups present on the MWCNT surface with hydroxyl functionalities of alginate, resulting in chemical bond formation or possible electrostatic interaction. Thus, alginate not only participates in an ionic cross-linking of the base hydrogel matrix, but also modifies the MWCNT surface resulting in its better dispersion in aqueous solution, keeping in mind that these interactions are rather weak (hydrogen bridging or Van der Waal forces), unless esterification between these components is considered. At this point of the study, we can only speculate on the nature of these interactions since the FTIR did not show a significant difference between pure hydrogel and MWCNT-containing samples (Figure 4).

3.4. Electrochemical Stability

An accelerated charge-discharge led to degradation of the hydrogel at the point of contact of materials with the FTO current collector (manifested by slightly darker color at the hydrogel/FTO contact). Furthermore, TGA analysis of as-prepared and degraded hydrogels showed no differences in the thermograms, leading to conclusions that the increase in impedance (observed only at lowest frequencies) are related to the passivation of the FTO when exposed to corrosive components such as 1 M KCl for extended periods of time. Both the thermal and spectroscopic analysis have revealed that there is no noticeable chemical change within the material after long-term charging at 0.5 mA.

The total conductivity of hydrogel decreases only slightly and within the same order of magnitude as demonstrated on the admittance spectra in Figure 7C,D. Since the bare MWCNTs did not show any changes when subjected to charging at various current loads, we concluded that the decrease in conductivity is primarily due to the passivation of the contacts. The analysis of the Warburg element did not show any changes in time constant, or if any, the values laid within the calculation error. Yet, the decrease in admittance is slightly smaller for pure hydrogel as compared to the MWCNT-containing electrodes when tested in the long-term charge-discharge. This indicates slightly faster degradation due to the presence of MWCNT. Since the MWCNT enhances electrical contact between hydrogels, current collectors, and within materials, the degradation due to charging is accelerated.

To further estimate the practical potential windows for hydrogels, samples were polarized at various regimes and corresponding cyclic voltammograms are presented in Figure 8. In general, all hydrogels are stable within −0.95 and +0.95 V regardless of the MWCNT content, with a slightly improved potential stability for samples containing 360 mg of MWCNT. The materials are pronounced unstable above ±1.0 V due to water evolution as revealed by strong redox signals in CVs in Figure 8. An important observation was made on the current magnitude recorded for pure (Figure 8B, insert) and MWCNT-embedded hydrogel (Figure 8A). Due to the high content of MWCNT that are well-known to be double-layer capacitors, the current observed for combined hydrogel is two orders of magnitude higher when compared to pure hydrogel. Both the current and the shape of the CV spectrum indicate the strong capacitive nature of the latter, making them a potentially good candidate as an electrode material for flexible capacitor devices.

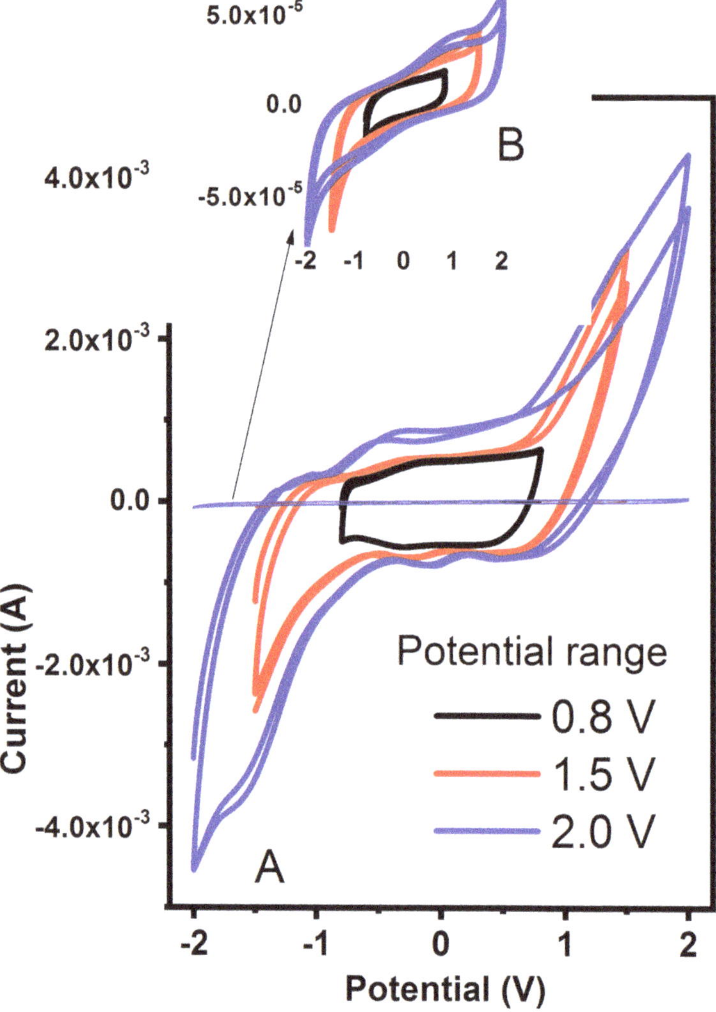

Figure 8. The cyclic voltammetry of the hydrogel containing 360 mg MWCNT (**A**) and pure hydrogel ((**B**), insert) scanned at various potential ranges at 0.2 V s^{-1}.

Cyclic voltammetry (Figure 9A) was utilized to define the area-specific interfacial capacitance of the material (C_s, F cm^{-2}), calculated through Equation (3):

$$C_s = \frac{1}{2}\frac{A_{CV}}{a\Delta V v}, \tag{3}$$

where A_{CV} is the area bound by the CV (0.00284 AV), a is the electrode surface area (12.0 mm × 12.0 mm), ΔV is the potential window (1.6 V), and v is the scan rate (0.2 V s^{-1}). For the MWCNT-hydrogel at 3.69 wt.% of carbon, the area-specific interfacial capacitance was 27 mF cm^{-2}. This interfacial double-layer capacitance of the carbon is within a range of the thin-film solid-state carbon electrodes (11 mF cm^{-2}; [35]). The impact of carbon content on a magnitude of charging current has been noticed only at higher MWCNT concentration (higher than 240 mg of MWCNT, that is 1.88 wt.%; Figure 9A). This leads to the conclusion that the minimum carbon content of 2.0 wt.% is required in order to create flexible electrodes with desired electrical characteristics. Figure 9B demonstrates a typical increase of current with increasing potential scan rate (for the electrode with 2.79 wt.% of MWCNT), showing the ability of this electrode to pass more charge. The bulk capacitance of these electrodes is expected to be much higher and will be tested in a three-electrode electrochemical cell in a future study. An increase in charging current for the carbon-containing hydrogel in comparison to the pure hydrogel electrolyte shown in Figure 9A (black scan) indicates the viability of this conducting platform and its possible implementation in stretchable electronics.

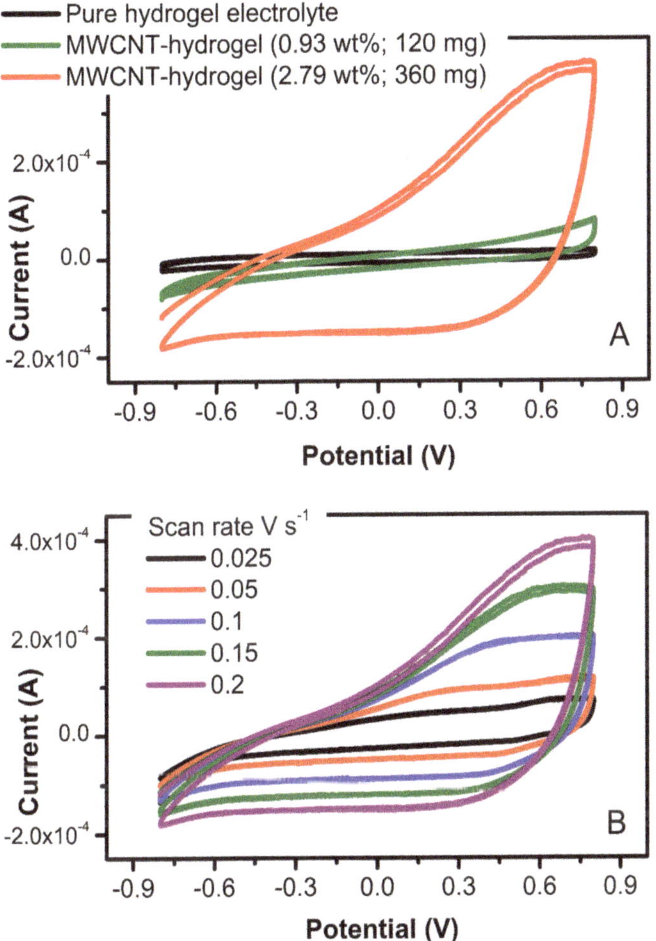

Figure 9. (**A**) CV scans recorded at 0.2 V s^{-1} of the pure hydrogel (black) and MWCNT-embedded hydrogel (green—0.93 wt.% and red—2.79 wt.% of carbon). (**B**) CVs of MWNT-embedded hydrogel (2.79 wt.%) at various scan rates.

4. Conclusions

We have investigated the effects of MWCNT addition within polyacrylamide-alginate hydrogels, on conductivity and mechanical characteristics in their steady state, and upon mechanical stretch. Based on the mechanical analysis, the tensile strength of pure hydrogel was similar to hydrogels reported in the literature, with increasing strength for MWCNT-embedded hydrogels synthesized in this work. The impedance spectroscopy revealed that the total resistance of electrodes decreased with increasing MWCNT content during their elongation, and that bending and twisting do not change their conductivity. The MWCNT-embedded hydrogels showed mixed ionic and electronic conductivities, both within the same range of 5–10 mS cm^{-1} in a steady state. In addition, the thermal stability of the electrode increased with increasing MWCNT content. This agreed with the long-term charge-discharge cycling that resulted in enhanced electrochemical stability of the MWCNT-hydrogel hybrid when compared to pure hydrogel electrolyte. These hydrogel films showed an interfacial double-layer capacitance at the high MWCNT content, which demonstrates their applicability in lightweight and yet flexible energy storage applications.

Author Contributions: J.T., all experimental work and writing—original draft preparation; A.I., impedance and TG analysis, writing—review and editing. All authors have read and agreed to the published version of the manuscript.

Funding: This work was financed by New Brunswick Innovation Foundation—Research Assistantship Initiative project number RAI2019-009 and Canada Foundation for Innovation John R. Evans Leaders Funds CFI JELF project number 36874.

Acknowledgments: We thank Steven McWilliams for his help with FTIR tests.

Conflicts of Interest: The authors declare no conflict of interest. The funders had no role in the design of the study; in the collection, analyses, or interpretation of data; in the writing of the manuscript, or in the decision to publish the results.

References

1. Honarvar, M.G.; Latifi, M. Overview of wearable electronics and smart textiles. *J. Text. Inst.* **2017**, *108*, 631–652. [CrossRef]
2. Vagott, J.; Parachuru, R. An Overview of Recent Developments in the Field of Wearable smart textiles. *J. Text. Sci. Eng.* **2018**, *08*, 1–10.
3. Stoppa, M.; Chiolerio, A. Wearable electronics and smart textiles: A critical review. *Sensors* **2014**, *14*, 11957–11992. [CrossRef] [PubMed]
4. Park, S.; Jayaraman, S. Enhancing the Quality of Life Through Wearable Technology. *IEEE Eng. Med. Biol.* **2003**, *22*, 41–48. [CrossRef] [PubMed]
5. Majumder, S.; Mondal, T.; Deen, M.J. Wearable sensors for remote health monitoring. *Sensors* **2017**, *17*, 130. [CrossRef]
6. Nia, A.M.; Mozaffari-Kermani, M.; Sur-Kolay, S.; Raghunathan, A.; Jha, N.K. Energy-Efficient Long-term Continuous Personal Health Monitoring. *IEEE Trans. Multi Scale Comput. Syst.* **2015**, *1*, 85–98. [CrossRef]
7. Lu, Y.; Biswas, M.C.; Guo, Z.; Jeon, J.W.; Wujcik, E.K. Recent developments in bio-monitoring via advanced polymer nanocomposite-based wearable strain sensors. *Biosens. Bioelectron.* **2019**, *123*, 167–177. [CrossRef]
8. Böttjer, R.; Grothe, T.; Ehrmann, A. Functional nanofiber mats for medical and biotechnological applications. *Narrow Smart Text.* **2017**, 203–214.
9. Persson, N.K.; Martinez, J.G.; Zhong, Y.; Maziz, A.; Jager, E.W.H. Actuating Textiles: Next Generation of Smart Textiles. *Adv. Mater. Technol.* **2018**, *3*, 1–12. [CrossRef]
10. Fernández-Caramés, T.M.; Fraga-Lamas, P. Towards the internet-of-smart-clothing: A review on IoT wearables and garments for creating intelligent connected E-textiles. *Electronics* **2018**, *7*, 405. [CrossRef]
11. Liu, H.; Zhong, J.; Lee, C.; Lee, S.W.; Lin, L. A comprehensive review on piezoelectric energy harvesting technology: Materials, mechanisms, and applications. *Appl. Phys. Rev.* **2018**, *5*, 1–35. [CrossRef]
12. Morshed, B.I.; Harmon, B.; Zaman, M.S.; Rahman, M.J.; Afroz, S.; Rahman, M. Inkjet printed fully-passive body-worn wireless sensors for smart and connected community (SCC). *J. Low Power Electron. Appl.* **2017**, *7*, 26. [CrossRef]

13. Yapici, M.K.; Alkhidir, T.E. Intelligent medical garments with graphene-functionalized smart-cloth ECG sensors. *Sensors* **2017**, *17*, 875. [CrossRef] [PubMed]
14. Ajami, S.; Teimouri, F. Features and application of wearable biosensors in medical care. *J. Res. Med. Sci.* **2015**, *20*, 1208–1215. [CrossRef]
15. Goforth, C.; Lisman, P.; Deuster, P. The Physiological Impact of Body Armor Cooling Devices in Hot Environments: A Systematic Review. *Mil. Med.* **2014**, *179*, 724–734. [CrossRef]
16. Kasliwal, M.H.; Patil, H.Y. Smart location tracking system for dementia patients. In Proceedings of the International Conference in Computing, Communication and Control (ICAC3), Mumbai, India, 20–21 December 2017; pp. 1–6.
17. Coughlin, T. Digital Storage in Smartphones and Wearables. *IEEE Consum. Electron. Mag.* **2018**, *7*, 108–120. [CrossRef]
18. Murray, S.A.; Farris, R.J.; Golfarb, M.; Hartigan, C.; Kandilakis, C.; Truex, D. FES Coupled with A Powered Exoskeleton for Cooperative Muscle Contribution in Persons with Paraplegia. In Proceedings of the 40th Annual International Conference of the IEEE Engineering in Medicine and Biology Society (EMBC), Honolulu, HI, USA, 17–21 July 2018; pp. 2788–2792.
19. Poppinga, S.; Zollfrank, C.; Prucker, O.; Rühe, J.; Menges, A.; Cheng, T.; Speck, T. Toward a New Generation of Smart Biomimetic Actuators for Architecture. *Adv. Mater.* **2018**, *30*, 1–10. [CrossRef]
20. Sekiguchi, A.; Tanaka, F.; Yamada, T.; Hata, K. Stretchable and robust transistor of single wall carbon nanotube, gel and elastomeric materials. In Proceedings of the International Conference on Electronics Packaging (ICEP), Yamagata, Japan, 19–22 April 2017; pp. 457–459.
21. Dam, N.; Ricketts, A.; Catlett, B.; Henriques, J. Wearable sensors for analyzing personal exposure to air pollution. In Proceedings of the Systems and Information Engineering Design Symposium (SIEDS 2017), Charlottesville, VA, USA, 28 April 2017; pp. 1–4.
22. Sun, J.Y.; Zhao, X.; Illeperuma, W.R.K.; Chaudhuri, O.; Oh, K.H.; Mooney, D.J.; Vlassak, J.J.; Suo, Z. Highly stretchable and tough hydrogels. *Nature* **2012**, *489*, 133–136. [CrossRef]
23. Gong, J.P.; Katsuyama, Y.; Kurokawa, T.; Osada, Y. Double-network hydrogels with extremely high mechanical strength. *Adv. Mater.* **2003**, *15*, 1155–1158. [CrossRef]
24. Henderson, K.J.; Zhou, T.C.; Otim, K.J.; Shull, K.R. Ionically cross-linked triblock copolymer hydrogels with high strength. *Macromolecules* **2010**, *43*, 6193–6201. [CrossRef]
25. Yang, C.H.; Wang, M.X.; Haider, H.; Yang, J.H.; Sun, J.Y.; Chen, Y.M.; Zhou, J.; Suo, Z. Strengthening alginate/polyacrylamide hydrogels using various multivalent cations. *ACS Appl. Mater. Interfaces* **2013**, *5*, 10418–10422. [CrossRef]
26. Demianenko, P.; Minisini, B. Stiff IPN Hydrogels of Poly (Acrylamide) and Alginate: Influence of the Crosslinking Ion's Valence on Hydrogel's Final Properties. *J. Chem. Eng. Process Technol.* **2016**, *7*, 1–6. [CrossRef]
27. Joddar, B.; Garcia, E.; Casas, A.; Stewart, C.M. Development of functionalized multi-walled carbon-nanotube-based alginate hydrogels for enabling biomimetic technologies. *Sci. Rep.* **2016**, *6*, 1–12. [CrossRef] [PubMed]
28. Hong, S.; Lee, J.; Do, K.; Lee, M.; Kim, J.H.; Lee, S.; Kim, D.H. Stretchable Electrode Based on Laterally Combed Carbon Nanotubes for Wearable Energy Harvesting and Storage Devices. *Adv. Funct. Mater.* **2017**, *27*, 1–7. [CrossRef]
29. Sudha; Mishra, B.M.; Kumar, D. Effect of multiwalled carbon nanotubes on the conductivity and swelling properties of porous polyacrylamide hydrogels. *Part. Sci. Technol.* **2014**, *32*, 624–631. [CrossRef]
30. Ji, C.; Yan, D.; Wang, Y.; Xiong, S.; Zhou, S.; Li, Y.; Sun, R.; Wong, S.-P. Thermal conductivity enhancement of CNT/MoS_2/graphene−epoxy nanocomposites based on structural synergistic effects and interpenetrating network. *Compos. Part B* **2019**, *163*, 363–370. [CrossRef]
31. Radtke, M.; Ignaszak, A. Stretchable current collectors based on carbon embedded in a poly (acrylamide)/poly (N, N-methylenebisacrylamide) hydrogel modified with Nafion 117®. *Mater. Renew. Sustain. Energy* **2018**, *7*, 1–13. [CrossRef]
32. Hu, H.; Yu, A.; Kim, E.; Zhao, B.; Itkis, M.E.; Bekyarova, E.; Haddon, R.C. Influence of the zeta potential on the dispersability and purification of single-walled carbon nanotubes. *J. Phys. Chem. B* **2005**, *109*, 11520–11524. [CrossRef]

33. Reinert, L.; Zeiger, M.; Suárez, S.; Presser, V.; Mücklich, F. Dispersion analysis of carbon nanotubes, carbon onions, and nanodiamonds for their application as reinforcement phase in nickel metal matrix composites. *RSC Adv.* **2015**, *5*, 95149–95159. [CrossRef]
34. Valente, A.J.M.; Polishchuk, A.Y.; Lobo, V.M.M.; Geuskens, G. Diffusion coefficients of lithium chloride and potassium chloride in hydrogel membranes derived from acrylamide. *Eur. Polym. J.* **2002**, *38*, 13–18. [CrossRef]
35. Patterson, N.; Ignaszak, A. Modification of glassy carbon with polypyrrole through an aminophenyl linker to create supercapacitive materials using bipolar electrochemistry. *Electrochem. Commun.* **2018**, *10*, 10–14. [CrossRef]

Article

Facile Fabrication of Conductive Graphene/Polyurethane Foam Composite and Its Application on Flexible Piezo-Resistive Sensors

Weibing Zhong [1], Xincheng Ding [2], Weixin Li [2], Chengyandan Shen [2], Ashish Yadav [2], Yuanli Chen [2], Mingze Bao [2], Haiqing Jiang [2,*] and Dong Wang [1,2,*]

1 College of Chemistry, Chemical Engineering and Biotechnology, Donghua University, Shanghai 201620, China

2 Hubei Key Laboratory of Advanced Textile Materials & Application, Wuhan Textile University, Wuhan 430200, China

* Correspondence: hqjiang@wtu.edu.cn (H.J.); wangdom08@126.com (D.W.)

Received: 16 July 2019; Accepted: 26 July 2019; Published: 1 August 2019

Abstract: Flexible pressure sensors have attracted tremendous research interests due to their wide applications in wearable electronics and smart robots. The easy-to-obtain fabrication and stable signal output are meaningful for the practical application of flexible pressure sensors. The graphene/polyurethane foam composites are prepared to develop a convenient method for piezo-resistive devices with simple structure and outstanding sensing performance. Graphene oxide was prepared through the modified Hummers method. Polyurethane foam was kept to soak in the obtained graphene oxide aqueous solution and then dried. After that, reduced graphene oxide/polyurethane composite foam has been fabricated under air phase reduction by hydrazine hydrate vapor. The chemical components and micro morphologies of the prepared samples have been observed by using FT-IR and scanning electron microscopy (SEM). The results predicted that the graphene is tightly adhered to the bare surface of the pores. The pressure sensing performance has been also evaluated by measuring the sensitivity, durability, and response time. The results indicate that the value of sensitivity under the range of 0–6 kPa and 6–25 kPa are 0.17 kPa^{-1} and 0.005 kPa^{-1}, respectively. Cycling stability test has been performed 30 times under three varying pressures. The signal output just exhibits slight fluctuations, which represents the good cycling stability of the pressure sensor. At the same stage, the response time of loading and unloading of 20 g weight turned out to be about 300 ms. These consequences showed the superiority of graphene/polyurethane composite foam while applied in piezo-resistive devices including wide sensitive pressure range, high sensitivity, outstanding durability, and fast response.

Keywords: reduced graphene oxide; polyurethane foam; flexible electronics; pressure sensing

1. Introduction

With the rapid development of the information technology, flexible pressure sensors have been extremely concerned due to their potential in human machine interface, flexible robots, wearing electronic equipment, and flexible electronic skin [1–15]. Normally, flexible pressure sensors are divided into three types: capacitive, piezoelectric, and piezo-resistive. Among them, piezo-resistive sensor is the most favorite because of its simple manufacturing techniques and easily collecting signal output [2,16–19].

A typical method to prepare the flexible pressure sensors is to construct a composite with elastomer material as the matrix and the conductive filling materials. The external force induced compression deformation would lead to the rearrangement of the conductive fillings, which finally changed the resistance of the elastic composite [13,20–23]. Generally, the piezo-resistive materials were mainly fabricated with conductive particles filled with elastomer composites. The deformation of the elastomer would force the conductive particles to form another electrical conductive network. Besides, thin conductive film such as carbon nanotubes (CNTs) and graphene were laminated on the elastomer film to improve the connecting efficiency of the conductive components. However, the deformations of these solid materials required overcoming strong energy to move the molecular chains, which seriously restricts the sensitivity of the manufactured piezo-resistive materials [24–26]. Obviously, substrate with lower Young's modulus possesses stronger deformation ability which significantly contributes the sensitivity. Polyurethane (PU) foam, which is an elastic material with large amount of through-pore structures and exhibiting low Young's modulus and high resilience, is an ideal candidate for high performance substrate for piezo-resistive sensors.

Graphene is a two-dimensional (2D) carbon material which has excellent physical properties [16,19,27]. It is a honeycomb shaped 2D planar crystal which is composed of 6 sp^2 hybrid carbon atoms connected by σ bond. As reported [28–31], graphene has exhibited excellent physical properties. For instance, specific surface area is as high as 2600 m^2g^{-1}, ultralow areal density of 0.77 mg/m^2, the optical transmittance can be as high as 98%, high electron mobility and thermal conductivity of 2.5×10^5 cm^2/Vs and 5000 $Wm^{-1}K^{-1}$, respectively. Therefore, graphene shows great potential in many fields.

In this paper, a convenient method of preparing the flexible piezo-resistive sensors is proposed. the reduce graphene oxide (rGO)/PU foam composite are fabricated via soaking the GO aqueous solution with pure PU foam and then reduced with hydrazine hydrate vapors. The chemical structure as well as the micro morphologies have been observed. Except that, the pressure sensing performance such as sensitivity, recognition ability to different pressures, cycling stability and response time have been systematically investigated. Due to the simple fabrication method, easily obtained raw materials and controllable preparation conditions, the composite materials can be manufactured with high reproducibility and very low cost. Except that, the chemically inert of the rGO enables it with good reusability over time as long as they were stored and used within the moderate environment.

2. Experimental

2.1. Materials

Fourier infrared tester (Tensor 27, Bruker, Ettlingen, Germany) and scanning electronic microscope (6510LV, JEOL, Tokyo, Japan) are applied for chemical structure and micro morphologies measurements. Electrochemical workstation (PGSTAT302N, Metrohm Autolab, Utrecht, Netherlands) and RLC digital electric bridge (TH2818, Tonghui, Changzhou, China) are used to detect and record the resistance of the composites foam under varying pressure. Digital display force gauge (Mark-10, ESM301, MARK-TEN, New York, NY, USA) and relative tension and compression testing bench used for applying the pressures. The microcrystal graphite used is purchased from Qingdao Jiaodong graphite Co., Ltd., Qingdao, China. The other chemicals used in the experiments are obtained from Sinopharm Chemical Reagent Co., Ltd., Shanghai, China. The deionized water is self-made in our lab.

2.2. The Preparation of Graphite

Modified Hummers method is used to synthesize the well delaminated GO. The oxidizing agents were concentrated with sulfuric acid and potassium permanganate. The acid and inorganic salts were removed from GO by centrifugation and dialysis. The purified GO aqueous solution was then configured to a fixed concentration for subsequent use.

2.3. Preparation of the Reduced Graphene Oxid/Polyurethane (rGO/PU) Foam Flexible Conductive Materials

As Figure 1 shows, the polyurethane (PU) foam was sliced into a fixed size of $40 \times 40 \times 30$ mm^3. They were pre-cleaned in the alcohol and then dried in oven at 60 °C for 2 h. After that, the PU foam were dipped into the GO aqueous solution with concentration of 3 mg/mL for 15 min. During this period, the foam was constantly squeezed in order to adsorb as much GO aqueous solution as possible. Then, the PU foam was transferred into vacuum oven at 55 °C for 4 h. After the drying process, the GO/PU foam was hanged in a sealed bottle with 8 mL hydrazine hydrate with concentration of 85 wt.%. The bottle was heated with oil bath with temperature of 100 °C for 90 min. The rGO/PU foam composite was obtained after three times cleaning and drying. The digital camera photographs of the pure PU foam and the rGO/PU composite foam was displayed in Figure S1.

Figure 1. Schematic figure of Preparation and assembling of the reduced graphene oxide/polyurethane (rGO/PU) foam piezo-resistive sensor.

2.4. Characterization and Methods

The quality of the GO and rGO were measured and analyzed by using X-ray diffraction (XRD). The chemical structure as well as the micro morphologies were observed with Fourier Transform infrared spectroscopy (FTIR) and Scanning electronic microscope (SEM). After that, the piezo-resistive sensor was assembled by adhering the copper electrodes to the two surfaces with 4×4 cm^2 of the conductive sponges. The pressure sensing performance such as sensitivity, recognition ability to different pressures, cycling stability, and response time were systematically investigated by using the electrochemical workstation, RLC digital electric bridge, digital display force gauge, and relative tension and compression testing bench.

3. Results and Discussions

3.1. X-ray Diffraction (XRD) Spectrum of Graphene Oxide (GO) and Reduced Graphene Oxide (rGO)

Figure 2 presents the XRD spectrum of the prepared GO and rGO. As can be seen, both spectra show an upward single peak. Being different, the peak position of the GO and rGO were 9.92° and 24.10°. Relying on the Bragg's law, the crystals distance of GO and rGO are observed 0.891 nm and 0.368 nm respectively. Due to the violent oxidation reaction, large amounts of oxygen-containing groups occur in layers of the graphite during the preparation, and bigger inter planar spacing forms. These oxygen-containing groups are progressively removed during the reduction, so that the layer spacing decreases to 0.368 nm. It is closed to the layer spacing of natural flake graphite, which is 0.34 nm. This is the main reason to showing XRD results. The GO and rGO were prepared accordingly previous reported literatures [32,33].

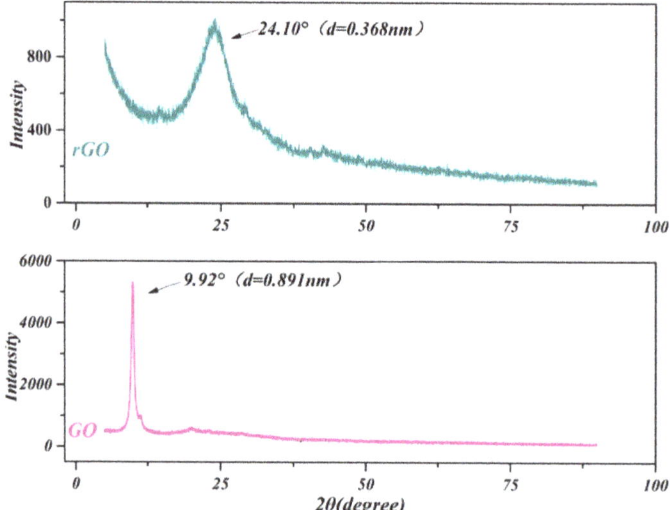

Figure 2. The X-ray diffraction (XRD) spectrum of graphene oxide (GO) and reduced graphene oxide (rGO) prepared with modified Hummers method.

3.2. Chemical Structure Analysis

Figure 3 shows the FTIR spectra of GO and rGO, blank PU foam obtained using Fourier Transform infrared spectroscopy. Infrared characteristic absorption peak of GO is shown in above figure. The stronger wide peak for free hydroxyl stretching resonance absorption at 3430 cm^{-1}. The tip of peak for telescopic resonance absorption peak of carboxyl-carbon-oxygen double bond in GO at 1725 cm^{-1}. The bending resonance absorption peak of C–OH is at 1630 cm^{-1}. Carbon oxide carbon epoxy bond absorption peak is at 1110 cm^{-1}. This indicates that the prepared GO have free hydroxyl and carboxyl groups and these groups are reduced by hydrazine hydrate vapor. The rGO has a little hydroxyl and epoxy groups. Comparing the change of the characteristic absorption peak in 5 infrared absorption curves, it could be known that after the compounding of PU foam and rGO, the peak of PU, itself has been weakened and the absorbing peaks of GO have occurred. In addition, the characteristic absorption peaks in the infrared spectra of rGO/PU foam are similar with the pure rGO. It can be inferred that the hydrazine hydrate vapor reduced rGO completely wrapped the PU foam because the specific absorption peaks of PU foam completely disappeared.

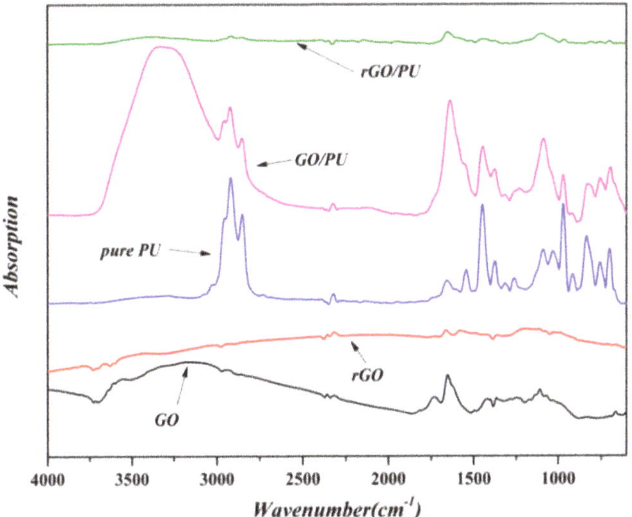

Figure 3. Fourier-transform infrared (FT-IR) of the graphene oxide (GO), reduced graphene oxide (rGO), pure polyurethane (PU), graphene oxide/polyurethane (GO/PU) and reduced graphene oxide/polyurethane (rGO/PU) composite foam.

3.3. Micro Morphology

Figure 4 shows the SEM images of a pure PU foam and rGO/PU foam, where Figure 4A, and B are images of pure PU foam at different magnifications, Figure 4C–F are rGO/PU foam at different magnifications. It can be seen from the images that the pores of the PU foam are regular, evenly distributed, and the surface is rough, which helps for the adhesion of GO.

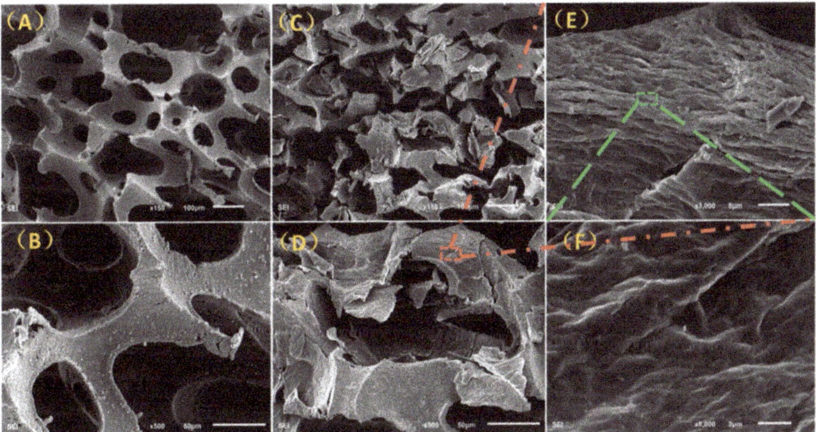

Figure 4. Scanning electron microscopy (SEM) images of pure polyurethane (PU) foam (**A**,**B**) and reduced graphene oxide/polyurethane (rGO/PU) foam (**C–F**) with varying magnifications.

From the Figure 4C–F, the rGO is sufficiently arranged on the wall of the foam pores. The surface of the rGO/PU foam exhibits lots of wrinkles and burrs. The directional arrangements of the rGO can be easily observed though the SEM image with high magnification (Figure 4E). The through-pore structure would be compressed while the external pressure was loaded on the composite foam. The wall of the foam pores that consisted of direactional arrangement rGOs would contact to each other forming electron transmission pathways. The wrinkles and burrs becomes "micro-switches" to adjust the contact area and quantity that dominated the equivalent circuit of the composite foam. As a result, the resistance would change rapidly when the rGO were rearranged under the pressure. This process was illustrated in Figure S2.

3.4. Sensitivity

To evaluate the sensing performance of the prepared rGO/PU foam composite, a series of different pressures were loaded on the assembled piezo-resistive sensor. The resistance and the compression amount were recorded. The original electrical resistance of the assembled piezo-resistive sensor was about 7.35 kΩ. Figure 5 represents the calculated relative resistance changes ($\Delta R/R_0$) and compression amount (ΔL) curves to the varying pressure. As can be seen, the relative resistance change shows similar trend with the compression. The relative resistance change and compression amount increased rapidly both in the pressure range of 0–6 kPa and then the rate of rise gradually decreases in the larger pressure range of 6–25 kPa. This may be caused by the variation of the pore structure during the compression process, which result in the different change rate of the contact area of the graphene sheets on the pore walls.

$$S = \frac{\partial(\Delta R/R_0)}{\partial P} \quad (1)$$

$$\Delta R = R - R_0 \quad (2)$$

where the R refers to the resistance under the pressure p and the R_0 refers to the original resistance of the piezo-resistive device.

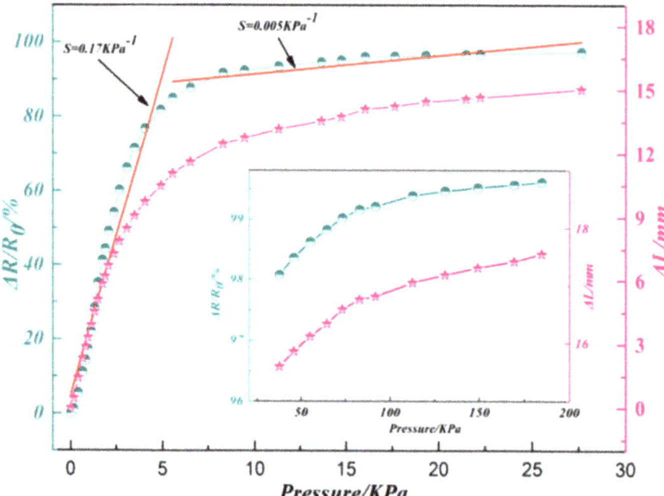

Figure 5. Relative resistance changes and compression amounts of reduced graphene oxide/polyurethane (rGO/PU) composite foam under pressure bellow 30 kPa.

According to the definition of equation of sensitivity (Equations (1) and (2)) [8,9] the $\Delta R/R_0$ curve can be divided into two linear sections and fitted. The slope of the two linear fitted line can be considered as the sensitivity in that pressure range. From the figure, the value of sensitivity under the range of 0–6 kPa and 6–25 kPa are 0.17 kPa^{-1} and 0.005 kPa^{-1}, respectively.

3.5. Pressure Discrimination

Figure 6 presents the real-time current response under constant voltage supply of 0.5 V when varying pressure (ranging from 0.625 kPa to 10.4 kPa) were loaded on the rGO/PU foam composite. As can be seen, lots of current stages existed in the very stable based line. What is worth noting is that the stage and the baseline are almost parallel to the horizontal axis. This indicates the high signal to noise ratio of the piezo-resistive sensor. The height of the current stage increased while the applied pressure increased which means the resistance of the device decreased. What is more, the different stable current stages induced by various pressures shows the excellent pressure discrimination of the piezo-resistive sensor which is practical useful to the application scenario requiring specific pressure value.

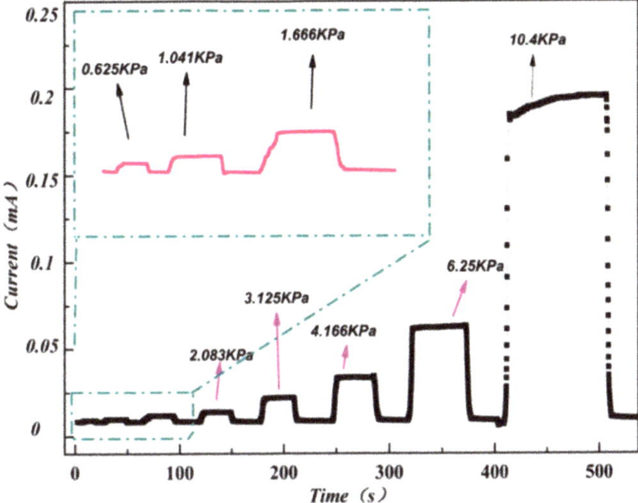

Figure 6. Real-time current response of reduced graphene oxide/polyurethane (rGO/PU) composite foam under different pressure.

3.6. Cycling Stability

In order to measure the cycling stability of the sensor, the different pressure of 0.625 kPa, 2.083 kPa, and 3.125 kPa were repeatedly applied and removed on the rGO/PU foam composite sensor for 30 times with time interval of 0.5 s. The results were displayed in Figure 7. As can be seen, there are several slight fluctuations for the cycling stability testing of each pressure. This may be caused by the little difference of the contacting area even under the same pressure. Actually, the rigid rGO plane fulfilled in the PU foam pore could contribute to the cycling stability by limiting the compression deformation trajectory.

In addition, comparing with different pressures, there is a little offset of the current change while removing the pressure. This may be caused by the inherent creep and stress relaxation properties of the PU foam. In other word, the PU foam structure changed themselves to resist the applied pressure but need a long time to recover, which is expressed as the offset of baseline.

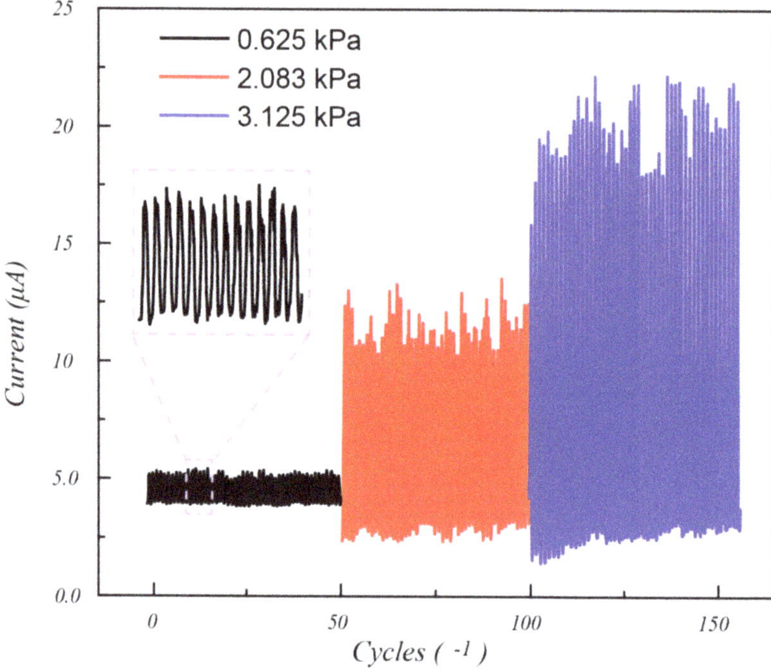

Figure 7. The 50 times cycling stability of the prepared reduced graphene oxide/polyurethane (rGO/PU) foam composite sensor under 0.625, 2.083 and 3.125 kPa.

3.7. Response Time

The response time was tested by loading and unloading a fixed pressure with 20 g weight on the piezo-resistive sensor with fast speed. The response time (τ) could match the formula $\tau = \tau_m + \tau_s$. Among them, τ_m is the time that the rGO/PU composite foam induct the pressure change when adding or removing the force; τ_s is the time which start from the rGO/PU composite foam inducting the pressure to the electrical parameters finishing the change. As τ_m is small and cannot be detected accurately, the value of τ_s is used to represent the completely response time of rGO/PU in this study. From Figure 8, we can find that the τ_s is about 0.3 s while using the 20 g weight as the pressure applying object.

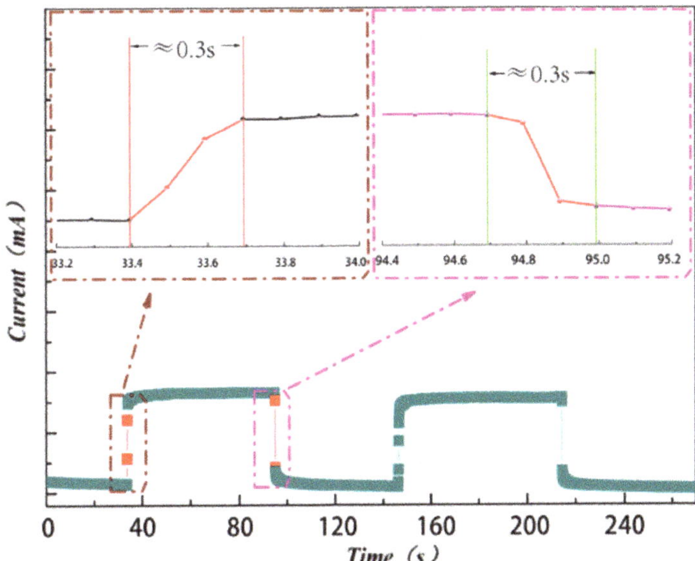

Figure 8. The response time and the recovery time of the reduced graphene oxide/polyurethane (rGO/PU) composite foam sensor.

4. Conclusions

In this present research, the graphene/polyurethane foam composites are prepared to develop a convenient method for piezo-resistive devices with simple structure and outstanding sensing performance. The chemical components and micro morphologies have been observed by using FT-IR and SEM. The results predicted that the graphene is tightly adhered to the bare surface of the pores. The pressure sensing performance has been also evaluated by measuring the sensitivity, durability, and response time. The results indicate that the value of sensitivity under the range of 0–6 kPa and 6–25 kPa are 0.17 kPa^{-1} and 0.005 kPa^{-1}, respectively. Cycling stability test has been performed 30 times under three varying pressures. The signal output just exhibits slight fluctuations, which represents the good cycling stability of the pressure sensor. At the same stage, the response time of loading and unloading of 20 g weight were turn out about 300 ms. These consequences showed the superiorities of graphene/polyurethane composite foam while applied in piezo-resistive devices including wide sensitive pressure range, high sensitivity, outstanding durability, and fast response.

Supplementary Materials: The following are available online at http://www.mdpi.com/2073-4360/11/8/1289/s1, Figure S1: The digital camera photographs of (a) pure PU foam, (b) prepared rGO/PU composite foam, (c) Digital display force gauge and relative tension and compression testing bench and (d) Electrochemical workstation; Figure S2: The pressure sensing mechanism of the rGO/PU composite foam sensor with pore structures and switchable rGO wrinkles and burrs.

Author Contributions: Conceptualization, H.J.; methodology, W.Z.; validation, Y.C.; formal analysis, W.Z. and W.L.; investigation, C.S.; resources, D.W.; data curation, X.D.; writing—original draft preparation, W.Z.; writing—review and editing, A.Y. and Y.C.; visualization, W.Z.; supervision, H.J. and D.W.; project administration, D.W.; funding acquisition, H.J., W.Z. and Y.C.

Funding: This study is supported by the University, Industry, Research and Innovation Fund from the Ministry of Education Science and Technology Development Center (No. 2018A02019) and National Natural Science Foundation of China (No. 51603155). The authors are also thankful for the financial support of the Fundamental Research Funds for the Central Universities (No. CUSF-DH-D-2018059).

Conflicts of Interest: The authors declare no conflict of interest.

References

1. Castro, H.F.; Correia, V.; Pereira, N.; Costab, P.; Oliveiraa, J.; Lanceros-Méndez, S. Printed Wheatstone bridge with embedded polymer based piezoresistive sensors for strain sensing applications. *Addit. Manuf.* **2018**, *20*, 119–125. [CrossRef]
2. Deng, J.; Zhuang, W.; Bao, L.; Wu, X.; Gao, J.; Wang, B.; Sun, X.; Peng, H. A tactile sensing textile with bending-independent pressure perception and spatial acuity. *Carbon* **2019**, *149*, 63–70. [CrossRef]
3. Domingues, M.F.; Rodriguez, C.A.; Martins, J.; Tavares, C.; Marques, C.; Alberto, N.; André, P.; Antunes, P. Cost-effective optical fiber pressure sensor based on intrinsic Fabry-Perot interferometric micro-cavities. *Opt. Fiber Technol.* **2018**, *42*, 56–62. [CrossRef]
4. Liu, K.; Zhou, Z.; Yan, X.; Meng, X.; Tang, H.; Qu, K.; Gao, Y.; Li, Y.; Yu, J.; Li, L. Polyaniline Nanofiber Wrapped Fabric for High Performance Flexible Pressure Sensors. *Polymers* **2019**, *11*, 1120. [CrossRef] [PubMed]
5. Liu, Y.; He, K.; Chen, G.; Leow, W.R.; Chen, X. Nature-Inspired Structural Materials for Flexible Electronic Devices. *Chem. Rev.* **2017**, *117*, 12893–12941. [CrossRef] [PubMed]
6. Nie, B.; Huang, R.; Zhang, Y.; Miao, Y.; Liu, C.; Liu, J.; Chen, X. Textile—Based Wireless Pressure Sensor Array for Human—Interactive Sensing. *Adv. Funct. Mater.* **2019**, *29*, 1808786. [CrossRef]
7. Shao, Q.; Niu, Z.; Hirtz, M.; Jiang, L.; Liu, Y.; Wang, Z.; Chen, X. High-Performance and Tailorable Pressure Sensor Based on Ultrathin Conductive Polymer Film. *Small* **2014**, *10*, 1466–1472. [CrossRef]
8. Zhong, W.; Liu, C.; Liu, Q.; Piao, L.; Jiang, H.; Wang, W.; Liu, K.; Li, M.; Sun, G.; Wang, D. Ultrasensitive Wearable Pressure Sensors Assembled by Surface-Patterned Polyolefin Elastomer Nanofiber Membrane Interpenetrated with Silver Nanowires. *ACS Appl. Mater. Interfaces* **2018**, *10*, 42706–42714. [CrossRef]
9. Zhong, W.; Liu, Q.; Wu, Y.; Wang, Y.; Qing, X.; Li, M.; Liu, K.; Wang, W.; Wang, D. A nanofiber based artificial electronic skin with high pressure sensitivity and 3D conformability. *Nanoscale* **2016**, *8*, 12105. [CrossRef]
10. Ma, Q.F.; Tou, Z.Q.; Ni, K.; Lim, Y.Y.; Lin, Y.F.; Wang, Y.R.; Zhou, M.H.; Shi, F.F.; Niu, L.; Dong, X.Y. Carbon-nanotube/Polyvinyl alcohol coated thin core fiber sensor for humidity measurement. *Sens. Actuators B Chem.* **2018**, *257*, 800–806. [CrossRef]
11. Wan, Y.; Wang, Y.; Guo, C.F. Recent progresses on flexible tactile sensors. *Mater. Today Phys.* **2017**, *1*, 61–73. [CrossRef]
12. Wang, T.; Zhang, Y.; Liu, Q.; Cheng, W.; Wang, X.; Pan, L.; Xu, B.; Xu, H. A Self-Healable, Highly Stretchable, and Solution Processable Conductive Polymer Composite for Ultrasensitive Strain and Pressure Sensing. *Adv. Funct. Mater.* **2018**, *28*, 1705551. [CrossRef]
13. Wu, X.; Han, Y.; Zhang, X.; Lu, C. Highly Sensitive, Stretchable, and Wash-Durable Strain Sensor Based on Ultrathin Conductive Layer@Polyurethane Yarn for Tiny Motion Monitoring. *ACS Appl. Mater. Interfaces* **2016**, *8*, 9936–9945. [CrossRef]
14. Yu, Z.; Tang, Y.; Cai, G.; Ren, R.; Tang, D. Paper Electrode-Based Flexible Pressure Sensor for Point-of-Care Immunoassay with Digital Multimeter. *Anal. Chem.* **2019**, *91*, 1222–1226. [CrossRef]
15. Zhong, W.; Liu, C.; Xiang, C.; Jin, Y.; Li, M.; Liu, K.; Liu, Q.; Wang, Y.; Sun, G.; Wang, D. Continuously Producible Ultrasensitive Wearable Strain Sensor Assembled with Three-Dimensional Interpenetrating Ag Nanowires/Polyolefin Elastomer Nanofibrous Composite Yarn. *ACS Appl. Mater. Interfaces* **2017**, *9*, 42058–42066. [CrossRef]
16. Zhu, Y.; Li, J.; Cai, H.; Wu, Y.; Ding, H.; Pan, N.; Wang, X. Highly sensitive and skin-like pressure sensor based on asymmetric double-layered structures of reduced graphite oxide. *Sens. Actuators B Chem.* **2018**, *255*, 1262–1267. [CrossRef]
17. Lee, J.; Kwon, H.; Seo, J.; Shin, S.; Koo, J.H.; Pang, C.; Son, S.; Kim, J.H.; Jang, Y.H.; Kim, D.E.; et al. Conductive fiber-based ultrasensitive textile pressure sensor for wearable electronics. *Adv. Mater.* **2015**, *27*, 2433–2439. [CrossRef]
18. Lee, Y.; Park, J.; Cho, S.; Shin, Y.E.; Lee, H.; Kim, J.; Myoung, J.; Cho, S.; Kang, S.; Baig, C.; et al. Flexible Ferroelectric Sensors with Ultrahigh Pressure Sensitivity and Linear Response over Exceptionally Broad Pressure Range. *ACS Nano* **2018**, *12*, 4045–4054. [CrossRef]
19. Ma, Y.; Yue, Y.; Zhang, H.; Cheng, F.; Zhao, W.; Rao, J.; Luo, S.; Wang, J.; Jiang, X.; Liu, Z.; et al. 3D Synergistical MXene/Reduced Graphene Oxide Aerogel for a Piezoresistive Sensor. *ACS Nano* **2018**, *12*, 3209–3216. [CrossRef]

20. Wang, D.; Li, H.; Li, M.; Jiang, H.; Xia, M.; Zhou, Z. Stretchable conductive polyurethane elastomer in situ polymerized with multi-walled carbon nanotubes. *J. Mater. Chem. C* **2013**, *1*, 2744. [CrossRef]
21. Wu, X.; Han, Y.; Zhang, X.; Zhou, Z.; Lu, C. Large-Area Compliant, Low-Cost, and Versatile Pressure-Sensing Platform Based on Microcrack-Designed Carbon Black@Polyurethane Sponge for Human-Machine Interfacing. *Adv. Funct. Mater.* **2016**, *26*, 6246–6256. [CrossRef]
22. Zhang, Q.; Liu, L.; Pan, C.; Li, D. Review of recent achievements in self-healing conductive materials and their applications. *J. Mater. Sci.* **2017**, *53*, 27–46. [CrossRef]
23. Shi, J.; Wang, L.; Dai, Z.; Zhao, L.; Du, M.; Li, H.; Fang, Y. Multiscale Hierarchical Design of a Flexible Piezoresistive Pressure Sensor with High Sensitivity and Wide Linearity Range. *Small* **2018**, *14*, e1800819. [CrossRef]
24. Gröttrup, J.; Paulowicz, I.; Schuchardt, A.; Kaidas, V.; Kaps, S.; Lupan, O.; Adelung, R.; KumarMishra, Y. Three-dimensional flexible ceramics based on interconnected network of highly porous pure and metal alloyed ZnO tetrapods. *Ceram. Int.* **2016**, *42*, 8664–8676. [CrossRef]
25. Kaps, S.; Bhowmick, S.; Gröttrup, J.; Hrkac, V.; Stauffer, D.; Guo, H.; Warren, O.L.; Adam, J.; Kienle, L.; Minor, A.M. Piezoresistive Response of Quasi-One-Dimensional ZnO Nanowires Using an in Situ Electromechanical Device. *ACS Omega* **2017**, *2*, 2985–2993. [CrossRef]
26. Schuchardt, A.; Braniste, T.; Mishra, Y.K.; Deng, M.; Mecklenburg, M.; Stevens-Kalceff, M.A.; Raevschi, S.; Schulte, K.; Kienle, L.; Adelung, R.; et al. Three-dimensional Aerographite-GaN hybrid networks: Single step fabrication of porous and mechanically flexible materials for multifunctional applications. *Sci. Rep.* **2015**, *5*, 8839. [CrossRef]
27. Liu, W.; Liu, N.; Yue, Y.; Rao, J.; Cheng, F.; Su, J.; Liu, Z.; Gao, Y. Piezoresistive Pressure Sensor Based on Synergistical Innerconnect Polyvinyl Alcohol Nanowires/Wrinkled Graphene Film. *Small* **2018**, *14*, e1704149. [CrossRef]
28. Lee, C.; Wei, X.; Kysar, J.W.; Hone, J. Measurement of the elastic properties and intrinsic strength of monolayer graphene. *Science* **2008**, *321*, 385–388. [CrossRef]
29. Geim, A.K. Graphene: Status and prospects. *Science* **2009**, *324*, 1530–1534. [CrossRef]
30. Geim, A.K.; Novoselov, K.S. The rise of graphene. *Nat. Mater.* **2007**, *6*, 183–191. [CrossRef]
31. Wu, J.; Becerril, H.A.; Bao, Z.; Liu, Z.; Chen, Y.; Peumans, P. Organic solar cells with solution-processed graphene transparent electrodes. *Appl. Phys. Lett.* **2008**, *92*, 237. [CrossRef]
32. Stobinski, L.; Lesiak, B.; Malolepszy, A.; Mazurkiewicz, M.; Mierzwa, B.; Zemekd, J.; Jiricek, P.; Bieloshapka, I. Graphene oxide and reduced graphene oxide studied by the XRD, TEM and electron spectroscopy methods. *J. Electron. Spectrosc. Relat. Phenom.* **2014**, *195*, 145–154. [CrossRef]
33. Wang, B.; Wu, X.; Shu, C.; Guo, Y.; Wang, C. Synthesis of CuO/graphene nanocomposite as a high-performance anode material for lithium-ion batteries. *J. Mater. Chem.* **2010**, *20*, 10661–10664. [CrossRef]

Article

Using an Ionic Liquid to Reduce the Electrical Percolation Threshold in Biobased Thermoplastic Polyurethane/Graphene Nanocomposites

Nora Aranburu, Itziar Otaegi and Gonzalo Guerrica-Echevarria *

POLYMAT and Polymer Science and Technology Department, Faculty of Chemistry, University of the Basque Country UPV/EHU, Paseo Manuel de Lardizabal 3, 20018 Donostia, Spain; nora.aramburu@ehu.eus (N.A.); itziar.otaegi@ehu.eus (I.O.)

* Correspondence: gonzalo.gerrika@ehu.eus; Tel.: +34-943-015443

Received: 11 February 2019; Accepted: 1 March 2019; Published: 6 March 2019

Abstract: Biobased thermoplastic polyurethane (bTPU)/unmodified graphene (GR) nanocomposites (NCs) were obtained by melt-mixing in a lab-scaled conventional twin-screw extruder. Alternatively, GR was also modified with an ionic liquid (GR-IL) using a simple preparation method with the aim of improving the dispersion level. XRD diffractograms indicated a minor presence of well-ordered structures in both bTPU/GR and bTPU/GR-IL NCs, which also showed, as observed by TEM, nonuniform dispersion. Electrical conductivity measurements pointed to an improved dispersion level when GR was modified with the IL, because the bTPU/GR-IL NCs showed a significantly lower electrical percolation threshold (1.99 wt%) than the bTPU/GR NCs (3.21 wt%), as well as higher conductivity values. Young's modulus increased upon the addition of the GR (by 65% with 4 wt%), as did the yield strength, while the ductile nature of the bTPU matrix maintained in all the compositions, with elongation at break values above 200%. This positive effect on the mechanical properties caused by the addition of GR maintained or slightly increased when GR-IL was used, pointing to the success of this method of modifying the nanofiller to obtain bTPU/GR NCs.

Keywords: nanocomposites; graphene; melt processing; mechanical properties; electrical conductivity

1. Introduction

Polymer nanocomposites (NCs) containing carbon-based electrically conductive nanoparticles have gained a certain advantage over other hybrid polymer systems because, in addition to the other usual improvements in mechanical and transport properties, they offer enhanced electrical and thermal conductivity. As a result, these NCs are particularly suitable for different applications such as low-cost, light-weight, EMI-shielded computer housing and cables, antistatic packaging, high-strength automotive and aerospace components, high-barrier packaging, and smart clothing/personal sensor systems [1].

Graphene (GR) is a two-dimensional carbon nanofiller with a one-atom-thick planar sheet of sp^2 bonded carbon atoms densely packed in a honeycomb crystal lattice [2]. The good thermal conductivity and excellent mechanical and electronic transport properties of graphene make it an ideal candidate for producing thermally and electrically-conductive reinforced NCs [3,4]. However, in order to achieve these goals, the graphene must be efficiently dispersed in the polymer matrix, a process which is hindered by the strong intrinsic van der Waals attraction and π–π stacking which tends to cause the reaggregation of the graphene sheets [5].

Among the different covalent and noncovalent modifications of graphene which are used in order to overcome the aforementioned restacking problem [6], the use of ionic liquids (IL) as dispersing agents has attracted the attention of the scientific community as their use is considered simple, ecofriendly, and highly efficient [7,8]. ILs are believed to interact with graphene by cation–π stacking and π–π interactions which prevent the reaggregation of graphene sheets while preserving the electronic structure of the graphene and its intrinsic properties [9,10]. As a result, ILs have been successfully used as dispersing agents for graphene in several thermoplastic matrices, including poly(vinyl alcohol) [11], a copolyamide [12], polylactide [9], polystyrene [13], poly(methyl methacrylate) [14], and polyvinylidene fluoride [7].

In the search for potential polymer matrices, thermoplastic polyurethanes (TPUs) have attracted much attention in the recent years [15,16]. This is because they are linear block copolymers with hard and soft segments and, thus, depending on the composition of the reactants used, such as isocyanates, polyols, and chain extenders, they can offer a variety of properties, rendering them suitable for a broad range of applications including automotive, electronics, sports goods, footwear, and medical devices. The addition of graphene-like structures to TPUs would further broaden the application potential of these polymers as it would make it possible to obtain electrically conductive TPUs with improved mechanical and other properties. This is why several papers on TPU-based NCs containing graphene [17–32] have been included in the bibliography. However, most of these studies focus on systems prepared by in situ polymerization [21,22,26] and solvent mixing [18–20,23,26,30–32]. Unfortunately, only a few studies are devoted to TPU/graphene NCs prepared by melt blending [24–29,31], which is the most practical, versatile, economical, and environmentally-friendly method as it can be easily adapted to existing conventional plastic processing equipment [33] and does not require the use of solvents or monomers [34]. Furthermore, in all these last cases, nonindustrial recirculating mixing equipment was used during processing. Even though the reinforcement effect of the graphene has been detailed [24–29,31] in several studies, only a few characterize the electrical conductivity of the NCs [26,28,31] in their analysis of the contribution of graphene as a conductive nanofiller. Moreover, the use of IL as a dispersing agent to improve the dispersion of graphene in TPU-based NCs has not been studied to the best of our knowledge.

Finally, the development and production of polymeric materials from renewable resources have become subjects of considerable interest in the academic and industrial sectors in recent years, due to their low cost and a growing concern about global warming, carbon emissions and the nonavailability of certain petrochemical raw materials [35,36]. In fact, biobased TPUs synthesized with polyols obtained from natural renewable resources are already commercially available [37,38]. The behavior of these commercial biobased TPUs as potential matrices for the preparation of NCs containing graphene has not, to our knowledge, been studied.

In this work, electrically conductive NCs with improved mechanical properties were obtained using conventional industrial-like melt-mixing equipment with a biobased TPU (bTPU) as the polymeric matrix. Neat GR and GR modified with an IL (GR-IL)—1-butyl-3-methylimidazolium tetrafluoroborate (BMITFB)—were used as nanofillers. The nanostructure, electrical conductivity, and thermal and mechanical properties of the bTPU/GR and bTPU/GR-IL NCs were studied and compared.

2. Materials and Methods

The polymer used in this work was a linear, biobased aromatic polyurethane, Pearlbond® Eco D590, based on specialty polyols from renewable sources and kindly supplied by Merquinsa (Barcelona, Spain). Its chemical structure, analyzed by NMR, consists of methylene diphenyl 4,4′-diisocyanate (MDI) (4 wt%) and polybutylene sebacate (PBSe) (96 wt%). Expanded graphene (GR), comprising 1–2 layers, with a purity > 98.5% and density < 0.2 g/cm^2, was purchased from Avanzare (Navarrete, Spain). The ionic liquid (IL) was 1-butyl-3-methylimidazolium tetrafluoroborate (purity $\geq$ 98%) from Aldrich (Saint Louis, MO, USA).

The modification of GR with IL was performed as follows: first, GR was sonicated in ethanol for 15 min after which the IL was added (in 1:1 ratio). Next, the mixture was sonicated for 1 h. Then it was dried in an oven at 60 °C for 24 h in order to remove the ethanol and obtain the GR-IL powder.

The composition of the prepared bTPU NCs, as well as their real GR content are listed on Table 1. The NCs were first melt-mixed by extrusion and then injection- or compression-molded to obtain standard test specimens.

Table 1. Prepared biobased thermoplastic polyurethane (bTPU) polymer nanocomposites (NCs) and their total nanofiller and graphene (GR) contents.

Composition	Nanofiller wt%	GR wt%
bTPU-0.5 GR-IL	1	0.5
bTPU-1 GR-IL	2	1
bTPU-1.5 GR-IL	3	1.5
bTPU-2 GR-IL	4	2
bTPU-3 GR-IL	6	3
bTPU-4 GR-IL	8	4
bTPU-1 GR	1	1
bTPU-2.5 GR	2.5	2.5
bTPU-3 GR	3	3
bTPU-3.5 GR	3.5	3.5
bTPU-4 GR	4	4
bTPU-5 GR	5	5
bTPU-6 GR	6	6
bTPU-7 GR	7	7

The extruder used was a Collin ZK25 corotating twin screw extruder-kneader (Collin, Maitenbeth, Germany) at a melt temperature of 130 °C and rotation speed of 400 rpm. The screw diameter and the L/D ratio were 25 mm and 30, respectively. The extrudates were cooled in a water bath and pelletized. Injection molding was carried out in a Battenfeld BA-230E reciprocating screw injection molding machine (Wittman, Wien, Germany) to obtain tensile (ASTM D638, type IV, thickness 1.84 mm) specimens. The screw of the plasticization unit was a standard screw with a diameter of 18 mm, L/D ratio of 17.8, and a compression ratio of 4. The melt temperature was 130 °C and the mold temperature was 15 °C. The injection speed was 10.2 $cm^3 \cdot s^{-1}$. The specimens were left to condition for 24 h in a desiccator before analysis or testing. The samples used for the electrical conductivity measurements, which were 70 mm in diameter and 1.1 mm thick, were obtained by hot-pressing at 130 °C using a Collin P200E press (Collin, Maitenbeth, Germany).

X-ray diffraction patterns were recorded in an X'pert X-ray diffractometer (PANalytical, Almelo, the Netherlands) at 40 kV and 40 mA, using a Ni-filtered Cu-K_α radiation source. The transmission electron microscopy (TEM) samples were ultrathin-sectioned at 30–40 nm using a cryo-ultramicrotome. The micrographs were obtained in a Tecnai G2 20 Twin microscope (FEI, Hillsboro, OR, USA) at an accelerating voltage of 200 kV.

For the electrical conductivity measurements, the electrical resistivity of the compression-molded sheets was determined and converted to conductivity values. Volume resistances were measured using a digital Keithley 6487 picoammeter (Keithley Instruments, Cleveland, OH, USA).

Tensile testing was carried out by means of an Instron 5569 machine (Instron, Norwood, MA, USA) at a cross-head speed of 50 mm/min and at 23 ± 2 °C and 50 ± 5% relative humidity. A minimum of five tensile specimens were tested for each reported value.

The melting and crystallization behavior of the NCs was studied by DSC in a DSC-7 calorimeter (PerkinElmer, Waltham, MA, USA) calibrated with an Indium standard. The samples were first heated from 0 to 100 °C at a heating rate of 20 °C/min, then cooled at the same rate and, subsequently, reheated in the same conditions. The melting temperature (T_m) and enthalpy (ΔH_m) of the bTPU were determined from the second heating scans using the peak maximum and area, respectively.

The crystallization temperature (T_c) was determined from the cooling scans. Dynamic mechanical analysis was carried out in a DMA Q800 apparatus (TA Instruments, New Castle, PA, USA) that provided the plots of the tanδ against temperature. The scans were carried out from −100 to 60 °C at a constant heating rate of 4 °C/min and a frequency of 1 Hz.

3. Results and Discussion

3.1. Nanostructure

The nanostructure of the NCs was analyzed by XRD and TEM with the aim of ascertaining the effect of the modification of the IL on the dispersion level of the GR. Figure 1 shows the XRD diffractograms of the bTPU-3 GR and bTPU-3 GR-IL NCs, as well as those of the neat components. As the section of the figure marked with a circle shows, the diffractograms of the neat GR and GR-IL do not show the characteristic strong, narrow peak of graphite ($2\theta \approx 26°$, not shown in the figure), which corresponds to the reflection of the 002 planes of well-ordered sheets [1,39,40]; instead, a small peak of very low intensity appears in the same position. This indicates the absence of ordered structures, and the exfoliated nature of GR and GR-IL [26,41]. In the case of the NCs, the intensity of this peak is slightly higher, which could be indicative of the formation of some ordered structures or the stacking of GR sheets in the NCs during melt mixing [26], caused by the considerable van der Waals forces and strong π–π interactions between the graphene sheets [41–44]. However, the low intensity of these peaks points to a minor presence of ordered graphene sheets in the NCs.

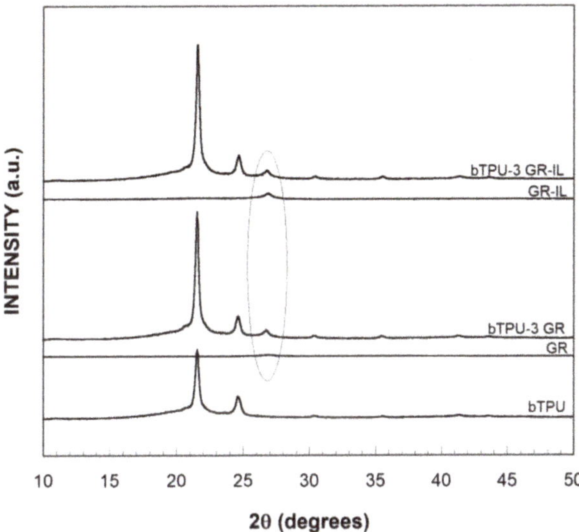

Figure 1. XRD diffractograms of the neat components and the bTPU-3 GR and bTPU-3 GR-IL NCs. The circle indicates the typical position of the corresponding peak of ordered graphene.

Additional information about the degree of dispersion of the nanofillers and the nanostructure of the NCs was obtained by TEM, and micrographs of the bTPU-3 GR and bTPU-3 GR-IL NCs are shown, as an example, in Figure 2. Other compositions showed similar nanostructures. As can be observed, the dispersion of both GR and GR-IL is nonuniform, presenting both relatively large stacks and stacks with only a few layers (Figure 2a,c), as well as individually dispersed graphene sheets (Figure 2b,d). The nanostructure of these bTPU NCs is similar to that observed for other NCs obtained by melt-mixing [26,33,45–47] and, taking into account that the samples for microscopy were prepared in perpendicular planes with respect to the flow direction, the GR platelets seem to have oriented in the flow direction during mold-filling [26,47,48]. Although complete exfoliation was not achieved, the dispersion of GR and GR-IL in the bTPU NCs was good enough to lead to significant improvements in the mechanical and electrical properties, as discussed below.

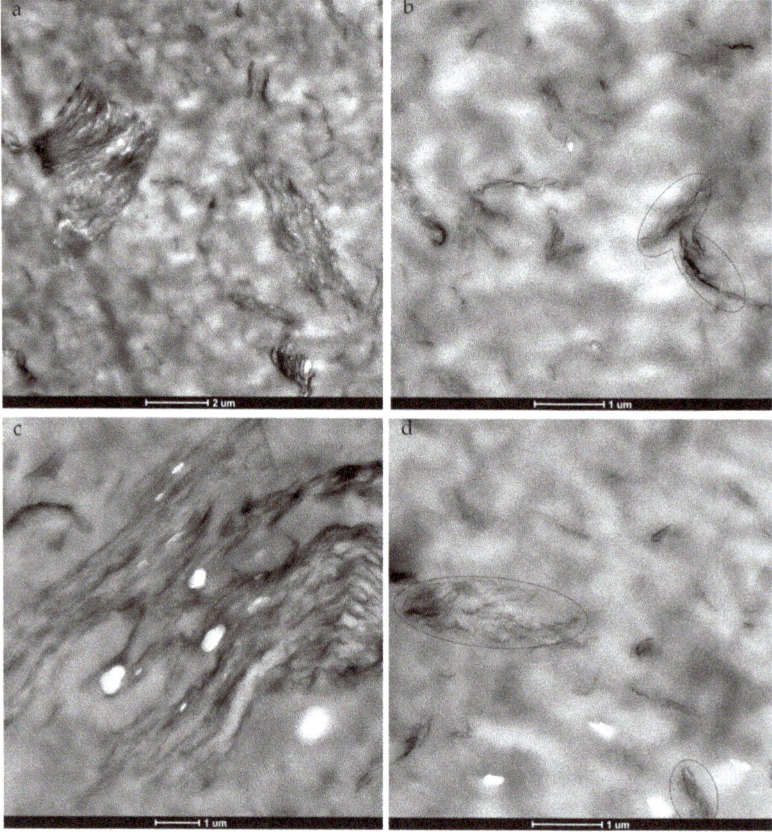

Figure 2. TEM micrographs of the bTPU-3 GR (**a**,**b**) and bTPU-3 GR-IL (**c**,**d**) NCs, obtained with low (left) and high (right) magnifications. Some graphene stacks have been highlighted with a circle in figures (b,d).

Regarding the effect of the IL-modified GR, the nonuniform dispersion makes it impossible to detect any significant difference in the degree of dispersion between the GR and the GR-IL. However, electrical conductivity is very sensitive to the dispersion level of nanofillers throughout the matrix and can be used as a qualitative measurement of this parameter.

Figure 3 shows the electrical conductivity of the bTPU NCs as a function of the GR content. Up to a graphene concentration of 2.5 wt% in bTPU/GR and 1 wt% in bTPU/GR-IL NCs, the electrical conductivity remained almost unchanged, but when the graphene concentration exceeded these contents, the electrical conductivity increased dramatically: from 5×10^{-14} S/cm for the neat bTPU to 2×10^{-7} S/cm for the bTPU-7 GR NC and to 5×10^{-7} S/cm for the bTPU-4 GR-IL, i.e., 7 orders of magnitude.

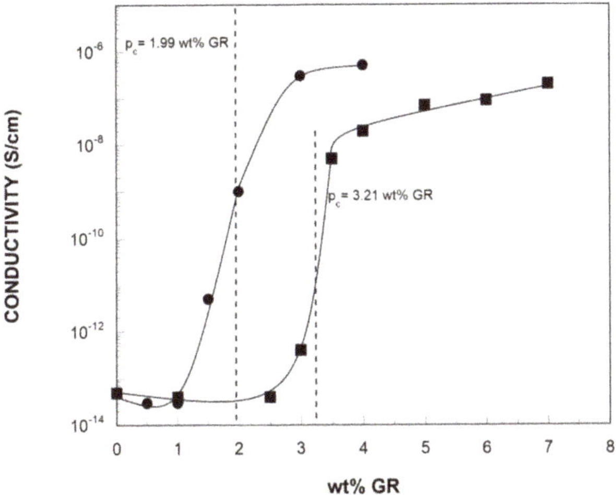

Figure 3. Electrical conductivity of the bTPU/GR (■) and bTPU/GR-IL (•) NCs as a function of the GR content. The dotted lines mark the calculated p_c.

The electrical percolation threshold (p_c) was fitted using the power law function for conductivity values near the p_c [49]: $\sigma(p) = B(p\text{-}p_c)^t$, where $\sigma(p)$ is the experimental conductivity value at concentrations $p > p_c$, B is the proportionality constant, and t is the critical exponent. The experimental results were fitted and the p_c was determined at 3.21 wt% GR for the bTPU/GR NCs and at 1.99 wt% GR for the bTPU/GR-IL NCs. Therefore, modifying the GR with the IL leads to a significant reduction in the percolation threshold of the bTPU NCs. In addition, when conductive NCs with the same amount of GR are compared (for example, bTPU-4 GR and bTPU-4 GR-IL), the latter showed a significantly higher electrical conductivity. ILs are known to interact with the graphene by cation–π stacking and π–π interactions which prevent reaggregation of graphene sheets [9,10]. Their ability to improve the dispersion of graphene in several thermoplastic matrices has been demonstrated [7,9,11–14]. Therefore, it can be concluded that, although not clearly detected by TEM, the modification of GR with IL leads to the enhanced dispersion of the nanofiller, which, in turn, gives rise to a significantly lower electrical percolation threshold in the bTPU/GR-IL NCs than in the bTPU/GR NCs.

In spite of the important contribution of graphene as a conductive nanofiller, relatively few research studies have analyzed its effects on the electrical conductivity of this type of polymer NCs. When compared with other TPU NCs obtained by melt blending, the p_cs obtained in this work are lower than those observed in a study on TPU NCs with expanded graphite [28]. In the same work, the p_c was reduced to 2 wt% upon optimization of the processing parameters; however, the NCs obtained presented an inhomogeneous nanostructure featuring large elongated aggregates, which is unsuitable from a mechanical point of view. Kim et al. [26] reported that the surface resistance of melt-processed TPU NCs with thermally-reduced graphite oxide (TRC) decreased at contents >0.5 vol% (~0.9 wt%). This finding is similar to the one obtained for the bTPU/GR-IL NCs in this work, where the conductivity increased at GR contents greater than 1 wt% (Figure 3). However, it should be mentioned that the value reported by Kim et al. refers to surface conductivity, and not volume conductivity as was measured in this work.

Higher [1,50–53] p_c values have been reported for melt-mixed NCs based on other polymer matrices and graphene. As in other NCs with carbonaceous nanofillers [54], the p_c is affected by various factors including the aspect ratio, the degree of dispersion or orientation of the nanofiller, the processing method and parameters, and the viscosity, molecular weight, and crystallinity of the polymer matrix.

In conclusion, the modification of GR with IL is an effective method of reducing considerably the electrical percolation threshold in polymer/graphene NCs, as the p_c determined in this study is one of the lowest electrical percolation thresholds reported in the literature for directly melt-blended NCs with a graphene-like nanofiller.

3.2. Mechanical Properties

As TPUs are used for a wide range of different applications, optimum control of their mechanical properties is critical. However, the stiffness of TPUs is low, so improving Young's modulus while retaining other desirable properties is an objective of particular interest. Therefore, the effects of the addition of GR and GR-IL on the mechanical properties of the bTPU were analyzed by tensile testing and the obtained results are summarized in Table 2. As the bTPU/GR NCs showed similar behavior, only the stress–strain curves of the neat bTPU and the bTPU/GR-IL NCs are shown in Figure 4. As can be observed, all the samples showed yielding and broke at high elongation values.

Table 2. Mechanical properties of the bTPU NCs.

Composition	Young's Modulus (MPa) (±10 MPa)	Elongation at Break (%) (±20%)	Tensile Strength (MPa) (±0.5 MPa)	Yield Strength (MPa) (±0.5 MPa)
bTPU	460	560	24.0	12.5
bTPU-0.5 GR-IL	490	500	20.5	13.5
bTPU-1 GR-IL	530	440	18.5	14.0
bTPU-1.5 GR-IL	550	420	18.0	14.0
bTPU-2 GR-IL	610	340	16.5	14.5
bTPU-3 GR-IL	700	300	15.0	15.0
bTPU-4 GR-IL	740	280	14.5	15.5
bTPU-1 GR	510	400	18.0	14.0
bTPU-2.5 GR	630	320	16.5	14.0
bTPU-3 GR	600	320	16.0	14.0
bTPU-3.5 GR	680	300	16.0	14.5
bTPU-4 GR	770	280	17.0	16.5
bTPU-5 GR	780	280	17.0	17.0
bTPU-6 GR	940	260	15.5	18.0
bTPU-7 GR	980	180	16.0	18.5

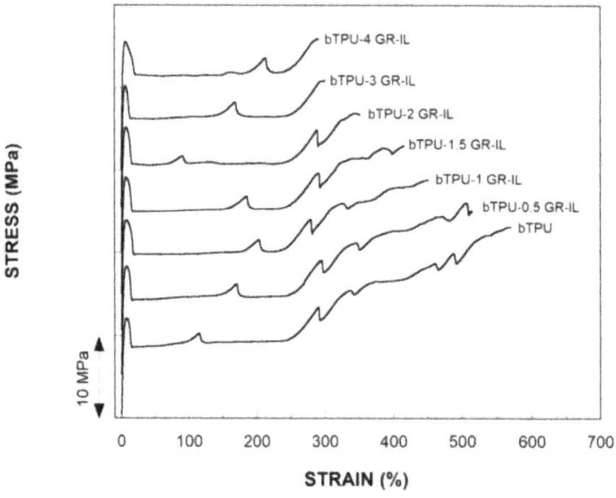

Figure 4. The stress–strain curves of the bTPU/GR-IL NCs. The curves are shifted along the vertical axis.

Figure 5 represents the Young modulus of the bTPU NCs as a function of the GR content. As can be seen, it increased almost linearly with the addition of GR and GR-IL. The linear modulus increase suggests that the dispersion and exfoliation levels of GR and GR-IL do not depend on their content in the NCs. As Figure 5 also reflects, very slight differences can be observed in the behavior of the GR and GR-IL NCs. This is a particularly positive result in the IL-modified NCs, because ionic liquids are known to act as plasticizers when added to polymeric matrices [55,56] and the IL used in this work did plasticize the bTPU matrix (see Table S1). Moreover, when the modulus increase per unit of added graphene was calculated, slightly higher values were obtained in the GR-IL NCs than in the GR NCs (average of 69 MPa vs. 65 MPa, respectively). Although the difference is in most cases close to the experimental error of measurement, it is deemed significant because the yield strength values showed a similar, even clearer trend, as discussed below. This enhanced mechanical behavior also suggests improved dispersion in the IL-modified GR, as already stated from the electrical conductivity measurements, which leads to greater reinforcing efficiency than the unmodified GR.

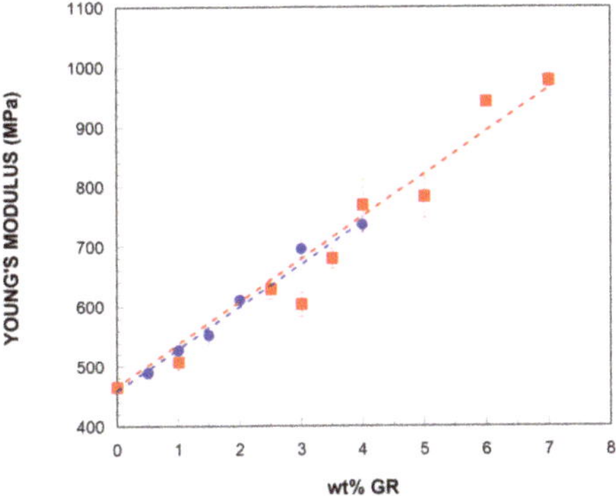

Figure 5. Young's modulus of the bTPU/GR (■) and bTPU/GR-IL (•) NCs as a function of the GR content.

The reinforcing effect of graphene in melt-mixed TPU NCs has been previously analyzed in the literature [24,26–29]; the extent of the increase in Young's modulus varies considerably in the studies. Maximum increases of 558% and 164% have been reported for TPU NCs with thermally-reduced graphite oxide [26] and expanded graphite [28], respectively. However, the modulus of the neat TPU in the referenced studies (6 and 27 MPa) was much lower than that of the bTPU used in this work (465 MPa), which would limit its applicability considerably. In addition, in the case of the TPU NCs with expanded graphite [28], the increase in Young's modulus was not due only to the reinforcing effect of the expanded graphite, but also to the greater crystallinity of the TPU in the NCs.

According to the bibliography, several factors may affect the stiffening effect of graphene-based nanofillers in polymer matrices, including the dispersion level, aspect ratio, the concentration, interface bonding, etc., [44], which helps to explain the broad range of modulus increases found in the literature. In the studies where graphene was melt-mixed with a variety of polymer matrices [46,47], the polarity match between the polymer and the filler, the nature of the polymer, and the processing conditions have all been observed to affect the degree of dispersion of the graphene and, thus, the enhancement in properties [46,47].

As can be observed in Figure 4, all the compositions broke during the strain hardening stage of the stress–strain curves, and the tensile strength value corresponds to the stress at break. However, the elongation at break of the NCs decreased, as discussed below, at increasing GR contents, which is reflected in the tensile strength values (Table 2). This has also been observed in other TPU/graphene NCs [28]. In order to analyze the effect of adding the GR, the yield strength of the bTPU NCs as a function of the GR content is presented in Figure 6, and the values are displayed in Table 2. Both NCs showed linear increases as the nanofiller content increased, but, as can be seen in the figure and was previously discussed, modifying GR with IL leads to enhanced yield strength compared with the unmodified GR. Young's modulus and yield strength usually show similar behaviors, as they do in the present work, but the positive effect of the GR-IL is even clearer in the latter than in the former. In fact, the average yield strength increase per unit of added graphene is 37% higher in the GR-IL NCs than in the GR NCs (1.18 MPa vs. 0.86 MPa, respectively).

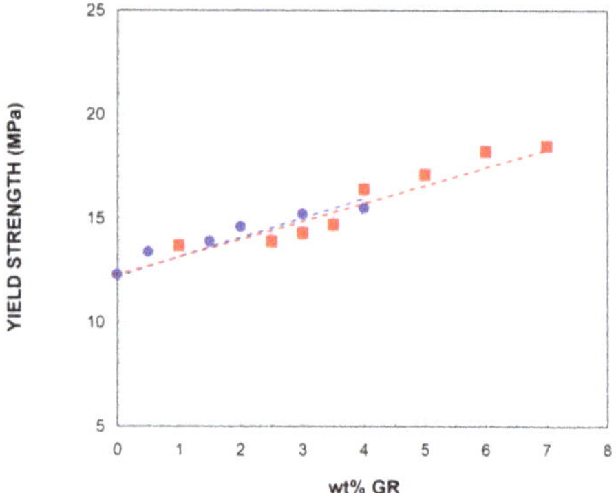

Figure 6. The yield strength of the bTPU/GR (■) and bTPU/GR-IL (•) NCs as a function of the GR content.

Finally, and as mentioned in the previous paragraph, the ductility, measured as the elongation at break, of the bTPU (Figure 7) decreased significantly with a minimum addition (1 wt%) of GR, and more slowly at higher GR contents. Similar behavior has been reported for other melt-mixed TPU/graphene NCs [24,28,29]. However, in this work, all the NCs showed a clearly ductile nature, with elongation at break values above 200%; however, the high standard deviation of the measurements does not allow for any conclusions to be drawn regarding the effect of modifying GR with IL on ductility.

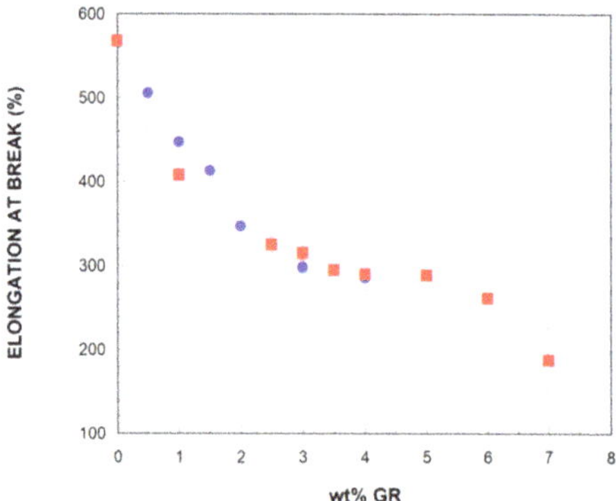

Figure 7. Elongation at break of the bTPU/GR (■) and bTPU/GR-IL (•) NCs as a function of the GR content.

3.3. Thermal Properties

Figure 8 shows the DSC cooling (a) and second heating (b) curves of the neat bTPU and the bTPU/GR-IL NCs. The curves corresponding to the bTPU/GR NCs are not presented as they show a similar trend. The crystallization temperatures (cooling scans), melting temperatures, and enthalpies (heating scans) of the bTPU NCs are shown in Table 3.

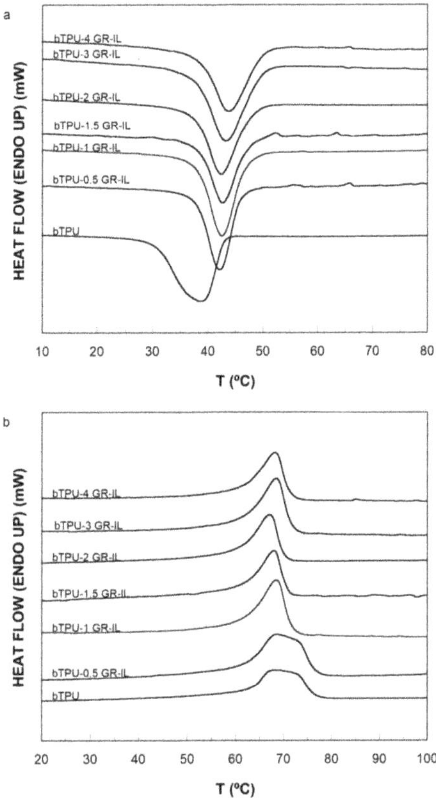

Figure 8. Cooling (**a**) and second heating (**b**) thermograms of the bTPU/GR-IL NCs.

Table 3. The crystallization temperature (T_c), melting temperature (T_m), melting enthalpy (ΔH_m), and glass transition temperature (T_g) of the bTPU NCs.

Composition	T_c (°C) [a] (±0.5 °C)	T_m (°C) [a] (±0.5 °C)	ΔH_m (J/g) [a] (±5 J/g)	T_g (°C) [b] (±0.5 °C)
bTPU	38.5	64.0–74.0	70	−16.0
bTPU-0.5 GR-IL	42.0	68.5	70	−16.0
bTPU-1 GR-IL	42.5	69.0	70	−15.5
bTPU-1.5 GR-IL	42.5	68.0	70	−15.5
bTPU-2 GR-IL	42.5	67.0	70	−15.5
bTPU-3 GR-IL	43.5	68.0	70	−15.0
bTPU-4 GR-IL	44.0	68.5	70	−15.0
bTPU-1 GR	42.0	66.5	75	−15.0
bTPU-2.5 GR	40.5	66.0	75	−15.0
bTPU-3 GR	44.5	66.5	70	−15.0
bTPU-3.5 GR	43.5	67.5	70	−14.5
bTPU-4 GR	44.0	67.5	75	−15.0
bTPU-5 GR	43.5	68.0	75	−16.0
bTPU-6 GR	44.0	66.0	75	−15.0
bTPU-7 GR	44.0	68.0	75	−14.5

[a] obtained by DSC. [b] obtained by DMTA.

As can be seen in Figure 8a, a broad crystallization peak appeared on cooling at between 30 and 40 °C in the neat bTPU, indicating a broad distribution of the perfection/size of the crystals, which is attributed to the hard segments of the bTPU. In both the bTPU/GR and the bTPU/GR-IL NCs, the crystallization exotherms narrowed and shifted to higher temperatures (Table 3) as a result of the nucleating effect of the GR [1,51]. The increase was pronounced at low GR contents, but T_c remained unchanged as greater amounts of GR were added.

With respect to the second heating scans (Figure 8b), the neat bTPU showed a very broad melting endothermic peak at ~64–74 °C. The characteristics of this melting endotherm indicate a broad distribution of crystal sizes, fully consistent with the corresponding exotherm in Figure 8a, attributed to the crystallization of the hard segments of the bTPU matrix. The melting endotherms were narrower for the NCs and the main peaks appeared at the lowest values of the melting interval of the neat bTPU endotherm. This indicates a more homogeneous population of crystal sizes in the NCs [51] and a smaller mean size of the crystals. As previously proposed for the cooling curves, this is probably related to the nucleating effect of the GR on the crystallization of the bTPU.

With respect to the melting (Table 3) or crystallization (not shown) enthalpies, they hardly changed when the GR content changed. Different behaviors have been observed in the crystallinity of the polymer matrix of NCs when graphene is added: (i) increases in crystallinity resulting from a strong nucleation effect [1,40] (although the changes were not significant enough to affect the mechanical properties [57]); (ii) decreases in crystallinity as graphene acts as a physical barrier due to its large surface area and causes interference in the crystallization process [47,51,58]; and (iii) no change [59]. The behavior observed in Table 3 seems to indicate that even though GR favors the formation of bTPU crystal sites, the degree of crystallinity of the bTPU is unaffected by its presence.

The glass transition temperatures of the soft segment phase of the bTPU were determined from the tanδ curves obtained by DMTA and the results are summarized in Table 3. The T_g of unfilled bTPU was located at −16.0 °C and the incorporation of either GR or GR-IL gave rise to a scarcely significant increase of 1 to 1.5 °C, close to the experimental error of the measurement. While the presence of graphene-like structures has, on occasion, been reported to have hindered the molecular mobility of the polymer matrix causing an upward shift in T_g [58,60], in most studies the T_g remained unchanged upon the addition of graphene [1,40,51,61].

4. Conclusions

Electrically and mechanically reinforced biobased thermoplastic polyurethane (bTPU)/graphene nanocomposites (NCs) were successfully obtained using conventional melt-mixing equipment. The modification of GR with an IL using a simple and rapid method in order to improve the dispersion level of the nanofiller was successful since it reduced the electrical percolation threshold and enhanced the mechanical behavior.

NCs with either GR or GR-IL presented nonuniform dispersion, featuring both relatively large stacks and stacks with a low number of layers, as well as individually dispersed graphene sheets. The modification of GR with IL led to an improved dispersion level, as the percolation threshold of the bTPU/GR-IL NCs was significantly lower (1.99 wt%) than that of the bTPU/GR NCs (3.21 wt%). In addition, the NCs with GR-IL showed higher electrical conductivity values than the bTPU/GR NCs.

The addition of either GR or GR-IL led to significant improvements in Young's modulus and yield strength. The superior mechanical behavior of the modified GR-IL NCs with respect to the unmodified GR NCs—which was only slightly noticeable in the case of Young's modulus but more significant in the yield strength—lends further support to the improved level of dispersion of the nanofiller. Although the ductility values of neat bTPU could not be maintained in the NCs, all NCs showed elongation at break values of over 200%.

The NCs showed a more homogeneous population of crystal sizes and a smaller mean size of crystal than the unfilled bTPU, due to the nucleating effect of the GR. However, the degree of crystallinity of the bTPU in the NCs remained unchanged. The glass transition temperature of the bTPU was also unaffected by the addition of GR. No differences in thermal properties were observed between the GR and the GR-IL NCs.

Supplementary Materials: The following are available online at http://www.mdpi.com/2073-4360/11/3/435/s1, Table S1: Young's modulus of the bTPU/IL blends.

Author Contributions: Investigation: N.A., I.O., and G.G.-E.; Writing—Original Draft: N.A., I.O., and G.G.-E.; Writing—Review and Editing: N.A., I.O., and G.G.-E.

Funding: This research was funded by MINECO (Spanish Government), project CTO 2013-4113-R; the Basque Government, project IT611-13; and the University of the Basque Country, UFI 11/56.

Conflicts of Interest: The authors declare no conflicts of interest.

References

1. Mayoral, B.; Harkin-Jones, E.; Khanam, P.N.; AlMaadeed, M.A.; Ouederni, M.; Hamilton, A.R.; Sun, D. Melt processing and characterization of polyamide 6/graphene nanoplatelet composites. *RSC Adv.* **2015**, *5*, 52395–52409. [CrossRef]
2. Kuilla, T.; Bhadra, S.; Yao, D.; Kim, N.H.; Bose, S.; Lee, J.H. Recent advances in graphene based polymer composites. *Prog. Polym. Sci.* **2010**, *35*, 1350–1375. [CrossRef]
3. Pan, Q.; Shim, E.; Pourdeyhimi, B.; Gao, W. Nylon-graphene composite nonwovens as monolithic conductive or capacitive fabrics. *ACS Appl. Mater. Interfaces* **2017**, *9*, 8308–8316. [CrossRef] [PubMed]
4. Pan, Q.; Shim, E.; Pourdeyhimi, B.; Gao, W. Highly conductive polypropylene-graphene nonwoven composite via interface engineering. *Langmuir* **2017**, *33*, 7452–7458. [CrossRef] [PubMed]
5. Phiri, J.; Gane, P.; Maloney, T.C. General overview of graphene: Production, properties and application in polymer composites. *Mater. Sci. Eng. B* **2017**, *215*, 9–28. [CrossRef]
6. Atif, R.; Inam, F. Reasons and remedies for the agglomeration of multilayered graphene and carbon nanotubes in polymers. *Beilstein J. Nanotechnol.* **2016**, *7*, 1174–1196. [CrossRef] [PubMed]
7. Xu, P.; Gui, H.; Wang, X.; Hu, Y.; Ding, Y. Improved dielectric properties of nanocomposites based on polyvinylidene fluoride and ionic liquid-functionalized graphene. *Compos. Sci. Technol.* **2015**, *117*, 282–288. [CrossRef]
8. Yin, B.; Zhang, X.; Zhang, X.; Wang, J.; Wen, Y.; Jia, H.; Ji, Q.; Ding, L. Ionic liquid functionalized graphene oxide for enhancement of styrene-butadiene rubber nanocomposites. *Polym. Adv. Technol.* **2017**, *28*, 293–302. [CrossRef]
9. Gui, H.; Xu, P.; Hu, Y.; Wang, J.; Yang, X.; Bahader, A.; Ding, Y. Synergistic effect of graphene and an ionic liquid containing phosphonium on the thermal stability and flame retardancy of polylactide. *RSC Adv.* **2015**, *5*, 27814–27822. [CrossRef]
10. Ma, W.-S.; Wu, L.; Yang, F.; Wang, S.-F. Non-covalently modified reduced graphene oxide/polyurethane nanocomposites with good mechanical and thermal properties. *J. Mater. Sci.* **2014**, *49*, 562–571. [CrossRef]
11. Wang, L.; Wang, W.; Fan, P.; Zhou, M.; Yang, J.; Chen, F.; Zhong, M. Ionic liquid modified graphene/poly(vinyl alcohol) composite with enhanced properties. *J. Appl. Polym. Sci.* **2017**, 134. [CrossRef]
12. Liu, X.; Wang, L.; Wan, Z.; Zhao, L.; He, H. Electrical conductivity and mechanical properties of ionic liquid modified shear exfoliation graphene/CO-PA nanocomposites at extremely low graphene loading. *Polym. Compos.* **2017**, *38*, E277–E284. [CrossRef]
13. Liu, N.; Luo, F.; Wu, H.; Liu, Y.; Zhang, C.; Chen, J. One-step ionic-liquid-assisted electrochemical synthesis of ionic-liquid-functionalized graphene sheets directly from graphite. *Adv. Funct. Mater.* **2008**, *18*, 1518–1525. [CrossRef]
14. Yang, Y.-K.; He, C.-E.; Peng, R.-G.; Baji, A.; Du, X.-S.; Huang, Y.-L.; Xie, X.-L.; Mai, Y.-W. Non-covalently modified graphene sheets by imidazolium ionic liquids for multifunctional polymer nanocomposites. *J. Mater. Chem.* **2012**, *22*, 5666–5675. [CrossRef]
15. Mohanty, A.K.; Misra, M.; Drzal, L.T. *Natural Fibers, Biopolymers, and Biocomposites*; CRC Press: Boca Raton, FL, USA, 2005.
16. Tayfun, U.; Dogan, M.; Bayramli, E. Effect of surface modification of rice straw on mechanical and flow properties of TPU-based green composites. *Polym. Compos.* **2016**, *37*, 1596–1602. [CrossRef]

17. Verma, M.; Chauhan, S.S.; Dhawan, S.K.; Choudhary, V. Graphene nanoplatelets/carbon nanotubes/polyurethane composites as efficient shield against electromagnetic polluting radiations. *Compos. Part B* **2017**, *120*, 118–127. [CrossRef]
18. Cataldi, P.; Ceseracciu, L.; Marras, S.; Athanassiou, A.; Bayer, I.S. Electrical conductivity enhancement in thermoplastic polyurethane-graphene nanoplatelet composites by stretch-release cycles. *Appl. Phys. Lett.* **2017**, *110*, 121904. [CrossRef]
19. Jan, R.; Habib, A.; Akram, M.A.; Ahmad, I.; Shah, A.; Sadiq, M.; Hussain, A. Flexible, thin films of graphene-polymer composites for EMI shielding. *Mater. Res. Express* **2017**, *4*, 035605. [CrossRef]
20. Li, Y.; Gao, F.; Xue, Z.; Luan, Y.; Yan, X.; Guo, Z.; Wang, Z. Synergistic effect of different graphene-CNT heterostructures on mechanical and self-healing properties of thermoplastic polyurethane composites. *Mater. Des.* **2018**, *137*, 438–445. [CrossRef]
21. Chen, K.; Tian, Q.; Tian, C.; Yan, G.; Cao, F.; Liang, S.; Wang, X. Mechanical reinforcement in thermoplastic polyurethane nanocomposite incorporated with polydopamine functionalized graphene nanoplatelet. *Ind. Eng. Chem. Res.* **2017**, *56*, 11827–11838. [CrossRef]
22. Bera, M.; Maji, P.K. Effect of structural disparity of graphene-based materials on thermo-mechanical and surface properties of thermoplastic polyurethane nanocomposites. *Polymer* **2017**, *119*, 118–133. [CrossRef]
23. Roy, S.; Srivastava, S.K.; Pionteck, J.; Mittal, V. Mechanically and thermally enhanced multiwalled carbon nanotube-graphene hybrid filled thermoplastic polyurethane nanocomposites. *Macromol. Mater. Eng.* **2015**, *300*, 346–357. [CrossRef]
24. Caradonna, A.; Colucci, G.; Giorcelli, M.; Frache, A.; Badini, C. Thermal behavior of thermoplastic polymer nanocomposites containing graphene nanoplatelets. *J. Appl. Polym. Sci.* **2017**, *134*. [CrossRef]
25. Strankowski, M.; Piszczyk, L.; Kosmela, P.; Korzeniewski, P. Morphology and the physical and thermal properties of thermoplastic polyurethane reinforced with thermally reduced graphene oxide. *Pol. J. Chem. Technol.* **2015**, *17*, 88–94. [CrossRef]
26. Kim, H.; Miura, Y.; Macosko, C.W. Graphene/polyurethane nanocomposites for improved gas barrier and electrical conductivity. *Chem. Mater.* **2010**, *22*, 3441–3450. [CrossRef]
27. Yuan, D.; Pedrazzoli, D.; Pircheraghi, G.; Manas-Zloczower, I. Melt compounding of thermoplastic polyurethanes incorporating 1D and 2D carbon nanofillers. *Polym.-Plast. Technol. Eng.* **2017**, *56*, 732–743. [CrossRef]
28. Piana, F.; Pionteck, J. Effect of the melt processing conditions on the conductive paths formation in thermoplastic polyurethane/expanded graphite (TPU/EG) composites. *Compos. Sci. Technol.* **2013**, *80*, 39–46. [CrossRef]
29. Yuan, D.; Pedrazzoli, D.; Manas-Zloczower, I. Synergistic effects in thermoplastic polyurethanes incorporating hybrid carbon nanofillers. *Int. Polym. Process.* **2016**, *31*, 554–561. [CrossRef]
30. Razeghi, M.; Pircheraghi, G. TPU/graphene nanocomposites: Effect of graphene functionality on the morphology of separated hard domains in thermoplastic polyurethane. *Polymer* **2018**, *148*, 169–180. [CrossRef]
31. Gomez, J.; Recio, I.; Navas, A.; Villaro, E.; Galindo, B.; Ortega-Murguialday, A. Processing influence on dielectric, mechanical, and electrical properties of reduced graphene oxide-TPU nanocomposites. *J. Appl. Polym. Sci.* **2019**, *136*. [CrossRef]
32. Strankowski, M.; Korzeniewski, P.; Strankowska, J.; Anu, A.S.; Thomas, S. Morphology, mechanical and thermal properties of thermoplastic polyurethane containing reduced graphene oxide and graphene nanoplatelets. *Materials* **2018**, *11*, 82. [CrossRef] [PubMed]
33. Inuwa, I.M.; Hassan, A.; Samsudin, S.A.; Haafiz, M.K.M.; Jawaid, M.; Majeed, K.; Abdul Razak, N.C. Characterization and mechanical properties of exfoliated graphite nanoplatelets reinforced polyethylene terephthalate/polypropylene composites. *J. Appl. Polym. Sci.* **2014**, *131*, 40582. [CrossRef]
34. Istrate, O.M.; Paton, K.R.; Khan, U.; O'Neill, A.; Bell, A.P.; Coleman, J.N. Reinforcement in melt-processed polymer-graphene composites at extremely low graphene loading level. *Carbon* **2014**, *78*, 243–249. [CrossRef]
35. Gandini, A. The irruption of polymers from renewable resources on the scene of macromolecular science and technology. *Green Chem.* **2011**, *13*, 1061–1083. [CrossRef]
36. Sinha Ray, S. *Environmentally Fribendly Polymer Nanocomposites: Types, Processing and Properties*; Woodhead Publishing: Cambridge, UK, 2013.
37. Babb, D.A. Polyurethanes from renewable resources. *Adv. Polym. Sci.* **2012**, *245*, 315–360.

38. Niaounakis, M. *Biopolymers: Reuse, Recycling and Disposal*; William Andrew Publishing: Norwich, NY, USA, 2013.
39. Pour, R.H.; Hassan, A.; Soheilmoghaddam, M.; Bidsorkhi, H.C. Mechanical, thermal, and morphological properties of graphene reinforced polycarbonate/acrylonitrile butadiene styrene nanocomposites. *Polym. Compos.* **2016**, *37*, 1633–1640. [CrossRef]
40. Sullivan, E.M.; Oh, Y.J.; Gerhardt, R.A.; Wang, B.; Kalaitzidou, K. Understanding the effect of polymer crystallinity on the electrical conductivity of exfoliated graphite nanoplatelet/polylactic acid composite films. *J. Polym. Res.* **2014**, *21*, 1–9. [CrossRef]
41. El Achaby, M.; Arrakhiz, F.-E.; Vaudreuil, S.; El Kacem Qaiss, A.; Bousmina, M.; Fassi-Fehri, O. Mechanical, thermal, and rheological properties of graphene-based polypropylene nanocomposites prepared by melt mixing. *Polym. Compos.* **2012**, *33*, 733–744. [CrossRef]
42. Shen, M.-Y.; Chang, T.-Y.; Hsieh, T.-H.; Li, Y.-L.; Chiang, C.-L.; Yang, H.; Yip, M.-C. Mechanical properties and tensile fatigue of graphene nanoplatelets reinforced polymer nanocomposites. *J. Nanomater.* **2013**, *2013*, 565401. [CrossRef]
43. Chieng, B.W.; Ibrahim, N.A.; Wan Yunus, W.M.Z.; Hussein, M.Z.; Loo, Y.Y. Effect of graphene nanoplatelets as nanofiller in plasticized poly(lactic acid) nanocomposites–Thermal properties and mechanical properties. *J. Therm. Anal. Calorim.* **2014**, *118*, 1551–1559. [CrossRef]
44. Fu, X.; Yao, C.; Yang, G. Recent advances in graphene/polyamide 6 composites: A review. *RSC Adv.* **2015**, *5*, 61688–61702. [CrossRef]
45. Inuwa, I.M.; Hassan, A.; Samsudin, S.A.; Mohamad Kassim, M.H.; Jawaid, M. Mechanical and thermal properties of exfoliated graphite nanoplatelets reinforced polyethylene terephthalate/polypropylene composites. *Polym. Compos.* **2014**, *35*, 2029–2035. [CrossRef]
46. Mittal, V.; Chaudhry, A.U.; Luckachan, G.E. Biopolymer–Thermally reduced graphene nanocomposites: Structural characterization and properties. *Mater. Chem. Phys.* **2014**, *147*, 319–332. [CrossRef]
47. Mittal, V.; Chaudhry, A.U. Polymer–graphene nanocomposites: Effect of polymer matrix and filler amount on properties. *Macromol. Mater. Eng.* **2015**, *300*, 510–521. [CrossRef]
48. Jin, J.; Rafiq, R.; Gill, Y.Q.; Song, M. Preparation and characterization of high performance of graphene/nylon nanocomposites. *Eur. Polym. J.* **2013**, *49*, 2617–2626. [CrossRef]
49. Stauffer, D.; Aharony, A. *Introduction in Percolation Theory*; Taylor and Francis: London, UK, 1994.
50. Kalaitzidou, K.; Fukushima, H.; Drzal, L.T. A new compounding method for exfoliated graphite-polypropylene nanocomposites with enhanced flexural properties and lower percolation threshold. *Compos. Sci. Technol.* **2007**, *67*, 2045–2051. [CrossRef]
51. Kumari Pallathadka, P.; Koh, X.Q.; Khatta, A.; Luckachan, G.E.; Mittal, V. Characteristics of biodegradable poly(butylene succinate) nanocomposites with thermally reduced graphene nanosheets. *Polym. Compos.* **2017**, *38*, E42–E48. [CrossRef]
52. Oyarzabal, A.; Cristiano-Tassi, A.; Laredo, E.; Newman, D.; Bello, A.; Etxeberria, A.; Eguiazabal, J.I.; Zubitur, M.; Mugica, A.; Mueller, A.J. Dielectric, mechanical and transport properties of bisphenol a polycarbonate/graphene nanocomposites prepared by melt blending. *J. Appl. Polym. Sci.* **2017**, *134*. [CrossRef]
53. Zhou, S.; Hrymak, A.; Kamal, M. Electrical and morphological properties of microinjection molded polypropylene/carbon nanocomposites. *J. Appl. Polym. Sci.* **2017**, *134*. [CrossRef]
54. Aranburu, N.; Eguiazabal, J.I. Electrically conductive multi-walled carbon nanotube-reinforced amorphous polyamide nanocomposites. *Polym. Compos.* **2014**, *35*, 587–595. [CrossRef]
55. Rynkowska, E.; Fatyeyeva, K.; Kujawa, J.; Dzieszkowski, K.; Wolan, A.; Kujawski, W. The effect of reactive ionic liquid or plasticizer incorporation on the physicochemical and transport properties of cellulose acetate propionate-based membranes. *Polymers* **2018**, *10*, 86. [CrossRef]
56. Chen, G.; Chen, N.; Li, L.; Wang, Q.; Duan, W. Ionic liquid modified poly(vinyl alcohol) with improved thermal processability and excellent electrical conductivity. *Ind. Eng. Chem. Res.* **2018**, *57*, 5472–5481. [CrossRef]
57. Chieng, B.W.; Ibrahim, N.A.; Yunus, W.M.Z.W.; Hussein, M.Z. Poly(lactic acid)/poly(ethylene glycol) polymer nanocomposites: Effects of graphene nanoplatelets. *Polymers* **2014**, *6*, 93–104. [CrossRef]
58. Yang, X.; Li, L.; Shang, S.; Tao, X.-M. Synthesis and characterization of layer-aligned poly(vinyl alcohol)/graphene nanocomposites. *Polymer* **2010**, *51*, 3431–3435. [CrossRef]

59. Cai, D.; Song, M. A simple route to enhance the interface between graphite oxide nanoplatelets and a semi-crystalline polymer for stress transfer. *Nanotechnology* **2009**, *2*, 315708. [CrossRef] [PubMed]
60. Liu, H.; Li, Y.; Dai, K.; Zheng, G.; Liu, C.; Shen, C.; Yan, X.; Guo, J.; Guo, Z. Electrically conductive thermoplastic elastomer nanocomposites at ultralow graphene loading levels for strain sensor applications. *J. Mater. Chem. C* **2016**, *4*, 157–166. [CrossRef]
61. Marini, J.; Pollet, E.; Averous, L.; Bretas, R.E.S. Elaboration and properties of novel biobased nanocomposites with halloysite nanotubes and thermoplastic polyurethane from dimerized fatty acids. *Polymer* **2014**, *55*, 5226–5234. [CrossRef]

Article

Mechanical Properties of Multi-Walled Carbon Nanotube/Waterborne Polyurethane Conductive Coatings Prepared by Electrostatic Spraying

Fangfang Wang [1], Lajun Feng [1,2,*] and Man Lu [1]

1 School of Materials Science and Engineering, Xi'an University of Technology, Xi'an 710048, China; wff1170111008@163.com (F.W.); luman569@163.com (M.L.)

2 Key Laboratory of Corrosion and Protection of Shaanxi Province, Xi'an 710048, China

* Correspondence: fenglajun@xaut.edu.cn; Tel./Fax: +86-(0)-29-8231-2733

Received: 26 March 2019; Accepted: 16 April 2019; Published: 19 April 2019

Abstract: Electrostatic spraying (ES) was used to prepare multi-walled carbon nanotube (MWCNT)/waterborne polyurethane (WPU) abrasion-proof, conductive coatings to improve the electrical conductivity and mechanical properties of WPU coatings. The dispersity of MWCNTs and the electrical conductivity, surface hardness, and wear resistance of the coating prepared by ES (ESC) were investigated. The ESC was further compared with coatings prepared by brushing (BrC). The results provide a theoretical basis for the preparation and application of conductive WPU coatings with excellent wear resistance. The dispersity of MWCNTs and the surface hardness and wear resistance of ESC were obviously better than those of BrC. With an increase in the MWCNT content, the surface hardness of both ESC and BrC went up. As the MWCNT content increased, the wear resistance of ESC first increased and then decreased, while the wear resistance of BrC decreased. It was evident that ESC with 0.3 wt% MWCNT was fully capable of conducting electricity, but BrC with 0.3 wt% MWCNT failed to conduct electricity. The best wear resistance was achieved for ESC with 0.3 wt% MWCNT. Its wear rate (1.18×10^{-10} cm^3/mm N) and friction coefficient (0.28) were the lowest, which were 50.21% and 20.00% lower, respectively, than those of pure WPU ESC.

Keywords: electrostatic spraying; multi-walled carbon nanotubes; waterborne polyurethane coating; dispersity; surface hardness; wear rate; friction coefficient

1. Introduction

Waterborne polyurethane (WPU) with water as the dispersion medium is a class of eco-friendly coatings [1–3]. It does not volatilize organic solvents into the air and is now widely being used in the industry to gradually replace solvent-based polyurethane owing to its environmentally friendly characteristic [4–7]. However, the poor mechanical strength and performance of WPU and its inability to conduct electricity may restrict its applications in some working conditions where relatively high antistatic property and wear resistance is required. Therefore, it is necessary for WPU to be modified to meet the requirements of harsh conditions [8]. Researchers have previously shown that adding nanoparticles to polymers to prepare organic/inorganic nanocomposites could strengthen the physical and mechanical properties of polymers [9,10]. As a kind of carbon materials, multi-walled carbon nanotubes (MWCNTs) holding the performance of conducting electricity are likely to maintain resistance to generally chemical corrosive media. The addition of MWCNTs in WPU could therefore effectively improve the electrical conductivity and other mechanical properties of WPU coatings. Moreover, some polar groups, such as –OH, may be adsorbed on the surface of the MWCNT due to the fibrous structure of the material and its outstanding surface activity. The crosslinking reaction that occurs between these polar groups and some polar groups in the molecular chain of WPU during the curing process of the coating could make the WPU coating form a crosslinked network structure [11], thus enhancing the mechanical property of the composite coating [12]. Khun et al. [13] prepared PU composite coatings with different MWCNT contents and found that the cathodic delamination of PU coatings was significantly lessened as the MWCNT content increased to 0.5 wt%. Manas-Zloczower et al. [14] obtained PU nanocomposites with the addition of MWCNTs via the in-situ polymerization of 1,4-phenyldiisocyanate (PPDI) and polycaprolactonediol, it was concluded that the dispersity of nano-fillers was well and properties of obtained nanocomposites were superior. Gao et al. [15] prepared a flexible conductive polymer nanofiber composite (FCPNC) with the addition of carbon nanotube (CNT) and found that the good electrical conductivity and interconnected porous structure of the FCPNC made it possible to be used as a chemical vapor sensor. However, MWCNTs are inclined to aggregate due to their characteristics of high aspect ratio and specific surface area [16]. When the MWCNT content is low, the number of agglomerated MWCNTs might be reduced to a certain extent. However, the electrical conductivity of the composite coating may be too poor to meet the requirement of antistatic performance, and its surface hardness and wear resistance may also be weak. When the MWCNT content is high, the antistatic property of the coating can satisfy the application requirements [17–19], but the composite coating structure would be loose, and the bond strength of the coating to the metal substrate may be poor. Thus, the coating will likely peel off in pieces once it is subjected to external forces during application.

In electrostatic spraying (ES), there is a nozzle with a spiculated edge in the head of the spray gun, which instantly generates high-voltage discharge as well as air ionization once the high-voltage negative electricity is connected. The advantage of high-voltage corona discharge is that it leads to the formation of electrostatic field between the spray gun and the metal substrate. Consequently, the liquid coating with negative charges ejected from the nozzle is attracted to the metal substrate with positive charges under the action of electrostatic attraction. If compressed air would be used as the driving force to transport the liquid coating, the coating could be more effectively atomized with the impact force of compressed air during spraying, it may prevent the agglomeration of conductive fillers to some extent. In addition, ES could overcome the problem of controlling the coating thickness, which is difficult for brushing (Br). In this work, a series of MWCNT/WPU nanocomposite coatings were prepared by ES to not only enhance the dispersity of conductive fillers to some extent but also promote the antistatic and mechanical properties of the WPU coating. The dispersity of MWCNTs and the electrical conductivity, surface hardness, and wear resistance of the coating were studied. The result was further investigated by comparing the coatings with those prepared by Br. This work provides a theoretical basis for the preparation and application [20,21] of MWCNT/WPU abrasion-proof, conductive coatings.

2. Materials and Methods

2.1. Experimental Materials

WPU was supplied by Jining Huakai Resin Co. Ltd., Jining, China. Its volatile organic compound (VOC) concentration, viscosity, and solid content were 253 g/L, 75 cps, and 35%, respectively. The MWCNTs (FloTube 9000 series) were supplied by Beijing Tiannai Technology Co., Ltd., Beijing, China. The purity, average diameter, average length, and tap density of MWCNTs were 95–97.5%, 10–15 nm, 10 μm, and 0.03–0.15 g/cm^3, respectively.

MWCNTs with different contents (0, 0.3, and 0.6 wt%) were each added in WPU by an 85-2 magnetic stirring device (Hangzhou Instrument Motor Co., Ltd., Hangzhou, China) at a low speed of 200–300 r/min for 30 min. Then, the mixtures were each treated using a KQ-50B ultrasonic dispersion device (Kunshan Ultrasonic Instrument Co., Ltd., Kunshan, China) for 30 min. The relatively evenly treated WPU dispersions with different MWCNT contents were obtained.

Q235 steel with a size of 50 × 20 × 3 mm was used as the metal substrate and was roughened by a YX-6050A sand blasting device (Anbangruiyuxin Machine Technology Development Co. Ltd., Wuhan, China). The process condition of sand blasting treatment was as follows. The air pressure was controlled at 0.6–0.8 MPa, the distance between the spray gun and the metal substrate was kept at 110–150 mm, and the time of sand blasting treatment was kept at 30–40 s.

2.2. *Preparation of Coatings*

2.2.1. Coating Prepared by ES (ESC)

Due to the characteristic of self-adjustment of coating thickness, the thickness of ES cannot be increased any further after reaching a certain thickness. Therefore, a method of multi-spraying was adopted to obtain a thicker coating. The obtained MWCNT/WPU dispersions were each sprayed on roughened metal substrates to form an underlayer coating using a NEW KCI-CU801 electrostatic spraying equipment (Shenzhen Honghaida Instrument Co., Ltd., Shenzhen, China). As the underlayer coating was in semidry and nonflowing conditions, the same MWCNT/WPU dispersion as the underlayer coating was sprayed again on the uncured underlayer coating to prepare an upper-layer coating. The average thickness of the multilayer coating was controlled at 80–87 μm. The samples (the Q235 steel substrate with a multilayer coating) were first cured at room temperature for 3 days and then at 70 °C for 24 h in an oven (Zhejiang YuyaoYuandong CNC Instrument Factory, Yuyao, China). The process condition of ES was as follows. The voltage of ES was set at 50–60 KV, the pressure of the compressed air was kept at 0.6–0.7 MPa, the distance between the spray gun and the Q235 steel substrate was controlled at 100–120 mm, the feedwell diameter was 1 mm, the liquid flow rate was 2 mL/min, and the spray time was 1–2 min.

2.2.2. Coating Prepared by Br (BrC)

The obtained WPU dispersions with different MWCNT contents were each brushed on roughened metal substrates to prepare MWCNT/WPU composite coatings. The average thickness of the coating was 80–87 μm. The samples were first cured at room temperature for 3 days and then at 70 °C for 24 h in an oven.

2.3. *Measurements*

2.3.1. Wear Resistance of the Coating

The wear resistance of the coating, which was tested according to ASTM G99-05, was evaluated by its wear rate and friction coefficient [22–24]. The experiment on the wear was conducted with an HT-1000 high-temperature scratch testing machine (Lanzhou Zhongke Kaihua Development Co., Ltd., Lanzhou, China) at room temperature using the Q235 steel substrate (50 × 20 × 3 mm) with the coating against a steel bearing ball (Φ2.5 mm) with a hardness level of HRC62. The applied load was 4 N, the rotation speed of the steel ball was 400 r/min, the sliding radius was 7 mm, and the wear time was 10 min. The wear rate was evaluated using Equation (1):

$$I = \Delta m / 2\pi rntF\rho \tag{1}$$

where I is the specific wear rate (cm^3/mm N), Δm is the loss weight (g), r is the sliding radius (mm), n is the rotation speed of the steel ball (r/min), t is the wear time (min), F is the applied load (N), and ρ is the density of the WPU coating (g/cm^3).

2.3.2. Surface Hardness of the Coating

The surface hardness of the coating was measured using a LX-A Shore durometer with a measurement range of 0–100 HA (Leqing Sanwen Metering and Detection Device, Wenzhou, China). The average value was calculated by five data points.

2.3.3. Electrical Conductivity of the Coating

The electrical conductivity of the coating was evaluated by its resistivity, which is equivalent to the multiplication of the thickness and the square resistance. The coating thickness was tested using a HCC-18 magnetoresistive thickness meter (Shanghai Huayang Testing Instrument Co., Ltd., Shanghai, China). The square resistance was tested at room temperature using a DY2101 digital multimeter (Duoyi Multimeter, Xi'an, China). The average value of each parameter was counted by six data points.

2.3.4. Micromorphology of the Coating

The morphology of the MWCNT in the uncured MWCNT/WPU coating that was just brushed or sprayed on the surface of the steel substrate was observed using a JEM-3010 high-resolution transmission electron microscope (TEM) (JEOL, Tokyo, Japan) to characterize the dispersion of MWCNTs in the WPU resins. The cross-sectional morphology of the coating was investigated using a Merlin Compact scanning electron microscope (SEM) (Zeiss, Oberkochen, Germany) to characterize the dispersion of MWCNTs in the coating.

When the test on the wear was finished, the surface morphology of the wear track was observed with a VEGA3 XMU SEM (TESCANSCAN, Brno, Czech) to characterize the effect of ES on the wear resistance of the water-based conductive coating.

3. Results and Discussion

3.1. *Dispersity of MWCNTs*

Figure 1 shows cross-sectional morphologies of WPU coatings with different MWCNT contents. When the MWCNT content was 0.3 wt% (Figure 1A), the MWCNTs in ESC were relatively evenly dispersed without obvious agglomeration and sedimentation. Although MWCNTs did not come into contact with each other, the generation of tunneling effect from the close average distance between them would enable the composite coating to conduct electricity [25]. As the MWCNT content increased to 0.6 wt% (Figure 1B), there were no apparently agglomerated MWCNTs in the ESC, and they intertwined with each other to form an infinite conductive network. The presence may be explained as follows. Due to the good atomization performance of ES, the still-agglomerated MWCNTs that had been treated by magnetic stirring and ultrasonic dispersion could be dispersed again. Thus, the agglomeration and sedimentation of MWCNTs could be weakened to some extent. Moreover, it was obvious that the morphology of BrC was different from that of ESC. When the MWCNT content was 0.3 wt%, there were few MWCNTs in WPU resins in the top area of BrC (Figure 1C), resulting in BrC being unable to conduct electricity. As the MWCNT content increased to 0.6 wt%, MWCNTs in BrC came into contact with each other to form a valid conductive network (Figure 1D). However, the MWCNTs were apparently agglomerated and deposited due to their uneven distributions, and the structure of BrC was thus less compact. The dispersion of MWCNTs in WPU resins was further studied by TEM. Figure 2A,B show the morphologies of MWCNTs in uncured ESC with 0.6 wt% MWCNT and BrC with 0.6 wt% MWCNT, respectively, on the surfaces of the steel substrates. Compared with the morphology of the MWCNT in ESC with 0.6 wt% MWCNT (Figure 2A), there were obvious agglomerated MWCNTs in BrC with 0.6 wt% MWCNT (Figure 2B), and the size of the agglomerated MWCNT particles was about 150–200 nm. The result was consistent with that of SEM.

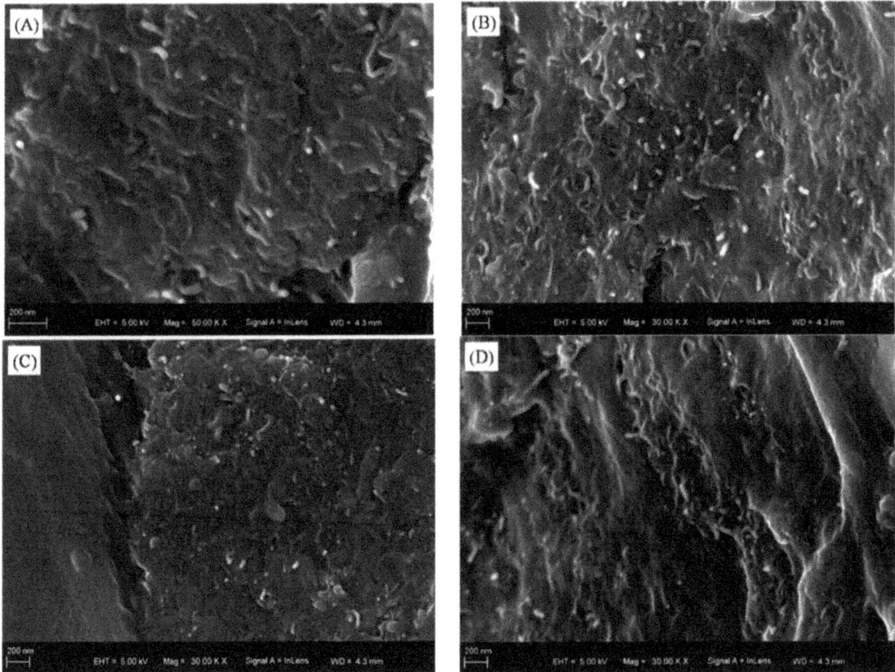

Figure 1. Cross-sectional morphologies of waterborne polyurethane (WPU) coatings with different multi-walled carbon nanotube (MWCNT) contents: (**A**) coating prepared by electrostatic spraying (ESC) with 0.3 wt% MWCNT, (**B**) ESC with 0.6 wt% MWCNT, (**C**) coating prepared by brushing (BrC) with 0.3 wt% MWCNT, and (**D**) BrC with 0.6 wt% MWCNT.

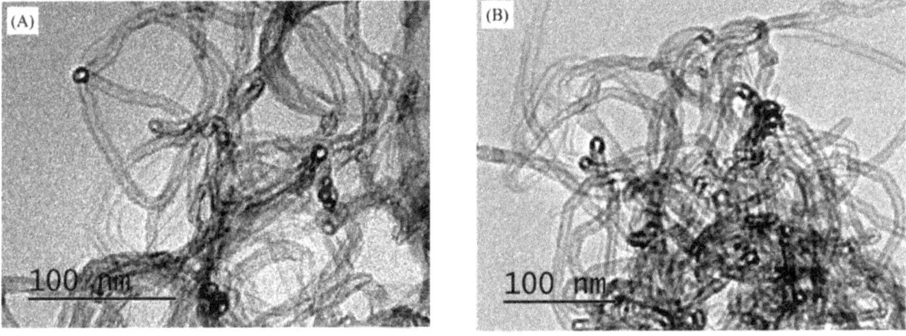

Figure 2. Morphologies of MWCNTs in uncured (**A**) ESC with 0.6 wt% MWCNT and (**B**) BrC with 0.6 wt% MWCNT on the steel substrates.

Table 1 summarizes the electrical conductivity of the coatings. It is evident that ESC with 0.3 wt% MWCNT was fully capable of conducting electricity, but BrC with 0.3 wt% MWCNT failed to conduct electricity. The electrical conductivity of ESC was better than that of BrC. Generally, the electrical conductivity results of the coatings were consistent with that of the dispersity of MWCNTs.

Table 1. Electrical conductivity of WPU coatings with different MWCNT contents.

Coatings	Properties	0 wt%	0.3 wt%	0.6 wt%
ESC	Thickness (μm)	81 ± 4	82 ± 4	83 ± 4
	Square resistance (MΩ)	0	156.2 ± 5	2.6 ± 0.2
	Resistivity (Ω m)	0	12,808.4	215.8
BrC	Thickness (μm)	82 ± 4	82 ± 4	83 ± 4
	Square resistance (MΩ)	0	0	155.7 ± 0.5
	Resistivity (Ω m)	0	0	12,923.1

3.2. Surface Hardness

The surface hardness of the WPU coating with different MWCNT content is shown in Figure 3. It is evident that the surface hardness of ESC was significantly higher than that of BrC with the same MWCNT content. As the MWCNT content rose, the surface hardness of both ESC and BrC went up. The difference was that the growth rate of the surface hardness of ESC was more rapid than that of BrC. The surface hardness of pure WPU ESC (79.5 HA) increased by 8.9% compared with that of pure WPU BrC (73.0 HA). The reason may be explained as follows. Generally, the reason the structure of ESC is more compact is that it is under the action of high-voltage electrostatic field. The defects that are generated in the coating during the former spray can be filled up by the atomized coating during the latter spray. However, when compressed air is used as the driving force to transport the liquid coating, the liquid coating may be attracted to the metal substrate by electrostatic attraction during spray, and it is also more effectively atomized by the impact force of compressed air, thus preventing agglomeration of conductive fillers. A denser coating structure is therefore formed after curing. Moreover, some of the moisture in the WPU coating is taken away during the spraying process, and the micropores caused by moisture volatilization during the curing process of WPU thus falls. Therefore, the coating structure becomes denser, and the surface hardness of the coating is strengthened. In our study, as the MWCNT content increased from 0.3 wt% to 0.6 wt%, the surface hardness of ESC increased by 6.6%, but the surface hardness of BrC merely increased by 4.0%. The reason the growth rate of the surface hardness of ESC was higher than that of BrC may be because the MWCNTs in the coating were relatively evenly dispersed with the help of ES, which effectively prevented the MWCNTs from depositing and agglomerating [26]. The MWCNTs with hardness greater than that of WPU resins filled up the micropores arising from the curing process of WPU, and the hardness of the WPU composite coating was thus greatly improved. In contrast, the ununiform dispersion of MWCNTs in BrC resulted in MWCNTs being more likely to agglomerate and deposit, and there might have been few or no MWCNTs in the WPU resins in the upper area of the coating. This intensified the formation of defects in the coating and thus weakened the effect additional MWCNTs might have had on improving the surface hardness of the coating.

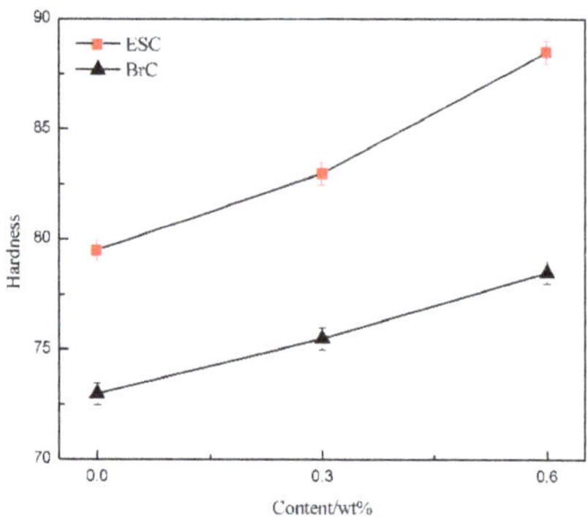

Figure 3. Surface hardness of ESC and BrC with different MWCNT content.

3.3. Wear Resistance

Figures 4 and 5 show the wear rates and friction coefficient–time curves of WPU coatings with different MWCNT contents. During the wear test, the applied load was 4 N, the rotation speed of the steel ball was 400 r/min, the sliding radius was 7 mm, and the wear time was 10 min. Figure 6 shows the wear track morphologies of WPU coatings with different MWCNT contents. It can be concluded from Figures 4 and 5 that, as the MWCNT content increased, the wear rate and friction coefficient of ESC first decreased and then increased, meaning that its wear resistance first increased and then decreased. However, the wear rate and friction coefficient of BrC accordingly increased, which meant that its wear resistance declined. It is apparent that the wear resistance of ESC was lower than that of pure WPU BrC when the MWCNT content was less than 0.6 wt%. The wear rate (2.37×10^{-10} cm^3/mm N) and friction coefficient (0.35) of pure WPU ESC decreased by 50.00% and 10.26%, respectively, compared with those of pure WPU BrC.

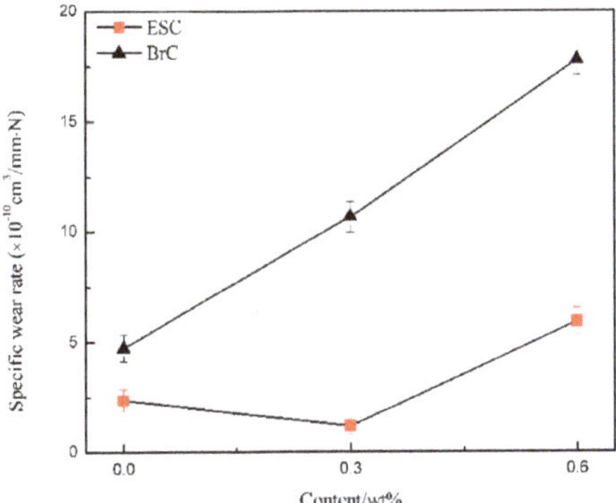

Figure 4. Wear rate curves of ESC and BrC with different MWCNT content.

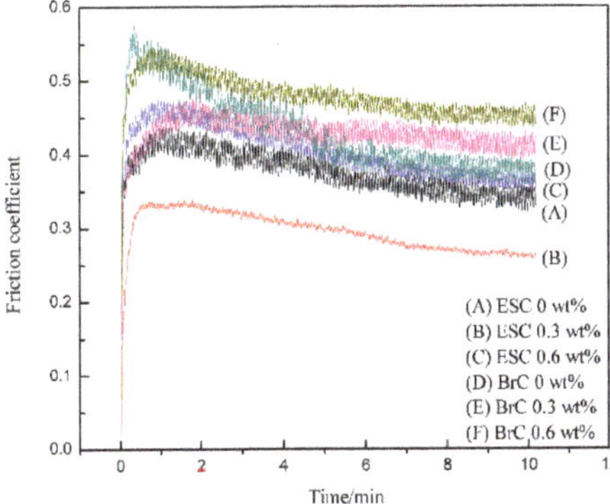

Figure 5. Friction coefficient—time curves of WPU coatings with different MWCNT contents.

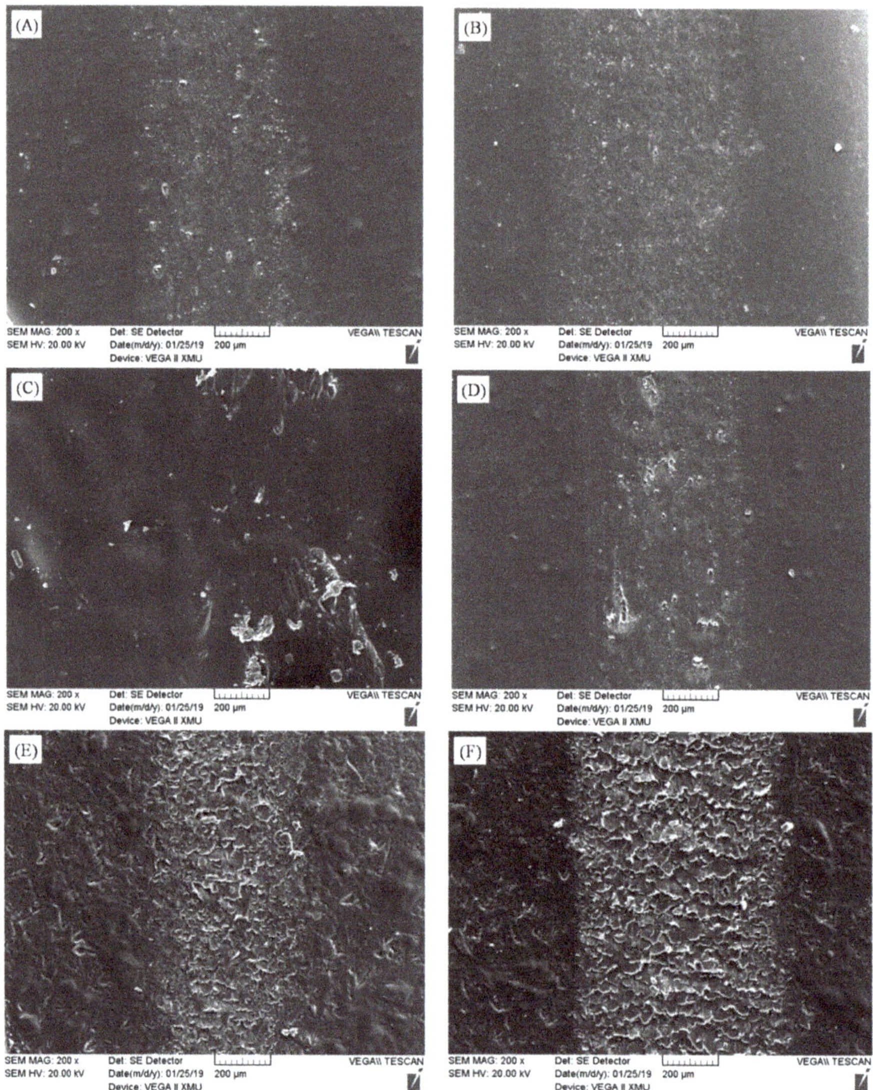

Figure 6. Wear track morphologies of coatings with different MWCNT contents: (**A**) ESC with 0 wt% MWCNT, (**B**) ESC with 0.3 wt% MWCNT, (**C**) ESC with 0.6 wt% MWCNT, (**D**) BrC with 0 wt% MWCNT, (**E**) BrC with 0.3 wt% MWCNT, and (**F**) BrC with 0.6 wt% MWCNT.

By analyzing the wear rate and friction coefficient, it was apparent that the best wear resistance was ESC with 0.3 wt% MWCNT (Figures 4 and 5). Its wear rate (1.18×10^{-10} cm^3/mm N) and friction coefficient (0.28), which were the lowest among all coatings, reduced by 50.21% and 20.00%, respectively, compared with those of pure WPU ESC. The main reason for this is that, when the MWCNT content was 0.3 wt%, the relatively evenly dispersed MWCNTs in WPU resins would have compensated micropores in ESC, thus enhancing the strength of the coating [27]. Therefore, it would be impossible to peel off the ESC from the steel substrate in pieces during the wear process, and its friction coefficient–time curve was relatively smooth. When the MWCNT content was lower than 0.3 wt%, the strength of the coating decreased, and the coating would be more inclined to wear. As the

MWCNT content increased to 0.6 wt%, the surface hardness of the composite coating was greater than that of ESC with 0.3 wt% MWCNT owing to the greater hardness of the MWCNT. However, the large addition of MWCNTs would cause a part of the MWCNT to agglomerate and the number of worn pieces may go up during the wear test. The formation of worn pieces may not only greatly aggravate the wear rate of the coating but also come into the wear track to become a sort of abrasive material that exacerbated the wear of the coating and generated a wavy friction coefficient–time curve. Figure 6A–C show wear track morphologies of ESC after the wear test, which may further explain the results of the wear experiment. It is obvious that there were fewer worn pieces on the wear track of pure WPU ESC. Its wear track seemed to be shallow, and there were almost no block pieces on the wear track of ESC with 0.3 wt% MWCNT. However, there were a large number of flake-like pieces on the wear track of ESC with 0.6 wt% MWCNT.

The analysis of the wear rate and friction coefficient curves of BrC (Figures 4 and 5) showed that the wear resistance of BrC was worse than that of ESC, and the friction coefficient–time curve of BrC was very wavy. This may be due to the uneven dispersion of MWCNTs and the poor compactness of BrC. In addition, the upward trend of agglomerated MWCNTs with the increase in the MWCNT content may have increased the microdefects in BrC. As the friction time increased, the number of worn pieces in the coating also increased. It came into the wear track as a kind of abrasive material, thus increasing the wear rate and friction coefficient of BrC. Figure 6D–F show wear track morphologies of BrC after the wear test. It is evident that the wear track of BrC was wider and deeper than that of ESC, and the number of worn pieces on the wear track of BrC was greater compared with that of ESC. Additionally, the degree of wear of BrC was aggravated with the increase in the MWCNT content.

4. Conclusions

It can be concluded that the dispersity of MWCNTs and the surface hardness and wear resistance of ESC were obviously better than those of BrC. When the MWCNT content increased, the surface hardness of both ESC and BrC went up. The wear resistance of ESC first increased and then decreased, while the wear resistance of BrC decreased. When the MWCNT content was only 0.3 wt%, the coating prepared by ES was capable of conducting electricity, but the coating prepared by Br failed to conduct electricity. The best wear resistance was achieved for ESC with 0.3 wt% MWCNT. Its wear rate (1.18×10^{-10} cm^3/mm N) and friction coefficient (0.28) were the lowest, which were 50.21% and 20.00% lower, respectively, than those of pure WPU ESC.

Author Contributions: Conceptualization, L.F. and F.W.; writing—original draft preparation, F.W.; writing—review and editing, L.F. and F.W.; supervision, M.L.

Funding: This work was supported by the Basic Research Program of Natural Science of Shaanxi Province, China, grant number 2018JM5053, and the Laboratory Project of the Education Department of Shaanxi Provincial Government, China, grant numbers 17JS084 and 18JS083.

Conflicts of Interest: The authors declare no conflict of interest.

References

1. Wang, C.; Mo, B.M.; He, Z.F.; Shao, Q.; Pan, D.; Wujick, E.; Guo, J.; Xie, X.L.; Xie, X.F.; Guo, Z.H. Crosslinked norbornene copolymer anion exchange membrane for fuel cells. *J. Membr. Sci.* **2018**, *556*, 118–125. [CrossRef]
2. Santamaria-Echart, A.; Fernandes, I.; Ugarte, L.; Barreiro, F.; Arbelaiz, A.; Corcuera, M.A.; Eceiza, A. Waterborne polyurethane-urea dispersion with chain extension step in homogeneous medium reinforced with cellulose nanocrystals. *Compos. Part B Eng.* **2018**, *137*, 31–38. [CrossRef]
3. Wang, C.; Mo, B.M.; He, Z.F.; Xie, X.F.; Zhao, C.X.X.; Zhang, L.Q.; Shao, Q.; Guo, X.K.; Wujcik, E.K.; Guo, Z.H. Hydroxide ions transportation in polynorbornene anion exchange membrane. *Polymer* **2018**, *138*, 363–368. [CrossRef]
4. Wu, Q.; Hu, J. Waterborne polyurethane based thermoelectric composites and their application potential in wearable thermoelectric textiles. *Compos. Part B Eng.* **2016**, *107*, 59–66. [CrossRef]

5. Kim, E.Y.; Lee, J.H.; Lee, D.J.; Lee, Y.H.; Lee, J.H.; Kim, H.D. Synthesis and properties of highly hydrophilic waterborne polyurethane-ureas containing various hardener content for waterproof breathable fabrics. *J. Appl. Polym. Sci.* **2013**, *129*, 1745–1751. [CrossRef]
6. Dong, M.Y.; Li, Q.; Liu, H.; Liu, C.T.; Wujcik, E.K.; Shao, Q.; Ding, T.; Mai, X.M.; Shen, C.Y.; Guo, Z.H. Thermoplastic polyurethane-carbon black nanocomposite coating: Fabrication and solid particle erosion resistance. *Polymer* **2018**, *158*, 381–390. [CrossRef]
7. Ashassisorkhabi, H.; Bagheri, R.; Rezaeimoghadam, B. Sonoelectrochemical synthesis of ppy-MWCNTs-chitosan nanocomposite coatings: Characterization and corrosion behavior. *J. Mater. Eng. Perform.* **2015**, *24*, 385–392. [CrossRef]
8. Santamaria-Echart, A.; Ugarte, L.; García-Astrain, C.; Arbelaiz, A.; Corcuera, M.A.; Eceiza, A. Cellulose nanocrystals reinforced environmentally-friendly waterborne polyurethane nanocomposites. *Carbohydr. Polym.* **2016**, *151*, 1203–1209. [CrossRef] [PubMed]
9. Kim, J.M.; Kim, J.H.; Ahn, J.H.; Kim, J.D.; Park, S.; Park, K.H.; Lee, J.M. Synthesis of nanoparticle-enhanced polyurethane foams and evaluation of mechanical characteristics. *Compos. Part B Eng.* **2018**, *136*, 28–38. [CrossRef]
10. Kamisho, T.; Takeshita, Y.; Sakata, S.; Sawada, T. Water absorption of water-based anticorrosive coatings and its effect on mechanical property and adhesive performance. *J. Coat. Technol. Res.* **2014**, *11*, 199–205. [CrossRef]
11. Sahoo, N.G.; Jung, Y.C.; So, H.H.; Cho, J.W. Polypyrrole coated carbon nanotubes: Synthesis, characterization, and enhanced electrical properties. *Synth. Met.* **2007**, *157*, 374–379. [CrossRef]
12. Wang, F.F.; Feng, L.J.; Li, G.Z. Properties of waterborne polyurethane conductive coating with low MWCNTs content by electrostatic spraying. *Polymers* **2018**, *10*, 1406. [CrossRef]
13. Khun, N.W.; Frankel, G.S. Cathodic delamination of polyurethane/multiwalled carbon nanotubecomposite coatings from steel substrates. *Prog. Org. Coat.* **2016**, *99*, 55–60. [CrossRef]
14. Bonab, V.S.; Manas-Zloczower, I. Chemorheology of thermoplastic polyurethane and thermoplastic polyurethane/carbon nanotube composite systems. *Polymer* **2016**, *99*, 513–520. [CrossRef]
15. Gao, J.F.; Wang, H.; Huang, X.W.; Hu, M.J.; Xue, H.G.; Li, R.K.Y. Electrically conductive polymer nanofiber composite with an ultralow percolation threshold for chemical vapor sensing. *Compos. Sci. Technol.* **2018**, *161*, 135–142. [CrossRef]
16. Chai, C.P.; Ma, Y.F.; Li, G.P.; Ge, Z.; Ma, S.Y.; Luo, Y.J. The preparation of high solid content waterborne polyurethane by special physical blending. *Prog. Org. Coat.* **2018**, *115*, 79–85. [CrossRef]
17. Bautista, Y.; Gonzalez, J.; Gilabert, J.; Ibañez, M.J.; Sanz, V. Correlation between the wear resistance, and the scratch resistance, for nanocomposite coatings. *Prog. Org. Coat.* **2011**, *70*, 178–185. [CrossRef]
18. Amerio, E.; Fabbri, P.; Malucelli, G.; Messori, M.; Sangermano, M.; Taurino, R. Scratch resistance of nano-silica reinforced acrylic coatings. *Prog. Org. Coat.* **2008**, *62*, 129–133. [CrossRef]
19. Qiu, F.X.; Xu, H.P.; Wang, Y.Y.; Xu, J.C.; Yang, D.Y. Preparation, characterization and properties of UV-curable waterborne polyurethane acrylate/SiO_2 coating. *J. Coat. Technol. Res.* **2012**, *9*, 503–514. [CrossRef]
20. Cai, G.F.; Wang, J.X.; Lin, M.F.; Chen, J.W.; Cui, M.Q.; Qian, K.; Li, S.H.; Cui, P.; Lee, P.S. A semitransparent snake-like tactile and olfactory bionic sensor with reversibly stretchable properties. *NPG Asia Mater.* **2017**, *9*, e437. [CrossRef]
21. Cai, G.F.; Wang, J.X.; Qian, K.; Chen, J.W.; Li, S.H.; Lee, P.S. Extremely stretchable strain sensors based on conductive self-healing dynamic cross-links hydrogels for human-motion detection. *Adv. Sci.* **2017**, *4*, 1600190. [CrossRef]
22. Abenojar, J.; Tutor, J.; Ballesteros, Y.; Real, J.C.; Martínez, M.A. Erosion-wear, mechanical and thermal properties of silica filled epoxy nanocomposites. *Compos. Part B Eng.* **2017**, *120*, 42–53. [CrossRef]
23. Tongyan, Y.; Magd, A.W. A numerical study on the effect of debris layer on fretting wear. *Materials* **2016**, *9*, 597.
24. Sawyer, W.G.; Freudenberg, K.D.; Bhimaraj, P.; Schadler, L.S. A study on the friction and wear behavior of PTFE filled with alumina nanoparticles. *Wear* **2003**, *254*, 573–580. [CrossRef]
25. Deng, H.; Skipa, T.; Zhang, R.; Lellinger, D.; Bilotti, E.; Alig, I.; Peijs, T. Effect of melting and crystallization on the conductive network in conductive polymer composites. *Polymer* **2009**, *50*, 3747–3754. [CrossRef]

26. Nezhad, H.Y.; Thakur, V.K. Effect of morphological changes due to increasing carbon nanoparticles content on the quasi-static mechanical response of epoxy resin. *Polymers* **2018**, *10*, 1106. [CrossRef]
27. Wang, F.F.; Feng, L.J.; Ma, H.N.; Zhai, Z.; Liu, Z. Influence of nano-SiO_2 on the bonding strength and wear resistance properties of polyurethane coating. *Sci. Eng. Compos. Mater.* **2018**, *26*, 77–83. [CrossRef]

Article

The Structure and Properties of Polyacrylonitrile Nascent Composite Fibers with Grafted Multi Walled Carbon Nanotubes Prepared by Wet Spinning Method

Hailong Zhang [1,3,*], Ling Quan [2], Aijun Gao [3], Yuping Tong [1], Fengjun Shi [1] and Lianghua Xu [3]

1 School of Materials Science and Engineering, North China University of Water Resources and Electric Power, Zhengzhou 450045, China; tongyuping@ncwu.edu.cn (Y.T.); shifengjun1962@126.com (F.S.)
2 School of Electric Power, North China University of Water Resources and Electric Power, Zhengzhou 450045, China; quanling@ncwu.edu.cn
3 Key Laboratory of Carbon Fiber and Functional Polymers Ministry of Education, Beijing University of Chemical Technology, Beijing 100029, China; bhgaoaijun@163.com (A.G.); xulh@mail.buct.edu.cn (L.X.)
* Correspondence: zhanghailongql@163.com; Tel.: +86-371-67120189

Received: 27 January 2019; Accepted: 28 February 2019; Published: 5 March 2019

Abstract: Polyacrylonitrile (PAN) grafted amino-functionalized multi walled carbon nanotubes (amino-MWCNTs) were synthesized by in situ polymerization under aqueous solvent. The grafted MWCNT/PAN nascent composite fibers were prepared by the wet spinning method. Fourier transform infrared spectroscopy and Raman spectroscopy indicated that the amino-MWCNTs and PAN macromolecular chains had interfacial interactions and formed chemical bonds. The grafting content of the PAN polymer on the amino-MWCNTs was up to 73.2% by thermo gravimetric analysis. The incorporation of the grafted MWCNTs improved the degree of crystallization and crystal size of PAN nascent fibers, and changed the thermal properties during exothermic processing in an air atmosphere. Morphology analysis and testing of mechanical properties showed that the grafted MWCNT/PAN nascent composite fibers with a more uniform diameter distribution and larger diameter had higher tensile strength and tensile modulus than the control PAN nascent fibers.

Keywords: multi walled carbon nanotubes; polyacrylonitrile; nascent fiber; thermal properties; morphological structure

1. Introduction

Owing to their unique structure and properties, such as high strength, low mass density, and large aspect ratio, carbon nanotubes (CNTs) have been used as an ideal reinforcing agent in composites [1,2]. However, CNTs are easily aggregated because of a strong van der Waals force between them and large specific surface area, which makes CNTs difficult to disperse into most solvents or polymer matrices. Moreover, it is difficult for strong interfacial interactions to form between the inert surface of the multi wall carbon nanotubes (MWCNTs) and the polymer matrix, which limits the excellent properties of the MWCNTs. Thus, various non-covalent and covalent methods are used to modify the surface of CNTs for improving the dispersion and interfacial interactions. The non-covalent method is based on the intermolecular interaction on the surface of CNTs, physical adsorption and/or wrapping [3], but the weak interfacial interaction between the CNTs and polymer matrix limits the effective transfer of stress. The covalent method is based on the formation of a chemical bond on the outer CNT wall, which can form a strong interfacial interaction between CNTs and the functionalized agent [4,5]. Katti et al. [6] reported that the poly(ether ether ketone) grafted on carboxyl functionalized MWCNTs improved the dispersion of MWCNTs in the epoxy resin by mechanical stirring and exhibited significant enhancement in mechanical properties. Konnola et al. [7] showed that hydroxyl terminated poly(ether sulfone) grafted on multi walled carbon nanotubes obviously increased the tensile strength and fracture toughness of the epoxy matrix. Therefore, the covalent method is considered, by many authors, an effective method to modify CNTs in order to obtain CNT/polymer matrix composites [8–11].

Polyacrylonitrile (PAN) is an important precursor for preparing high performance carbon fibers. The structure and properties of nascent fibers prepared by wet spinning are formed by liquid solid phase separation under stretching, which influences the evolution of subsequent properties during the preparation of PAN precursor fibers, and affects the final properties of carbon fibers. Up to now, CNT/PAN composite precursor fibers have been reported by many researchers [12–16], but the CNTs were often directly dispersed in the organic solvent and needed to be removed in the last processing, which polluted the environment. However, in situ polymerization provides an effective way to solve this problem, and could improve the interfacial interaction and dispersion of the CNTs. Zhou et al. [17] reported that multi walled carbon nanotube/polyacrylonitrile composite fibers were obtained by in situ polymerization, which exhibited better dispersion of CNTs and interfacial interaction between PAN and CNTs compared with fibers prepared by mechanical mixing. Poochai et al. [18] reported that PAN grafted CNTs by admicellar polymerization would improve the dispersion of CNTs in the PAN matrix. However, the effects of grafted CNTs on the structure and mechanical properties for the PAN nascent fibers were rarely reported. Although our research teams have reported the orientation, crystal structure and thermal properties of CNT/PAN nascent composite fibers [19], grafted CNT/PAN nascent composite fibers prepared by the wet spinning method need to be further researched and discussed.

In this paper, PAN grafted amino-functionalization MWCNT (amino-MWCNT) nanocomposites were synthesized by in situ polymerization under aqueous solvent with ultra-sonication. The nanocomposites, acrylonitrile (AN) and itaconic acid (IA) monomer, were dissolved in dimethylsulphoxide (DMSO) and initiated by solution polymerization to obtain the composite spinning solutions. The grafted MWCNT/PAN nascent composite fibers were prepared by the wet spinning method. The interfacial interaction between amino-MWCNTs and PAN polymers was characterized by Fourier transform infrared (FTIR) spectroscopy and Raman spectroscopy. The grafting amount of PAN polymer was performed by thermo gravimetric (TG) analysis. The effects of the grafted MWCNTs on the crystal structure, thermal properties, cross-sectional structure and surface morphology, as well as mechanical properties of the PAN nascent fibers were analyzed by X-ray diffraction (XRD), differential scanning calorimetry (DSC), transmission electron microscopy (TEM), scanning electron microscopy (SEM), and a monofilament tensile testing machine, respectively.

2. Experiment

2.1. Materials

The pristine MWCNTs ($\geq$95 wt %) with 10–15 nm diameter and 1–10 μm length were purchased from Shenzhen Nanotech Port Co., Ltd. (Shenzhen, China), synthesized using the chemical vapor deposition method. Acrylonitrile (AN) was provided by Beijing Xingjin Chemical Factory (Beijing, China) and purified before polymerization. DMSO and IA were supplied by the Beijing Yili Reagent Corporation (Beijing, China). Hydrochloric acid (HCl), hydrogen peroxide (H_2O_2), thionyl chloride ($SOCl_2$), tetrahydrofuran (THF) and triethylenetetramine (TETA) were obtained from Sinopharm Chemical Reagent Co., Ltd. (Shanghai, China).

2.2. Preparation of the Amino-MWCNTs

The pristine MWCNTs were first purified by HCl to remove the metallic catalysts. The purified MWCNTs were dispersed in H_2O_2 solution with a concentration of 30% under ultrasonic bath for 2 h, and then stirred at 70 °C for 24 h to obtain the MWCNTs with a carboxyl group. The obtained MWCNTs were refluxed with $SOCl_2$ solution under ultra-sonication at 23 °C for 2 h and stirred at

75 °C for 24 h. The MWCNTs with a Cl element were reacted with excessive TETA by ultrasound for 2 h and stirred at 120 °C for 24 h to obtain the amino-functionalized MWCNTs (amino-MWCNTs). The schematic route is shown in Figure 1.

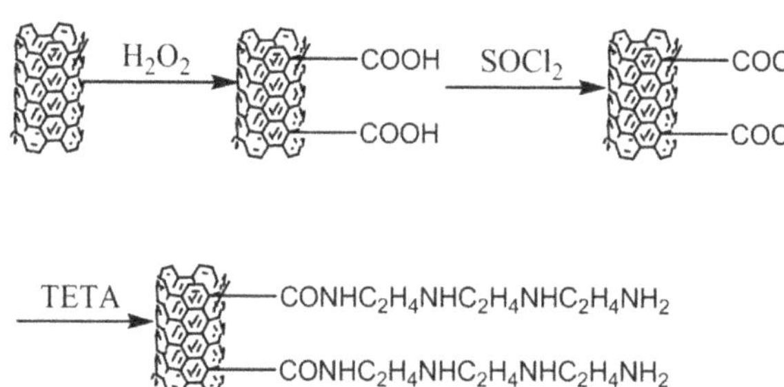

Figure 1. Schematic representation of the formation of amino-MWCNTs.

2.3. *PAN Grafted Amino-MWCNT Nanocomposites*

The PAN grafted amino-MWCNT nanocomposites were synthesized in aqueous solvent by in situ polymerization under ultra-sonication. The AN monomer was dispersed in aqueous solution by ultrasonic treatment for 10 min under nitrogen atmosphere. The certain amounts of amino-MWCNTs were placed into the mixed solution by ultrasound at 50 °C for 30 min. The initiator was added and the mixture solution begun to polymerize for 2 h. Finally, the mixture was filtered and dried under vacuum at 40 °C. The control PAN polymer was synthesized under the same conditions without amino-MWCNTs.

2.4. *The Grafted MWCNT/PAN Nascent Composite Fibers*

The nanocomposites were dissolved in DMSO solvent under an ultra-sonication bath in a nitrogen atmosphere for 5 min. The AN and IA monomers were added into solvent, and then initiated by the initiator after 10 min to obtain the grafted MWCNT/PAN composite spinning solution with a concentration of about 22%. The grafted MWCNT/PAN nascent composite fibers were prepared by the wet spinning method. The mass concentration of MWCNTs in the nascent composite fibers was about 1%. The control PAN nascent fibers were prepared under the same conditions without the grafted MWCNTs.

2.5. *Characterization*

The chemical bonds were characterized by FTIR spectroscopy (Nicolet 5700 spectrometer, Thermo Nicolet Company, Madison, WI, USA), and the samples were pressed into a pellet with potassium bromide (KBr). The spectra obtained ranged from 400 to 4000 cm^{-1} at a resolution of 4 cm^{-1}. The Raman spectrum was recorded using a Microscopic Confocal Raman Spectrometer (RM2000, Renishaw, London, UK) with a He-Ne laser (Spectra-Physics) excitation at 632.8 nm. The amino-MWCNTs and the PAN grafted amino-MWCNTs were deposited on glass slides, and the laser beam was focused with a 5 μm diameter spot on the samples. Morphologies of amino-MWCNTs and the PAN grafted amino-MWCNT nanocomposites were directly observed by transmission electron microscopy (H-800, Hitachi, Tokyo, Japan) at an operating voltage of 200 kV. The samples were dispersed into alcohol solvent under ultrasonic bath at 40 kHz for 5 min, and then dripped onto the copper grid covered with a perforated carbon film. The cross-sectional structure and surface morphology of the PAN nascent fibers and the grafted MWCNT/PAN nascent composite fibers were examined by SEM (S 4700, Hitachi, Tokyo, Japan) at an operating voltage of 20 kV. The nascent fibers were cured in epoxy resin, and then fractured in liquid nitrogen to observe the cross-sectional structure after sputtering gold. The samples were fixed on the surface of a sample table and sputtered with gold to examine the surface morphology. The diameter distribution was carried out using ImageJ software (National Institutes of Health, Berhesda, Rockville, MD, USA), and 20 positions in a SEM image were chosen for analyzing. The thermal stabilities of the PAN polymer and the different MWCNTs were evaluated using a thermo gravimetric instrument (TG 209, Netzsch, Bavarian State, Selb, Germany). Specimens were heated from 50 to 800 °C at a heating rate of 20 °C/min in a nitrogen atmosphere. DSC was carried out in an air atmosphere using a Q 100 instrument (TA Company, Boston, PA, USA). Samples of about 3.0 mg were used to evaluate the exothermic properties of the different nascent fibers at 10 °C heating rate from 50 to 400 °C. The mechanical properties were measured by a monofilament tensile machine (YG001N, Nantong Hongda Company, Nantong, Jiangsu, China). The specimens were tested at a crosshead speed of 5 mm/min with gage length of 50 mm at room temperature. The number of tensile tests performed was 25. XRD was recorded using a D/max 2500VB2+/PC diffractometer (Rigaku, Japan) with a range of 2θ from 5° to 55° at the scanning speed of 10°/min and a step of 1.0 s. The operation was used at 40 kV and 50 mA with Cu Kα radiation (wavelength λ = 0.154056 nm). The degree of crystallization and the crystal size were calculated according to the formula $C(\%) = A_c/(A_a + A_c) \times 100\%$ and the Scherrer equation $L_c = K\lambda/(\beta \cos\theta)$, respectively. Where C % is the degree of crystallization, A_c is the area of the crystalline region, A_a is the area of the amorphous region; L_c is the crystal size, K is 0.89, λ = 0.154056 nm is the wavelength of the XRD, and β is the full width at half maximum at $2\theta \approx 17°$ [20,21].

3. Results and Discussion

The FTIR spectra of amino-MWCNTs, PAN polymer and PAN grafted amino-MWCNT nanocomposites are shown in Figure 2. The FTIR spectrum of amino-MWCNTs, as can be seen in Figure 2a, showed a very strong absorption peak at 1661 cm^{-1}, which is assigned to the stretching vibration of amide carbonyl groups, and the band at 1086 cm^{-1} is attributed to the stretching vibration of the C–N bond. These results are consistent with previous reports [22,23], indicating that the amino groups are grafted onto the surface of MWCNTs. The FTIR spectrum of the control PAN polymer showed a very strong absorption peak at about 2243 cm^{-1}, which is attributed to the nitrile group vibration, and the peak at 1736 cm^{-1} corresponds to the carboxylic acid groups of IA on the PAN macromolecular chains [24]. The peaks at 1454 and 1077 cm^{-1} are assigned to the bending vibration of C–H groups. The weak peak at 1628 cm^{-1} belongs to the NH_2 bending that was likely caused by the hydrolysis of nitrile groups [25]. For the PAN grafted amino-MWCNT nanocomposites, the peak of the carboxylic acid group shifted from 1736 to 1724 cm^{-1}, and the peak of C–H bending vibration shifted from 1454 to 1451 cm^{-1} and from 1077 to 1069 cm^{-1}, respectively. Moreover, the strong peak of amino-MWCNTs shifted from 1661 to 1632 cm^{-1} of the PAN grafted amino-MWCNT nanocomposites. The results coming from these shifted peaks suggest that the amino-functionalized groups on the surface of MWCNTs reacted with the carboxylic acid groups on the control PAN macromolecule chains, and a chemical bond was formed between them.

Raman spectroscopy was used to characterize the structural changes of the surface of MWCNTs, and the interfacial interaction between the MWCNTs and polymer matrix according to the shift of peaks about MWCNTs. The purified MWCNTs, as shown in Figure 3a, had two significant peaks. The strong peak at about 1579.0 cm^{-1} (G-band) is assigned to the sp^2 graphite carbon of the MWCNTs, and the broad weak peak at about 1323.1 cm^{-1} (D-band) is attributed to the sp^3 hybridized carbon of the MWCNTs [26]. After being modified by amino-functionalized groups, as shown in Figure 3b, the amino-MWCNTs also had two typical peaks. However, the peaks about the G-band and D-band shifted from 1579.0 to 1583.7 cm^{-1} and from 1323.1 to 1325.0 cm^{-1}, respectively. Compared with the amino-MWCNTs, the peak of the G-band for the PAN grafted amino-MWCNT nanocomposites shifted from 1583.7 to 1585.6 cm^{-1}, and the peak of the D-band shifted from 1325.0 to 1327.8 cm^{-1}. These results indicate that a strong chemical interfacial interaction exists between the PAN and the amino-MWCNTs after in situ polymerization [27]. The value of the G-band/D-band intensity ratio (I_G/I_D) was used to characterize the surface structure of MWCNTs.

The value of I_G/I_D for amino-MWCNTs is 5.98, which is lower than that of purified MWCNTs at 9.84. This suggests that the surface of the amino-MWCNTs has more relative content of sp^3 hybridized carbon than the purified MWCNTs owing to grafting of the amino-functionalized groups. However, the value of I_G/I_D for the PAN grafted amino-MWCNT nanocomposites is 4.87, indicating that the relative number of sp^3 hybridized carbon attaching onto the amino-MWCNTs increased after AN polymerization [28]. Thus, the Raman results suggest that the PAN polymer covalently attached to the surface of the amino-MWCNTs.

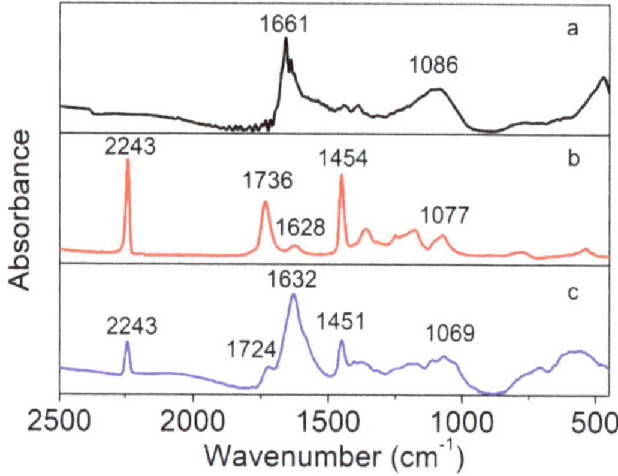

Figure 2. FTIR spectra of (**a**) amino-MWCNTs, (**b**) PAN polymer and (**c**) PAN grafted amino-MWCNT nanocomposites.

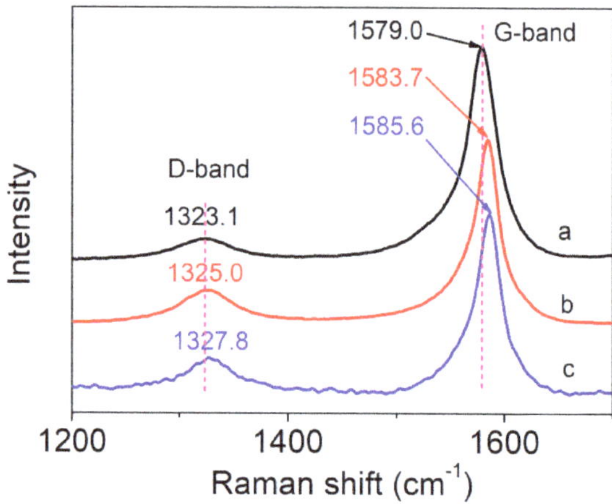

Figure 3. Raman spectra of (**a**) purified MWCNTs, (**b**) amino-MWCNTs and (**c**) PAN grafted amino-MWCNT nanocomposites.

The relative amount of the grafted PAN can be calculated by TG curves through thermal decomposition. The TG curves of purified MWCNTs, amino-MWCNTs, the PAN polymer and the PAN grafted amino-MWCNT nanocomposites in a nitrogen atmosphere are shown in Figure 4. As can be seen in Figure 4a, the weight loss of purified MWCNTs was 9.6% at 800 °C. For the amino-MWCNTs, as shown in Figure 4b, rapid weight loss occurred between 250 and 400 °C, which is ascribed to some functional groups on the surface of MWCNTs. The weight loss of the amino-MWCNTs at 800 °C was 15.2%, which is larger than that of purified MWCNTs. The PAN polymer had similar weight loss to

PAN grafted amino-MWCNTs at lower temperatures between 80 and 300 °C, as shown in Figure 4c,d, which is attributed to the loss of the residual water and small molecules. At a temperature above 300 °C, the PAN polymer begun to lose weight quickly up to 470 °C, and then the rate of weight loss slowed down. During this process, the PAN polymer underwent a cyclized reaction, dehydrogenation reaction and oxidative reaction to form a ring structure. The weight loss for the PAN polymer at 800 °C was 50.5%. However, the rate of weight loss for the PAN grafted amino-MWCNTs was lower than that of the PAN polymer for a temperature from 300 to 470 °C, and had the same parallel line as the PAN polymer between 470 and 800 °C. The weight loss of the PAN grafted amino-MWCNTs was 38.9% at 800 °C. The higher residual weight suggests that the amino-MNCNTs changed the complex chemical reaction and promoted formation of a more stable ring structure because of the interfacial interaction between them [29], which was beneficial for obtaining a high yield of the carbon materials. According to the weight loss at 800 °C for the amino-MWCNTs, PAN polymer and PAN grafted amino-MWCNTs, the content of grafted PAN on the surface of amino-MWCNTs was about 73.2%.

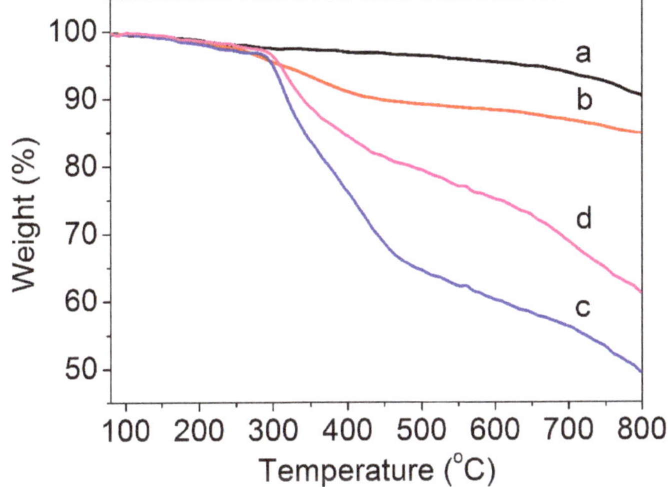

Figure 4. TG curves of (**a**) purified MWCNTs, (**b**) amino-MWCNTs, (**c**) PAN polymer and (**d**) PAN grafted amino-MWCNT nanocomposites.

To confirm the PAN polymer grafted onto the surface of the amino-MWCNTs, the TEM images of amino-MWCNTs and the PAN grafted amino-MWCNTs nanocomposites are shown in Figure 5. The amino-MWCNTs, having a smooth surface, are individually dispersed with a diameter of about 10 nm, as shown in Figure 5a, and have a hollow structure with an internal diameter of about 5 nm. However, the PAN grafted amino-MWCNTs, as can be seen in Figure 5b, exhibit a non-hollow structure owing to the grafted PAN polymer layer, with a diameter of about 20–25 nm and a rough surface. These results suggest that the PAN polymer was grafted onto the surface of the amino-MWCNTs by in situ polymerization.

XRD curves of the PAN nascent fibers and the grafted MWCNT/PAN nascent composite fibers are shown in Figure 6. A PAN macromolecule is considered as a semi-crystalline polymer because of the strong interaction of the nitrile groups. PAN nascent fibers show a strong absorption peak at around 16.8°, which is ascribed to the (100) crystal plane. The weak peak at about 25.6° is assigned to the amorphous region [30]. Because the mass concentration of the grafted MWCNTs in the composite fibers is small, the grafted MWCNT/PAN nascent composite fibers show a similar shape to the PAN nascent fibers, indicating that the incorporation of grafted MWCNTs would not change the crystal structure of PAN nascent fibers. From the XRD curves, the degree of crystallization and the crystal size of different nascent fibers are listed in Table 1.

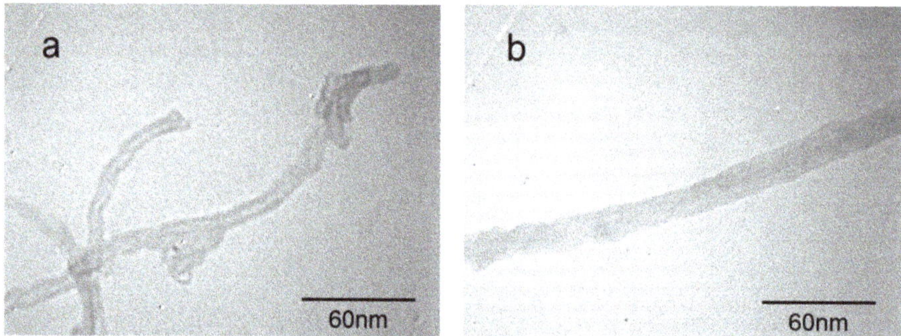

Figure 5. TEM images of (**a**) amino-MWCNTs and (**b**) PAN grafted amino-MWCNTs nanocomposites.

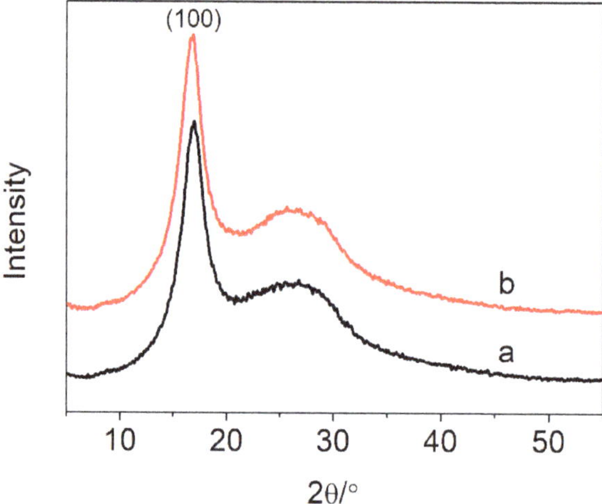

Figure 6. XRD curves of (**a**) the PAN nascent fibers and (**b**) the grafted MWCNT/PAN nascent composite fibers.

Table 1. XRD data for the PAN nascent fibers and the grafted MWCNT/PAN nascent composite fibers.

Samples	Degree of Crystallization (%)	Crystal Size (nm)
PAN	35.23	3.01
grafted MWCNT/PAN	38.73	3.42

From Table 1, it can be seen that the degree of crystallization of the PAN nascent fibers was 35.23%, which was calculated from the (100) crystal plane. The crystal size of the PAN nascent fibers was 3.01 nm, which is assigned to the rod-like structure due to the intermolecular repulsion of the nitrile dipoles [31]. The degree of crystallization of the grafted MWCNT/PAN nascent composite fibers was 38.73%, which is higher than that of the PAN nascent fibers. Meanwhile, the addition of grafted MWCNTs into the PAN nascent fibers improved the crystal size from 3.01 to 3.42 nm. This is attributed to the MWCNTs with a small diameter as the nucleating agent, which could induce growth of PAN crystallites, and transform the amorphous regions around the MWCNTs into a crystal region, resulting in an improvement in the degree of crystallization [32].

In order to investigate the effect of grafted MWCNTs on the thermal properties of PAN nascent fibers during the thermal stabilization, the DSC curves of the PAN nascent fibers and grafted MWCNT/PAN nascent composite fibers under an air atmosphere are shown in Figure 7. It was clearly seen that the DSC curves displayed two exothermic peaks. The first strong peak is attributed to the cyclization reaction of the PAN polymer, and the second weak peak is assigned to the oxygen reaction. The initial temperature (T_i), the first peak temperature (T_1), the second temperature (T_2), the finish temperature (T_f) and the evolved heat (ΔH) are shown in Table 2.

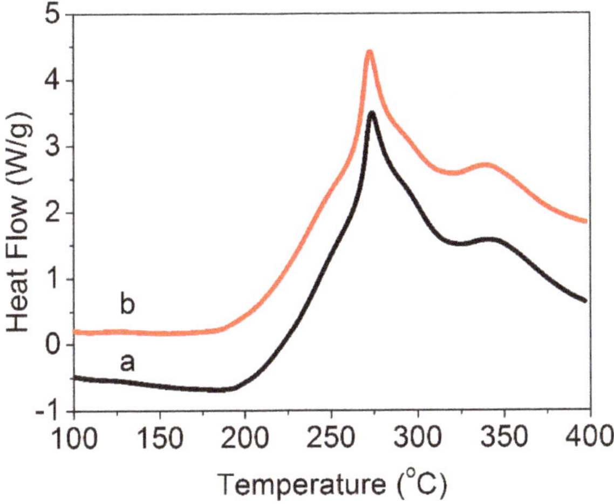

Figure 7. DSC curves of (**a**) the PAN nascent fibers and (**b**) the grafted MWCNT/PAN nascent composite fibers.

Table 2. DSC data for the PAN nascent fibers and the grafted MWCNT/PAN nascent composite fibers in an air atmosphere.

Samples	T_i (°C)	T_1 (°C)	T_2 (°C)	T_f (°C)	ΔH ($J \cdot g^{-1}$)	ΔT (°C)	$\Delta H/\Delta T$ ($J \cdot g^{-1} \cdot °C^{-1}$)
PAN	194.0	273.9	340.9	390.3	1546	196.3	7.9
grafted MWCNT/PAN	188.1	273.0	339.5	388.9	1470	200.8	7.3

According to the data in Table 2, the grafted MWCNT/PAN nascent composite fibers have a much lower value of T_i than the PAN nascent fibers, indicating a greater ease of initiation of stabilization in an air atmosphere. This phenomenon may be ascribed to the grafted MWCNTs inducing the onset temperature of the cyclization reaction. The grafted MWCNT/PAN nascent composite fibers had a lower T_1, T_2 and T_f than the PAN nascent fibers, which could be assigned to the MWCNTs having high heat conductivity. However, the grafted MWCNT/PAN nascent composite fibers with a high degree of crystallization had lower ΔH and broader ΔT than the PAN nascent fibers, suggesting that the nascent composite fibers were more easily controlled during the thermal stabilization process.

Figure 8 shows the cross-sectional structure of the PAN nascent fibers and the grafted MWCNT/PAN nascent composite fibers. The cross-sectional structure of the PAN nascent fibers in Figure 8a shows some porous structures between the skin and core, which is attributed to the exchange of water and DMSO solvent when the spinning solution passes through the spinneret holes and spins into the coagulation bath. Furthermore, there are some nanopores on the cross-sectional morphology in Figure 8b. However, the addition of grafted MWCNTs into the PAN matrix reduces the number of porous structures and nanopores in the grafted MWCNT/PAN composite fibers, as shown in Figure 8c,d. Especially as shown by the arrow in Figure 8d, the individual grafted MWCNTs with a diameter about 20 nm was exhibited in the cross-section of the nascent composite fibers. Moreover, the morphology of cross-section for the PAN nascent fibers shows a lamellar structure in Figure 8b. However, the composite fibers in Figure 8d exhibit some wire drawing on the cross-section owing to the PAN macromolecular chains. These structures indicate that the grafted MWCNT/PAN nascent composite fibers have better tensile properties than the PAN nascent fibers. Meanwhile, a less porous structure means higher compactness, and higher compactness suggests that the nascent fibers have higher mechanical properties, as shown in Table 3. The addition of the grafted MWCNTs increases the tensile strength from 3.83 to 4.37 cN/dtex and the tensile modulus from 101.73 to 121.82 cN/dtex, and decreases the breaking elongation from 50.63% to 32.68%, respectively. Moreover, this porous structure would continue to be retained in the PAN precursor fibers through the drawing process, which reduced the tensile strength of the final carbon fibers. Therefore, the grafted MWCNT/PAN nascent composite fibers would achieve high performance carbon fibers compared with the PAN nascent fibers.

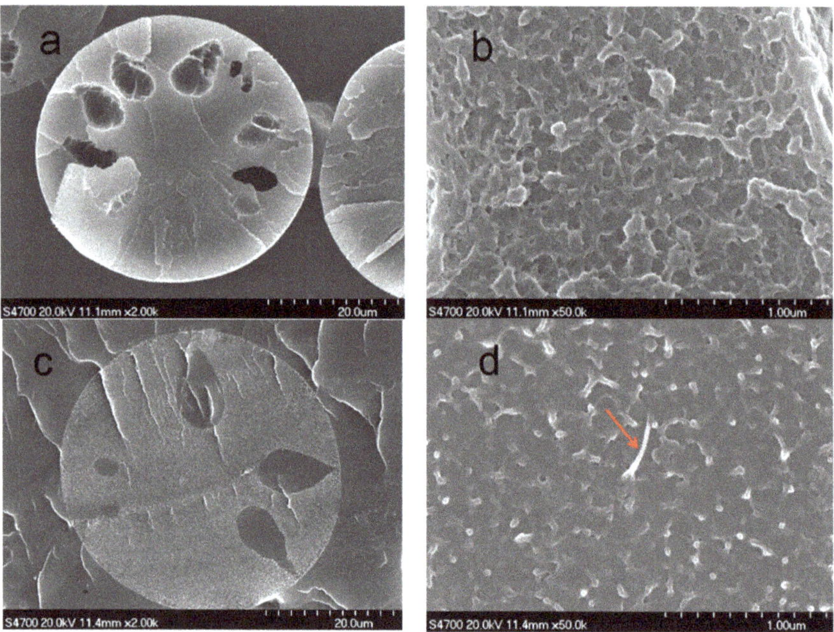

Figure 8. SEM images of cross sectional structure: (**a**,**b**) the PAN nascent fibers and (**c**,**d**) the grafted MWCNT/PAN nascent composite fibers.

Table 3. The mechanical properties of the PAN nascent fibers and the grafted MWCNT/PAN nascent composite fibers.

Samples	Tensile Strength (cN/dtex)	Tensile Modulus (cN/dtex)	Breaking Elongation (%)
PAN	3.83 ± 0.34	101.73 ± 9.32	50.63 ± 4.24
grafted MWCNT/PAN	4.37 ± 0.28	121.82 ± 8.13	32.68 ± 2.33

Figure 9 shows the surface morphology of the PAN nascent fibers and the grafted MWCNT/PAN nascent composite fibers. Compared with the PAN nascent fibers in Figure 9a,d, the grafted MWCNT/PAN nascent composite fibers in Figure 9e,h had more uniform diameter distribution than that of the PAN nascent fibers. Moreover, at the same wet spinning conditions, the average diameter of the grafted MWCNT/PAN nascent composite fibers was about 44 μm, which is larger than that of the PAN nascent fibers with a diameter of about 41 μm. These may be ascribed to the effects of the grafted MWCNT on the rheological properties of spinning solutions [33], and indicated that network structures were formed between the grafted MWCNTs and the PAN polymer during polymerization. These results are beneficial for improving the mechanical properties of the nascent composite fibers, which can be proved in Table 3. The surface morphology of the PAN nascent fibers in Figure 9b,c shows an amount of relatively parallel grooves along the axes of the fibers, which is a typical surface structure of PAN fibers by the wet spinning method. These grooves would last until the final carbon fibers, and increase the surface area and improve the interfacial interaction between the carbon fibers and polymer matrix. However, the grafted MWCNT/PAN nascent composite fibers in Figure 9f,g have a relatively smooth surface, and the grooves of which are not obvious compared to that of the PAN nascent fibers. These results indicate that the grafted MWCNT affected the phase separation and solidification behaviors and changed the parameters of crystal structure.

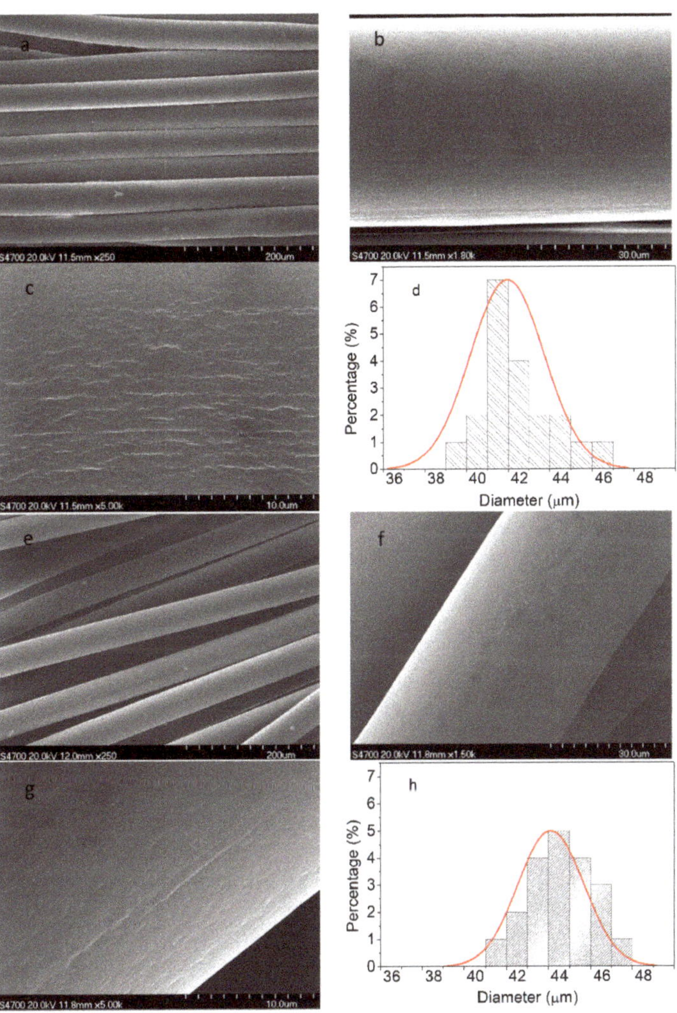

Figure 9. SEM images of surface morphology and corresponding diameter distribution: (**a–d**) the PAN nascent fibers and (**e–h**) the grafted MWCNT/PAN nascent composite fibers.

4. Conclusions

In this paper, amino-MWCNTs were obtained by chemical modification. In aqueous solvent, the PAN polymer synthesized by an AN monomer under initiator, was grafted onto the surface of the amino-MWCNTs by in situ polymerization. The grafted MWCNT/PAN nascent composite fibers with parallel grooves on the surface were prepared by the wet spinning method, which would increase the surface area of fibers and improve the interfacial interaction between the fibers and polymer matrix. The results from the FTIR spectroscopy suggest that a chemical bond was formed between the amino-MWCNTs and the PAN macromolecular chains. The shift of the G-band for amino-MWCNTs coming from Raman spectroscopy indicates that a strong interfacial interaction existed between the amino-MWCNTs and the PAN polymer. TEM images showed that the grafted MWCNTs after in situ polymerization had larger diameters, of about 20–25 nm, compared to the control amino-MWCNTs with a diameter of 10 nm. TG curves showed that the amino-MWCNTs decreased the rate of weight loss at temperatures between 300 and 470 °C, and the grafted content of the PAN polymer on the surface of amino-MWCNTs was 73.2%. The addition of the grafted MWNCTs into the PAN nascent fibers improved the degree of crystallization from 35.23 to 38.73% and the crystal size from 3.01 to 3.42 nm. The DSC data indicated that the grafted MWCNTs in the nascent composite fibers at an air atmosphere decreased the initial exothermal temperature about 5.9 °C, and broadened the range of exothermic temperature from 196.3 to 200.8 °C. The cross-sectional structure coming from SEM images showed that the PAN grafted amino-MWCNTs nascent composite fibers had more compactness structure than the PAN nascent fibers, and the surface morphology showed that the composite fibers with more uniform diameter distribution had a higher diameter than the control PAN nascent fibers. Furthermore, the incorporation of the grafted MWCNTs improved the tensile strength by 14.1% and the tensile modulus by 19.7%. These results suggest that the PAN grafted MWCNT/PAN nascent composite fibers are beneficial for preparing a high performance carbon fiber.

Author Contributions: Conceptualization, H.Z., F.S. and L.X.; Data curation, L.Q., A.G. and Y.T.; Funding acquisition, A.G. and Y.T.; Methodology, H.Z., A.G., F.S. and L.X.; Resources, Y.T. and L.X.; Writing-original draft, H.Z. and L.Q.

Funding: This research was funded by the Natural Science Foundation of Henan Province of China (grant number: 162300410193) and the Joint Fund of the Equipment Pre-research Ministry of Education (grant number: 6141A02033233).

Conflicts of Interest: The authors declare no conflict of interest.

References

1. Ebbesen, T.W.; Lezec, H.J.; Hiura, H.; Bennett, J.W.; Ghaemi, H.F.; Thio, T. Electrical conductivity of individual carbon nanotubes. *Nature* **1996**, *382*, 54–56. [CrossRef]
2. Liu, Y.; Kumar, S. Polymer/carbon nanotube nano composite fibers-a review. *ACS Appl. Mater. Interface* **2014**, *6*, 6069–6087. [CrossRef] [PubMed]
3. Islam, M.F.; Rojas, E.; Bergey, D.M.; Johnson, A.T.; Yodh, A.G. High weight fraction surfactant solubilization of single-wall carbon nanotubes in water. *Nano Lett.* **2003**, *3*, 269–273. [CrossRef]
4. Qin, S.; Qin, D.; Ford, W.T.; Resasco, D.E.; Herrera, J.E. Polymers brushes on single-walled carbon nanotubes by atom transfer radical polymerization of n-butyl methacrylate. *J. Am. Chem. Soc.* **2003**, *126*, 170–176. [CrossRef] [PubMed]
5. Yao, Z.; Braidy, N.; Botton, G.A.; Adronov, A. Polymerization from the surface of single-walled carbon nanotubes-preparation and characterization of nanocomposites. *J. Am. Chem. Soc.* **2003**, *125*, 16015–16024. [CrossRef] [PubMed]
6. Katti, P.; Bose, S.; Kumar, S. Tailored interface resulting in improvement in mechanical properties of epoxy composited containing poly(ether ether ketone) grafted multiwall carbon nanotubes. *Polymer* **2016**, *102*, 43–53. [CrossRef]
7. Konnola, R.; Nair, C.P.R.; Joseph, K. High strength toughened epoxy nanocomposite based on poly(ether sulfone)-grafted multiwalled carbon nanotube. *Polym. Adv. Technol.* **2016**, *27*, 82–89. [CrossRef]

8. Patole, A.S.; Patole, S.P.; Yoo, J.B.; An, J.H.; Kim, T.H. Fabrication of polystyrene/multiwalled carbon nanotube composite films synthesized by in situ microemulsion polymerization. *Polym. Eng. Sci.* **2013**, *53*, 1327–1336. [CrossRef]
9. Mahmood, N.; Islam, M.; Hameed, A.; Saeed, S. Polyamide 6/multiwalled carbon nanotubes nanocomposites with modified morphology and thermal properties. *Polymers* **2013**, *5*, 1380–1391. [CrossRef]
10. Khan, M.U.; Reddy, K.R.; Snguanwongchai, T.; Haque, E.; Gomes, V.G. Polymer brush synthesis on surface modified carbon nanotubes via in situ emulsion polymerization. *Colloid Polym. Sci.* **2016**, *294*, 1599–1610. [CrossRef]
11. Wang, C.; Xu, J.; Yang, J.; Qian, Y.; Liu, H. In-situ polymerization and multifunctional properties of surface-modified multiwalled carbon nanotube-reinforced polyimide nanocomposites. *High Perform. Polym.* **2017**, *29*, 797–807. [CrossRef]
12. Yan, D.; Yang, G. A novel approach of in situ grafting polyamide 6 to the surface of multi-walled carbon nanotubes. *Mater. Lett.* **2009**, *63*, 298–300. [CrossRef]
13. Hwang, G.L.; Shieh, Y.T.; Hwang, K.C. Efficient load transfer to polymer-grafted multiwalled carbon nanotubes in polymer composites. *Adv. Funct. Mater.* **2004**, *14*, 487–491. [CrossRef]
14. Sreekumar, T.V.; Liu, T.; Min, B.G.; Guo, H.; Kumar, S.; Hauge, R.H.; Smalley, R.E. Polyacrylonitrile single-walled carbon nanotube composite fibers. *Adv. Mater.* **2004**, *16*, 58–61. [CrossRef]
15. Majeed, S.; Fierro, D.; Buhr, K.; Wind, J.; Du, B.; Boschetti-de-Fierro, A.; Abetz, V. Multi-walled carbon nanotubes (MWCNTs) mixed polyacrylonitrile (PAN) ultrafiltration membranes. *J. Membr. Sci.* **2012**, *403*, 101–109. [CrossRef]
16. Kim, S.H.; Min, B.G.; Lee, S.C. Properties of polyacrylonitrile/single wall carbon nanotube composite films prepared in nitric acid. *Fibers Polym.* **2005**, *6*, 108–112. [CrossRef]
17. Zhou, H.; Tang, X.; Dong, Y.; Chen, L.; Zhang, L.; Wang, W.; Xiong, X. Multiwalled carbon nanotube/polyacrylonitrile composite fibers prepared by in situ polymerization. *J. Appl. Polym. Sci.* **2011**, *120*, 1385–1389. [CrossRef]
18. Poochai, C.; Pongprayoon, T. Enhancing dispersion of carbon nanotube in polyacrylonitrile matrix using admicellar polymerization. *Colloids Surf. A* **2014**, *456*, 67–74. [CrossRef]
19. Quan, L.; Zhang, H.; Xu, L. Orientation and thermal properties of carbon nanotube/polyacrylonitrile nascent composite fibers. *J. Polym. Res.* **2015**, *22*, 1–6. [CrossRef]
20. Bell, J.P.; Dumbleton, J.H. Changes in the structure of wet-spun acrylic fibers during processing. *Text. Res. J.* **1971**, *41*, 196–203. [CrossRef]
21. Bashir, Z. Thermoreversible gelation and plasticization of polyacrylonitrile. *Polymer* **1992**, *33*, 4304–4313. [CrossRef]
22. Vuković, G.; Marinković, A.; Obradović, M.; Radmilović, V.; Čolić, M.; Aleksić, R.; Uskoković, P.S. Synthesis, characterization and cytotoxicity of surface amino-functionalized water-dispersible multi-walled carbon nanotubes. *Appl. Surf. Sci.* **2009**, *255*, 8067–8075. [CrossRef]
23. Valentini, L.; Macan, J.; Armentano, I.; Mengoni, F.; Kenny, J.M. Modification of fluorinated single-walled carbon nanotubes with aminosilane molecules. *Carbon* **2006**, *44*, 2196–2201. [CrossRef]
24. Zhao, Y.Q.; Wang, C.G.; Bai, Y.J.; Chen, G.W.; Jing, M.; Zhu, B. Property changes of powdery polyacrylonitrile synthesized by aqueous suspension polymerization during heat treatment process under air atmosphere. *J. Colloid Interface Sci.* **2009**, *329*, 48–53. [CrossRef] [PubMed]
25. Bajaj, P.; Sen, K.; Bahrami, S.H. Solution polymerization of acrylonitrile with vinyl acids in dimethylformamide. *J. Appl. Polym. Sci.* **1996**, *59*, 1539–1550. [CrossRef]
26. Ma, H.Y.; Tong, L.F.; Xu, Z.B.; Fang, Z.P. Functionalizing carbon nanotubes by grafting on intumescent flame ratardant: Nanocomposite synthesis, morphology, rheology, and flammability. *Adv. Funct. Mater.* **2008**, *18*, 414–421. [CrossRef]
27. Owens, F.J. Raman and mechanical properties measurements of single walled carbon nanotube composites of polyisobutylene. *J. Mater. Chem.* **2006**, *16*, 505–508. [CrossRef]
28. Liu, M.; Zhu, T.; Li, Z.; Liu, Z. One-step in situ synthesis of poly(methyl mechacrylate)-grafted single-walled carbon nanotube composites. *J. Phys. Chem. C* **2009**, *113*, 9670–9675. [CrossRef]
29. Koganemaru, A.; Bin, Y.; Agari, Y.; Matsuo, M. Composites of polyacrylonitrile and multiwalled carbon nanotubes prepared by gelation/crystallization from solution. *Adv. Funct. Mater.* **2004**, *14*, 842–850. [CrossRef]

30. Dong, X.G.; Wang, C.G.; Chen, J.; Cao, W.W. Crystallinity development in polyacrylonitrile nascent fibers during coagulation. *J. Polym. Res.* **2008**, *15*, 125–130. [CrossRef]
31. Zhang, H.; Xu, L.; Yang, F.; Geng, L. The synthesis of polyacrylonitrile/carbon nanotube microspheres by aqueous deposition polymerization under ultrasonication. *Carbon* **2010**, *48*, 688–695. [CrossRef]
32. Youssefi, M.; Safaie, B. Effect of multi walled carbon nanotube on the crystalline structure of polypropylene fibers. *Fibers Polym.* **2013**, *14*, 1602–1607. [CrossRef]
33. Zhang, H.; Quan, L.; Shi, F.; Li, C.; Liu, H.; Xu, L. Multi-walled carbon nanotube/polyacrylonitrile concentrated solutions and crystal structure of composite fibers. *Polymers* **2018**, *10*, 186. [CrossRef]

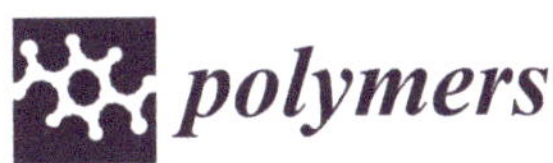

Article

Modulating Carrier Type for Enhanced Thermoelectric Performance of Single-Walled Carbon Nanotubes/Polyethyleneimine Composites

Xiao-Xi Peng, Xuan Qiao, Shuai Luo, Jun-An Yao, Yun-Fei Zhang * and Fei-Peng Du *

School of Materials Science and Engineering, Wuhan Institute of Technology, Wuhan 430205, China
* Correspondence: zyf3006@126.com (Y.-F.Z.); hsdfp@163.com (F.-P.D.)

Received: 1 July 2019; Accepted: 17 July 2019; Published: 2 August 2019

Abstract: Thermoelectric (TE) generators consisting of flexible and lightweight *p*- and *n*-type single-walled carbon nanotube (SWCNT)-based composites have potential applications in powering wearable electronics using the temperature difference between the human body and the environment. Tuning the TE properties of SWCNTs, particularly *p*- versus *n*-type control, is currently of significant interest. Herein, the TE properties of SWCNT-based flexible films consisting of SWCNTs doped with polyethyleneimine (PEI) were evaluated. The carrier type of the SWCNT/PEI composites was modulated by regulating the proportion of SWCNTs and PEI using simple mixing techniques. The as-prepared SWCNT/PEI composite films were switched from *p*- to *n*-type by the addition of a high amount of PEI (>13.0 wt.%). Moreover, interconnected SWCNTs networks were formed due to the excellent SWNT dispersion and film formation. These parameters were improved by the addition of PEI and Nafion, which facilitated effective carrier transport. A TE generator with three thermocouples of *p*- and *n*-type SWCNT/PEI flexible composite films delivered an open circuit voltage of 17 mV and a maximum output power of 224 nW at the temperature gradient of 50 K. These promising results showed that the flexible SWCNT/PEI composites have potential applications in wearable and autonomous devices.

Keywords: single-walled carbon nanotubes; polyethyleneimine; thermoelectric properties; carrier type

---

1. Introduction

Thermoelectric (TE) energy-harvesting generators, composed of multiple *p*- and *n*-type TE materials that are electrically and thermally connected in series and in parallel, respectively, can directly convert heat energy into electric energy and vice versa under a temperature gradient without moving parts. These generators are also quiet, exhibit long lifetimes, and are environmentally friendly which gives them wide applicability in power generation for frontier electronic devices [1–3]. The performance of TE materials is evaluated by a dimensionless figure of merit, ZT, through the equation $ZT = S^2\sigma T/\kappa$ (where S is the Seebeck coefficient, σ is the electrical conductivity, κ is the thermal conductivity, and T is the absolute temperature), wherein $S^2\sigma$ is defined as power factor (PF) [4–6]. The Seebeck coefficient can be positive (holes, *p*-type) or negative (electron, *n*-type) depending on the main charge carrier type [7–9]. Inorganic semiconductors, such as Bi_2Te_3, PbTe, and Sb_2Te_3, are often used as TE materials due to their high ZT values. However, their relatively high cost, high toxicity, and poor processability impede their application in flexible electronics [10,11]. Therefore, it is necessary to develop advanced *p*- and *n*-type TE materials for flexible electronic devices.

Recently, carbon nanotubes (CNTs) and their composites have been intensively studied and utilized as TE materials for energy harvesting due to their mechanical flexibility, light-weight characteristics, and facile processability [11–15]. By improving the dispersion of CNTs in CNT/polymer composites, the TE performance of the composite could be effectively enhanced. For example, through non-covalent modification of the surface of single-walled carbon nanotubes (SWCNTs) with polymer, He et al. prepared a substrate-free and well-dispersed polythiophene/SWCNT composite, which exhibited a ZT of 3.1×10^{-2} with a PF of 28.8 μW/(m·K^2) at room temperature [16]. Kim et al. improved the dispersion state of SWCNTs in composite fibers via simple wet-spinning of SWCNT/poly(vinylidene fluoride) (PVDF) pastes and achieved optimized PFs for the *p*- and *n*-type SWCNT/PVDF composite fibers of about 378 and 289 μW/(m·K^2), respectively [17]. It was noted that the simultaneous increase in electrical conductivity and the absolute value of the Seebeck coefficient is a useful way to enhance the PF, as observed for CNT-based TE materials, such as polyaniline (PANI)/CNTs [18] and paper-based CNT composites [19]. Yao et al. reported that the carrier mobility, which contributes to the increase in electrical conductivity, of PANI/SWCNT composites increased with increasing SWCNT content, which was ascribed to the ordering of PANI chains induced by π–π interactions between SWCNTs and PANI [18,20]. Most of the CNT-based materials reported in the past few decades have been *p*-type, due to oxygen impurities [21]. Recently, unstable *n*-type CNT composites have attracted great attention due to their role in fabricating complete TE generators [22–24]. Doping CNTs with *n*-type polymers or molecules containing electron donor groups, such as poly (vinyl pyrrolidone) [25], poly (vinyl chloride), PVDF [26], polyethyleneimine (PEI) [23,27], $NaBH_4$ [23], and N_2H_4 [28], has been considered a promising strategy for obtaining *n*-type CNT-based TE materials. Yu et al. reported that treatment with the strong reducing agent $NaBH_4$ followed by lamination improved the air stability of *n*-type PEI-doped CNT composites, yielding Seebeck coefficients as large as −80 μV/K [23]. PEI is a relatively simple, effective, and low-cost *n*-type doping agent for CNTs. Also, branched PEI can be physically adsorbed onto the sidewalls of CNTs by facile techniques such as solution mixing [22,25], spin coating [28], or layer-by-layer (LbL) assembly [29]. Therefore, PEI-doped CNT composites have been widely used as *n*-type legs for flexible TE generators [22,23,30]. While most studies only report the instability of the composite in the air and methods to improve the air stability of PEI-doped CNTs [21,23,30], little work has been focused on the effect of PEI on the dispersion of SWCNTs and the comprehensive TE performance of SWCNT/PEI composites. Our recent findings demonstrated that a small amount of PEI improves the dispersion and film-forming properties of SWCNTs, causing a simultaneous increase in the electrical conductivity and Seebeck coefficient of *p*-type SWCNT/PEI composites as compared to pure SWCNTs.

In this work, both *p*- and *n*-type SWCNT/PEI composite TEs were developed by modulating the proportion of SWCNTs to PEI using simple mixing techniques. Through controlling the carrier type of SWCNT/PEI composites, the prepared composites exhibited maximum Seebeck coefficients of 49 and −40 μV/K with electrical conductivities of 210 and 170 S/cm for *p*- and *n*-type legs, respectively. Furthermore, a flexible TE generator comprising *p*- and *n*-type SWCNT/PEI composites has been assembled to demonstrate the TE energy conversion ability of these composites.

2. Experimental

2.1. Materials

SWCNTs were purchased from the Shenzhen Nanotech Port Corporation (Shenzhen, China) at >83% purity, > 15 μm in length, and 1–3 nm in diameter. Branched PEI (molecular weight: 1800, 99%) was purchased from Aladdin Industrial Corporation (Shanghai, China). The Nafion solution (5.2 wt.%) was bought from DuPont Corporation (Wilmington, DE, USA). All other analytical reagents were purchased from the Sinopharm Chemical Reagent Corporation (Shanghai, China). All of the chemicals were used as received without further purification.

2.2. Preparation of Flexible p- and n-Type Single-Walled Carbon Nanotube/Polyethyleneimine (SWCNT/PEI) Films

In order to obtain a SWCNT/PEI composite film, 12.0 mg SWCNTs and certain amount of PEI were dispersed in 6.0 mL ethanol to obtain a suspension. The PEI content was varied from 0.0 to 20.0 wt.% to adjust the TE performance. Accordingly, the concentration of PEI was selected as 0.0 wt.%, 2.0 wt.%, 7.0 wt.%, 13.0 wt.%, 15.0 wt.% and 20.0 wt.%, named as SWCNT/PEI-0, SWCNT/PEI-2, SWCNT/PEI-7, SWCNT/PEI-13, SWCNT/PEI-15 and SWCNT/PEI-20, respectively. Then, 0.05 mL of a 1.0 wt.% Nafion solution was added to improve the dispersion of the mixture. The mixture was then stirred at room temperature for 12 h to form a homogeneous, well-distributed solution. Finally, the SWCNT/PEI composite film was prepared via vacuum filtration of the SWCNT-PEI dispersion onto a PVDF membrane (0.22 μm) and subsequent drying at 55 °C overnight in a vacuum oven (Shanghai Soxpec Instrument,, Shanghai, China). The thickness of the obtained SWCNTs/PEI film was approximately 30 μm, which was measured with a thickness gauge (Resolution: 1 μm).

2.3. Thermoelectric Device Fabrication

The as-prepared *p*- and *n*-type SWCNT/PEI films were cut into rectangular shapes (5 × 20 mm) and pasted onto a polypropylene (PP) substrate. Then, *p*- and *n*-type SWCNT/PEI films were alternately linked with conductive silver paste to fabricate a flexible TE generator consisting of 3 *p*-*n* junctions. Finally, the TE generator was sealed with a transparent PP film. A temperature gradient was used to measure the output voltage and current of the TE generator. These parameters were also measured under different load resistances.

2.4. Characterization

Raman spectra were recorded using a Raman spectrometer (DXR, USA) at a wavelength of 532 nm. The surface microstructures of the SWCNT/PEI composite films were characterized using scanning electron microscopy (SEM, Phillips XL30). Transmission electron microscopy (TEM) images were recorded on a JEOL 2010 electron microscope at an accelerating voltage of 200 kV. X-ray diffraction (XRD) patterns (from 10° to 90° of 2θ, Cu-K_α radiation, λ = 1.54 Å) were obtained using a New D8-Advance/Bruker-AXS (Karlsruhe, Germany) powder X-ray diffractometer operating at 40 kV and 30 mA, with a scanning rate of 6°/min and a scanning step of 0.02. The electrical conductivity and Seebeck coefficients were measured using a TE parameter test system (Namicro-III, Wuhan Schwab Instruments, Wuhan, China). When measuring the Seebeck coefficient, all of the samples had similar dimensions of 14.0 mm × 14.0 mm × 30 μm. The temperature gradient was set from 0.3 to 3.5 °C along the longitudinal direction of the samples. The PF was obtained via the formula PF = $S^2\sigma$. The Hall effect measurement was recorded under a magnetic field of 0.68 T using a Hall effect test system (HET-RT, Wuhan Schwab Instruments, Wuhan, China). The voltage and power generated by the TE generator were measured with a home-built system in an ambient environment and calculated by the temperature difference (ΔT, provided by Peltier devices and determined by T-type thermocouples) and the change in thermal electrical voltage (ΔV, measured with Keithley 2000 multimeter). Then, the output power of the TE generator was obtained as a function of the load resistance (R) using the equation $P = V^2/R$ [31].

3. Results and Discussion

Figure 1a indicates that the dispersion ability of SWCNTs was improved by the addition of PEI and Nafion. Also, all of the prepared films were smooth and compacted and showed great flexibility (Figure 1b), wherein they could endure bending without destruction, which is beneficial for the assembly of these films into TE devices.

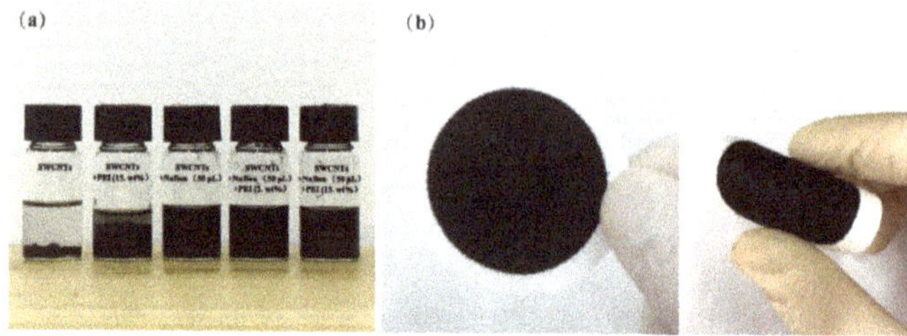

Figure 1. Photograph of SWCNT dispersions containing PEI or Nafion (**a**) and the SWCNT/PEI composite containing 2.0 wt.% PEI (**b**).

Figure 2a–c displays typical SEM images of the SWCNT/PEI composites containing 0, 2, and 15 wt.% PEI, respectively. In these images, it is clearly seen that numerous interconnected tubes form a porous network due to the excellent flexibility of the individual SWCNTs. The TEM images of the composites (Figure 2e–f) show that the SWCNTs' surface was mainly coated with PEI while some parts were naked (indicated by arrows), implying strong binding between the SWCNTs and PEI. The black patches visible in the TEM images might be ascribed to Nafion adsorption on the surface of the tubes.

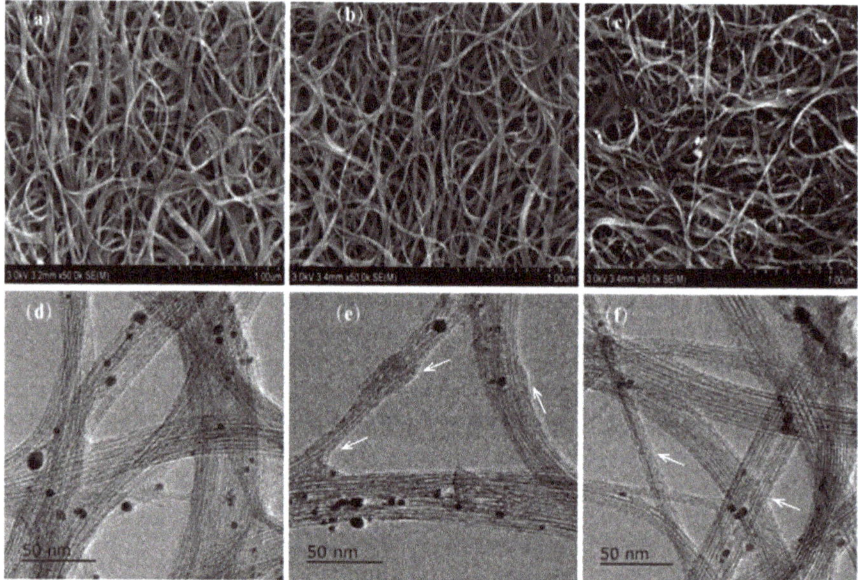

Figure 2. Transmission electron microscope (TEM) and scanning electron microscope (SEM) images of single-walled carbon nanotubes (SWCNTs) doped with different amount of polyethyleneimine (PEI). (**a**,**d**) 0 wt.%, (**b**,**e**) 2 wt.%, (**c**,**f**) 15 wt.%.

Raman spectrometry was used to examine the defects and disordered structures of the carbon-based materials. Raman spectra of the pristine SWCNTs and the PEI-doped SWCNTs are shown in Figure 3. The G band (in-plane stretching, E_{2g} mode) can be attributed to the vibration of sp^2 bonded carbon atoms while the D band corresponds to the vibration of carbon atoms with dangling bonds, which reflects the defects and the disordered structures of the SWCNTs [23,31–33]. The D bands of both the pristine and PEI-doped SWCNTs were found at approximately 1337 and 2670 cm^{-1} (D_1 and D_2, respectively), while the G bands were located at about 1587 cm^{-1}. Usually, the intensity ratio of D:G (I_D:I_G) qualitatively indicates the amount of PEI that is attached to the carbon nanotubes. In this work, as the PEI content increased from 0 to 20 wt.%, the intensity ratio was similar and around 0.015. These values are quite small compared with the reported results [23,34]. The SWCNTs employed here were not functionalized for the purpose of high electrical conductivity; therefore, we believe that the interaction between PEI and the SWCNTs was non-covalent.

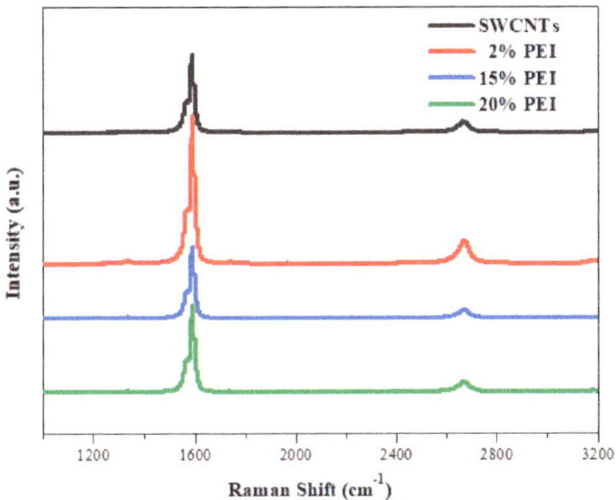

Figure 3. Raman spectra of pristine SWCNTs and PEI-doped SWCNTs.

XRD measurements were employed to identify the crystalline phase and structure of pristine SWCNTs and SWCNT/PEI composites (Figure 4). The XRD patterns of the pure SWCNTs exhibit a typical diffraction peak at $2\theta = 26.5°$, which corresponds to graphite reflection [18,20]. The diffraction peak at $2\theta = 26.4°$ of the SWCNTs was observed in all four samples, indicating that their crystalline structure remains intact and was not destroyed during the composite preparation process. A SWCNTs bundle diffraction peak below 16° was not found, which might be ascribed to low quality of SWCNTs. The peaks located at 18.1, 21.6, 23.9, and 36.8° (2θ) were assigned to Nafion [35,36] and these peaks were observed in all four samples. Compared with pristine Nafion, the crystallinity of Nafion mixed with SWCNTs was increased, suggesting that strong interfacial interactions between SWCNTs and Nafion induced the crystallization of Nafion [29]. The Nafion present in the composites might act as a barrier to permeating oxygen, imparting long-term air stability to the composites.

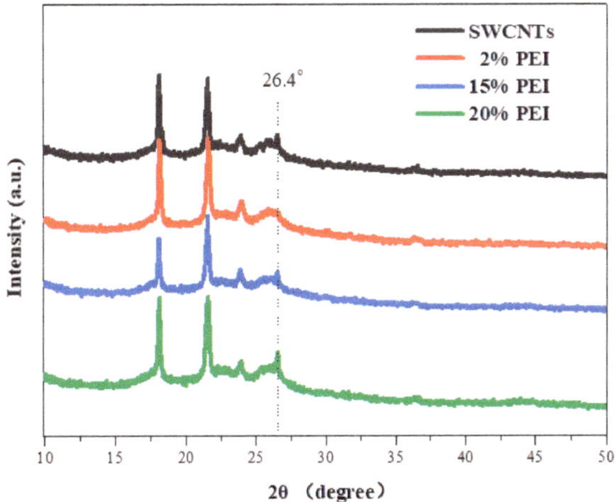

Figure 4. X-ray diffraction (XRD) patterns of pristine SWCNTs and SWCNT/PEI composites.

Figure 5a shows the influence of PEI content on the electrical conductivity, Seebeck coefficient, and PF of SWCNT/PEI composite films. With an increase in PEI content, the electrical conductivity of the composite first shows an increasing trend, reaching maximum conductivity at 13 wt.% PEI content, followed by a rapid decrease. The increase in electrical conductivity might be attributed to the formation of an interconnected SWCNT network, due to improved dispersion and film formation prompted by the addition of PEI and Nafion, which facilitates effective carrier transport. Since PEI is an electrical insulator, when a large amount of PEI is coated on the surface of the SWCNTs, it may prevent carrier transport across the junctions between the SWCNTs, resulting in a reduction in electrical conductivity [37]. Amine-rich PEI was adsorbed onto the surface of the SWCNTs where it acts as an electron donor, which changed the carrier type of the SWCNT/PEI composites from *p*- to *n*-type when high amounts of PEI were used (Figure 5b) [22,38]. The composite films exhibited *p*-type TE behavior when the PEI content was below 13 wt.%, and *n*-type behavior when the PEI content was above 13 wt.%, as shown in Figure 5a, where the maximum Seebeck coefficients of the *p*- and *n*-type composites are 43 μV/K (SWCNT/PEI-2) and −40 μV/K (SWCNT/PEI-15), respectively. Consequently, the maximum PFs of the *p*- and *n*-type composites reached 40.5 and 25.5 μW/(m·K^2), respectively.

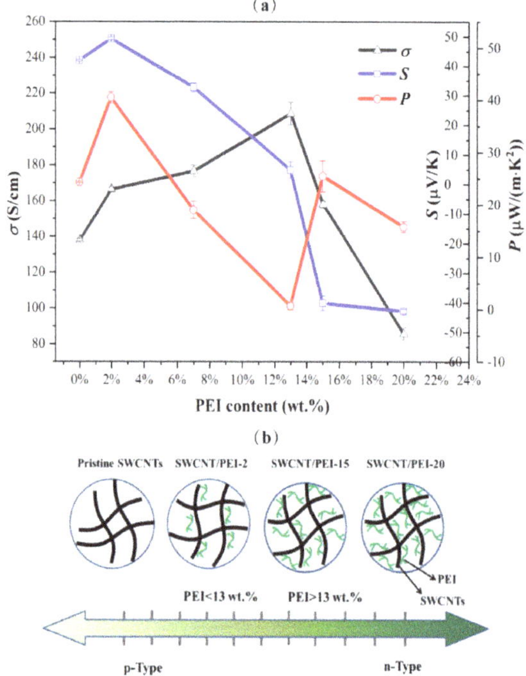

Figure 5. (**a**) The influence of PEI content on the thermoelectric properties of SWCNT/PEI composites. (**b**) Schematic representation of the carrier type of SWCNT/PEI composites tuned by PEI content.

Air stability is a critical issue for *n*-type PEI-doped CNT TE materials. In previous works, negative Seebeck coefficient values of PEI-doped CNTs became positive after 10 h [21] or 9 days [22]. Song et al. [22] and Yu et al. [23] reported that an oxygen adsorption and oxidation reaction may be taking place in the *n*-type PEI-doped CNTs, which changes the sample from n to *p*-type. In our work, the prepared *n*-type SWCNTs/PEI sample is also suffering the problem of being oxidized to *p*-type. For example, the Seebeck coefficient of SWCNT/PEI-15 as a function of exposure time in an ambient environment has been examined and the results are shown in Figure 6. It can be observed that the

Seebeck coefficient value of SWCNT/PEI-15 decreases dramatically after it is exposed to air for the given period of time. The Seebeck coefficient value began to change from negative to positive (1.3 μV/K) after 12 days of exposure and became stable (21.9 μV/K) after 16 days of exposure. Therefore, for the practical application of PEI-doped CNTs, strategies such as providing a protective coating, sealing or lamination should be considered to preserve the *n*-type characteristics of the composites [22,23].

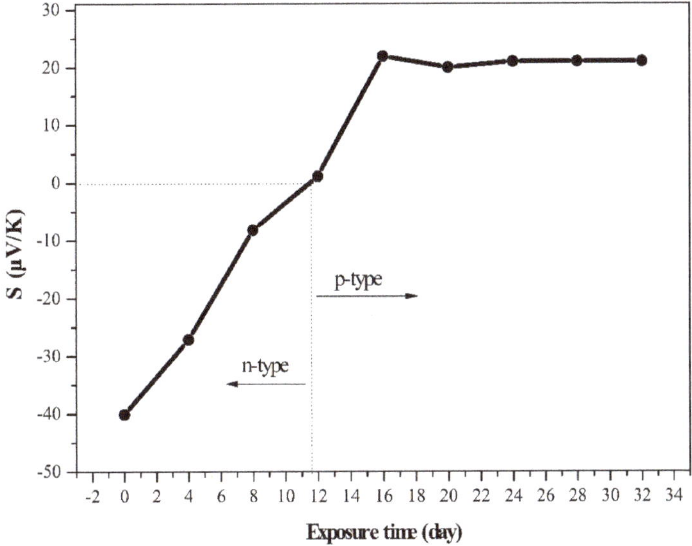

Figure 6. Seebeck coefficient of the SWCNT/PEI-15 composite as a function of exposure time in an ambient environment.

The Hall effect measurement was used to study the carrier concentration and carrier mobility of SWCNT/PEI samples (see Table 1). Compared with pristine SWCNTs, the carrier concentration of SWCNT/PEI composites increased with the addition of PEI. This is most likely due to the action of the amine-rich PEI, which acts as an electron donor, wherein a large amount of lone pair electrons might be donated to the PEI-doped SWCNTs. However, the carrier mobility of SWCNT/PEI composites decreased with the addition of PEI compared with pristine SWCNTs. This could be due to the high density of amine groups with lone pairs of electrons acting as scattering points for the carrier [25,38,39]. Therefore, the electrical conductivity of heavily doped SWCNT/PEI composites was reduced because of the low carrier mobility.

Table 1. Results of Hall effect test of SWCNT/PEI films at room temperature.

Sample	Carrier Type	Carrier Concentration (n/(cm^3))	Carrier Mobility (μ/(cm^2/V·s))
SWCNT/PEI-0	*p*	4.6×10^{20}	5.314
SWCNT/PEI-2	*p*	3.7×10^{21}	0.252
SWCNT/PEI-15	*n*	9.3×10^{20}	0.814
SWCNT/PEI-20	*n*	2.6×10^{22}	0.057

A flexible TE generator was assembled by electrically connecting the *p*-type SWCNT/PEI-2 film with the *n*-type SWCNT/PEI-15 film using silver paste. Figure 7a,b depict the schematic representation and demonstration of the prepared flexible TE generator consisting of three *p*-*n* junctions pasted on a PP substrate, respectively. The temperature gradient (50 K) provided an output voltage of ~6 mV for one *p*-*n* junction couple in our experiment. Figure 7c shows the relationship between the total output voltage and current versus temperature for the TE generator with three *p-n* junctions. As the temperature gradient increased, both output voltage and current increased linearly, which is ascribed to the Seebeck effect. When temperature gradients of 26 and 50 K were separately applied to the prepared TE generator, the voltage was approximately 8.8 and 17 mV, respectively. Power measurements were conducted for several different load resistances and the results are shown in Figure 7d. It was seen that the power reached maximum output (224 nW) when the load resistance was 330 Ω, which is close to the resistance of the TE generator. Despite the fact that the prepared TE generator was composed of only three *p-n* junctions, the power output is higher than that of generators composed of six *p-n* junctions or *p*- and *n*-type SWCNTs [22,38], which can be attributed to the high PF of both the *p*- and *n*-type SWCNT/PEI composites that were synthesized and investigated in this work.

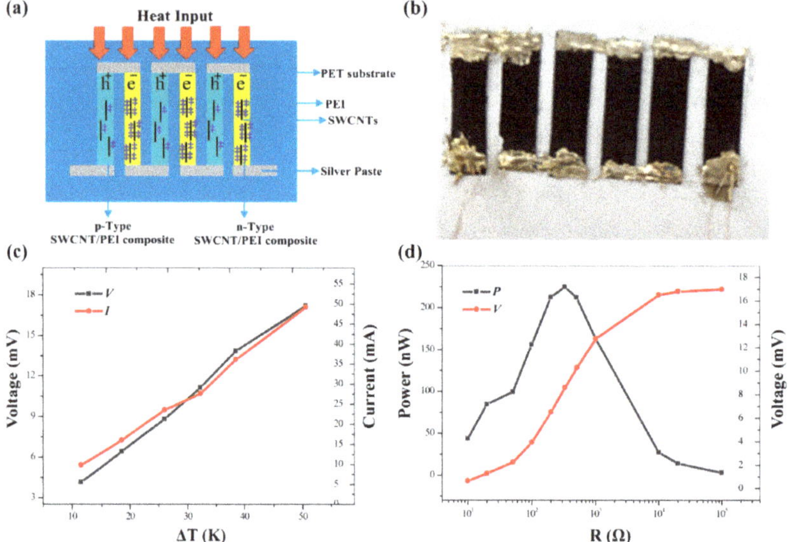

Figure 7. (**a**) Scheme and (**b**) demonstration of the TE generator comprised of three thermocouples. (**c**) Output voltage generated by a three *p-n* junction couples as a function of temperature gradient. (**d**) Output voltage and power as functions of different load resistances for the prepared device with a temperature gradient of 50 K.

The output voltage (8.6 mV) is lower than that of the open circuit voltage (17 mV) of the generator at the load resistance of 330 Ω and the temperature gradient of 50 K. This is mainly due to the relatively high internal resistance of the TE generator [22]. Although the output voltage here is low, it can be scaled up by increasing the number of *p-n* junctions, making the fabrication of flexible TE generators with much higher power outputs feasible.

Table 2 shows a comparison of the thermoelectric performance obtained from this work as compared with the results of some typical flexible thermoelectric devices. It is found that the output voltage and power of SWCNTs/PEI devices with small dimensions prepared in this work are superior to most of the reported low toxic flexible composite devices, which exhibit high potential for producing flexible thermoelectric devices.

Table 2. A summary of some high-performance flexible thermoelectric devices.

TE Units	Device Dimensions	ΔT	Output Voltage	Output Power	Comment	Ref.
p-type: poly(3,4-ethylenedioxythiophene): polystyrenesulfonate (PEDOT:PSS)	14 thermocouples	50	12 mV	16.8 μW	Parallel connected single-leg structure; screening printing; paper substrate.	[40]
p-type: PEDOT:PSS *n*-type: N-doped graphene	8 cm × 10 cm	10 K	3 mV	≈0.24 $mW\cdot m^{-2}$	Roll-to roll printing.	[41]
p-type: carbon nanotubes (CNTs)/poly(3-hexyl thiophene)	41 *p*-type stripes (0.1 cm × 1.5 cm)	10 K	≈32 mV	32.7 nW	Spray-printing; polyimide substrate.	[42]
p-type: CNTs/PEDOT:PSS *n*-type: CNTs/PEDOT:PSS treated by hydrazine	14 thermocouples	10 K	8 mV	0.43 μW	Wet-spinning process.	[43]
p-type: PEDOT-coated textiles *n*-type: *n*-type CNTs	5 thermocouples	100 K	≈22 mV	62 nW	Vapor phase polymerization; wearable thermoelectric strain sensor.	[44]
p-type: $Bi_{0.5}Sb_{1.5}Te_3$ pastes infiltrated paper; *n*-type: $Bi_2Se_{0.3}Te_{2.7}$ pastes infiltrated paper;	10 thermocouples	35 K	8.3 mV	10 nW	Impregnating method; transparent paper-based thermoelectric generator.	[45]
n-type: Bi_2Te_3/polyvinyl alcohol hybrid composites	10 stripes (0.2 cm × 2.5 cm)	46 K	24 mV	≈9 $\mu W\cdot cm^{-2}$	Solid-state reaction and ball milling method.	[46]
p-type: polypyrrole film *n*-type: copper tape	7 thermocouples	80 K	336 μV	6.4 pW	Flexible scaleable free-standing polypyrrole films; interfacial chemical polymerization.	[47]
p-type: PEDOT film	6 stripes (0.7 cm × 3 cm)	51.6 K	≈5.6 mV	157.2 nW	Simple self-assembled micellar soft-template method; vacuum-assisted filtration; ultrafine PEDOT nanowires.	[48]
p-type: poly(acrylic acid)/CNTs *n*-type: PEI/CNTs	30 thermocouples	10 K	21 mV	131 nW	A flexible thermoelectric generator prepared by printing the *p*-type and *n*-type CNTs ink on a curved surface.	[49]
p-type: PEI/SWCNTs *n*-type: PEI/SWCNTs	6 stripes (0.5 cm × 2 cm)	50 K	8.6 mV	224 nW	Flexible scaleable PEI/SWCNTs films; simple mixing techniques.	Our work

In general, regulating the proportion of SWCNTs and PEI can modulate the carrier type and balance the thermal conductivity and electrical conductivity of the SWCNTs/PEI composites. The SWCNTs/ PEI composites and the preparation method demonstrated in this study are environmentally friendly, cost effective, and easy for large scale production. The prepared SWCNTs/PEI composites exhibited high Seebeck coefficient and relative high power factor, demonstrating a great potential in the preparation of lightweight and cheap thermoelectric devices, such as wearable thermoelectric devices.

4. Conclusions

In summary, our work demonstrates for the first time the feasibility of fabricating both *p*- and *n*-type SWCNT/PEI composites with enhanced thermoelectrical properties by modulating carrier type via simple mixing techniques. The addition of PEI and Nafion increased the dispersion of the SWCNTs and improved the electrically percolated SWCNT network as a carrier transport pathway. The addition of 13 wt.% PEI modulates the carrier type from *p*- to *n*-type. The Seebeck coefficient of the SWCNT/PEI composite was changed from 49 to −40 μV/K under the experimental range of PEI content. High PFs were obtained with values of 40.5 and 25.5 μW/(m·K^2) for *p*- and *n*-type SWCNT/PEI composites, respectively. A TE generator consisting of three thermocouples of *p*- and *n*-type SWCNT/PEI composite films delivered an open circuit voltage of 17 mV and a maximum output power of 224 nW at the temperature gradient of 50 K.

Author Contributions: F.-P.D. and Y.-F.Z. conceived and designed the experiments; X.-X.P. and X.Q. performed the experiments and analyzed the data; S.L. and J.-A.Y. contributed reagents/materials/analysis tools; X.-X.P., Y.-F.Z. and F.-P.D. wrote the paper.

Funding: This work was funded by the Natural National Science Foundation of China (51803157), the Scientific Research Fund Project of Wuhan Institute of Technology (K201807) and the Youth Fund Research Project of Wuhan Institute of Technology (Q201701).

Conflicts of Interest: The authors declare no conflict of interest.

References

1. Meng, Q.; Cai, K.; Du, Y.; Chen, L. Preparation and thermoelectric properties of SWCNT/PEDOT: PSS coated tellurium nanorod composite films. *J. Alloys Compd.* **2019**, *778*, 163–169. [CrossRef]
2. Zhang, Y.H.; Park, S.J. Flexible organic thermoelectric materials and devices for wearable green energy harvesting. *Polymers* **2019**, *11*. [CrossRef] [PubMed]
3. Du, F.P.; Cao, N.N.; Zhang, Y.F.; Fu, P.; Wu, Y.G.; Lin, Z.D.; Shi, R.; Amini, A.; Cheng, C. PEDOT:PSS/graphene quantum dots films with enhanced thermoelectric properties via strong interfacial interaction and phase separation. *Sci. Rep.* **2018**, *8*, 6441. [CrossRef] [PubMed]
4. Fan, Z.; Du, D.; Guan, X.; Ouyang, J. Polymer films with ultrahigh thermoelectric properties arising from significant seebeck coefficient enhancement by ion accumulation on surface. *Nano Energy* **2018**, *51*, 481–488. [CrossRef]
5. Zhu, Z.; Liu, C.; Jiang, F.; Xu, J.; Liu, E. Effective treatment methods on PEDOT: PSS to enhance its thermoelectric performance. *Synth. Met.* **2017**, *225*, 31–40. [CrossRef]
6. Ding, Y.; Qiu, Y.; Cai, K.; Yao, Q.; Chen, S.; Chen, L.; He, J. High performance n-type ag2se film on nylon membrane for flexible thermoelectric power generator. *Nat. Commun.* **2019**, *10*, 841. [CrossRef]
7. Luo, J.; Cerretti, G.; Krause, B.; Zhang, L.; Otto, T.; Jenschke, W.; Ullrich, M.; Tremel, W.; Voit, B.; Poetschke, P. Polypropylene-based melt mixed composites with singlewalled carbon nanotubes for thermoelectric applications: Switching from p-type to n-type by the addition of polyethylene glycol. *Polymer* **2017**, *108*, 513–520. [CrossRef]
8. Dörling, B.; Ryan, J.; Craddock, J. Photoinduced p- to n- type switching in thermoelectric polymer-carbon nanotube composites. *Adv. Mater.* **2016**, *28*, 2782–2789. [CrossRef]
9. Cheng, X.L.; Zhang, Y.F.; Wu, Y.G.; Fu, P.; Lin, Z.D.; Du, F.P.; Cheng, C. Thermally sensitive n-type thermoelectric aniline oligomer-block-polyethylene glycol-block-aniline oligomer aba triblock copolymers. *Macromol. Chem. Phys.* **2018**, *219*. [CrossRef]

10. Yao, C.J.; Zhang, H.L.; Zhang, Q.C. Recent progress in thermoelectric materials based on conjugated polymers. *Polymers* **2019**, *11*. [CrossRef]
11. Xue, Y.F.; Gao, C.M.; Liang, L.R.; Wang, X.; Chen, G.M. Nanostructure controlled construction of high-performance thermoelectric materials of polymers and their composites. *J. Mater. Chem. A* **2018**, *6*, 22381–22390. [CrossRef]
12. Du, F.P.; Qiao, X.; Wu, Y.G.; Fu, P.; Liu, S.P.; Zhang, Y.F.; Wang, Q.Y. Fabrication of porous polyvinylidene fluoride/multi-walled carbon nanotube nanocomposites and their enhanced thermoelectric performance. *Polymers* **2018**, *10*. [CrossRef]
13. Zhang, Y.H.; Heo, Y.J.; Park, M.; Park, S.J. Recent advances in organic thermoelectric materials: Principle mechanisms and emerging carbon-based green energy materials. *Polymers* **2019**, *11*. [CrossRef]
14. Pan, C.J.; Wang, L.H.; Zhou, W.Q.; Cai, L.R.; Xie, D.X.; Chen, Z.M.; Wang, L. Preparation and thermoelectric properties study of bipyridine-containing polyfluorene derivative/swcnt composites. *Polymers* **2019**, *11*. [CrossRef]
15. Aghelinejad, M.; Zhang, Y.C.; Leung, S.N. Processing parameters to enhance the electrical conductivity and thermoelectric power factor of polypyrrole/multi-walled carbon nanotubes nanocomposites. *Synth. Met.* **2019**, *247*, 59–66. [CrossRef]
16. He, P.; Shimano, S.; Salikolimi, K.; Isoshima, T.; Kakefuda, Y.; Mori, T.; Taguchi, Y.; Ito, Y.; Kawamoto, M. Noncovalent modification of single-walled carbon nanotubes using thermally cleavable polythiophenes for solution-processed thermoelectric films. *ACS Appl. Mater. Inter.* **2019**, *11*, 4211–4218. [CrossRef]
17. Kim, J.Y.; Mo, J.H.; Kang, Y.H.; Cho, S.Y.; Jang, K.S. Thermoelectric fibers from well-dispersed carbon nanotube/poly (vinyliedene fluoride) pastes for fiber-based thermoelectric generators. *Nanoscale* **2018**, *10*, 19766–19773. [CrossRef]
18. Yao, Q.; Wang, Q.; Wang, L.; Chen, L. Abnormally enhanced thermoelectric transport properties of SWNT/PANI hybrid films by the strengthened PANI molecular ordering. *Energy Environ. Sci.* **2014**, *7*, 3801–3807. [CrossRef]
19. Okhay, O.; Gonçalves, G.; Dias, C.; Ventura, J.; Vieira, E.M.F.; Gonçalves, L.M.V.; Tkach, A. Tuning electrical and thermoelectric properties of freestanding graphene oxide papers by carbon nanotubes and heat treatment. *J. Alloys Compd.* **2019**, *781*, 196–200. [CrossRef]
20. Yao, Q.; Chen, L.; Zhang, W.; Liufu, S.; Chen, X. Enhanced thermoelectric performance of single-walled carbon nanotubes/polyaniline hybrid nanocomposites. *ACS Nano* **2010**, *4*, 2445–2451. [CrossRef]
21. Nonoguchi, Y.; Nakano, M.; Murayama, T.; Hagino, H.; Hama, S.; Miyazaki, K.; Matsubara, R.; Nakamura, M.; Kawai, T. Simple salt-coordinated n-type nanocarbon materials stable in air. *Adv. Funct. Mater.* **2016**, *26*, 3021–3028. [CrossRef]
22. Song, H.; Qiu, Y.; Wang, Y.; Cai, K.; Li, D.; Deng, Y.; He, J. Polymer/carbon nanotube composite materials for flexible thermoelectric power generator. *Compos. Sci. Technol.* **2017**, *153*, 71–83. [CrossRef]
23. Yu, C.; Murali, A.; Choi, K.; Ryu, Y. Air-stable fabric thermoelectric modules made of n- and p-type carbon nanotubes. *Energy Environ. Sci.* **2012**, *5*, 9481–9486. [CrossRef]
24. Wang, H.; Hsu, J.H.; Yi, S.I.; Kim, S.L.; Choi, K.; Yang, G.; Yu, C. Thermally driven large n-type voltage responses from hybrids of carbon nanotubes and poly(3,4-ethylenedioxythiophene) with tetrakis(dimethylamino) ethylene. *Adv. Mater.* **2015**, *27*, 6855–6861. [CrossRef]
25. Sarabia-Riquelme, R.; Craddock, J.; Morris, E.A.; Eaton, D.; Andrews, R.; Anthony, J.; Weisenberger, M.C. Simple, low-cost, water-processable n-type thermoelectric composite films from multiwall carbon nanotubes in polyvinylpyrrolidone. *Synth. Met.* **2017**, *225*, 86–92. [CrossRef]
26. Oshima, K.; Yanagawa, Y.; Asano, H.; Shiraishi, Y.; Toshima, N. Improvement of stability of n-type super growth cnts by hybridization with polymer for organic hybrid thermoelectrics. *Synth. Met.* **2017**, *225*, 81–85. [CrossRef]
27. Liu, J.; Jia, Y.; Jiang, Q.; Jiang, F.; Li, C.; Wang, X.; Liu, P.; Liu, P.; Hu, F.; Du, Y.; et al. Highly conductive hydrogel polymer fibers toward promising wearable thermoelectric energy harvesting. *ACS Appl. Mater. Inter.* **2018**, *10*, 44033–44040. [CrossRef]

28. Wang, P.C.; Liao, Y.C.; Lai, Y.L.; Lin, Y.C.; Su, C.Y.; Tsai, C.H.; Hsu, Y.J. Conversion of pristine and p-doped sulfuric-acid-treated single-walled carbon nanotubes to n-type materials by a facile hydrazine vapor exposure process. *Mater. Chem. Phys.* **2012**, *134*, 325–332. [CrossRef]
29. Cho, C.; Culebras, M.; Wallace, K.L.; Song, Y.X.; Holder, K.; Hsu, J.H.; Yu, C.; Grunlan, J.C. Stable n-type thermoelectric multilayer thin films with high power factor from carbonaceous nanofillers. *Nano Energy* **2016**, *28*, 426–432. [CrossRef]
30. Wang, X.; Liu, P.; Jiang, Q.; Zhou, W.; Xu, J.; Liu, J.; Jia, Y.; Duan, X.; Liu, Y.; Du, Y.; et al. Efficient dmso-vapor annealing for enhancing thermoelectric performance of PEDOT:PSS-based aerogel. *ACS Appl. Mater. Inter.* **2019**, *11*, 2408–2417. [CrossRef]
31. Li, H.; Liang, Y.; Liu, S.; Qiao, F.; Li, P.; He, C. Modulating carrier transport for the enhanced thermoelectric performance of carbon nanotubes/polyaniline composites. *Org. Electron.* **2019**, *69*, 62–68. [CrossRef]
32. Gebhardt, B.; Hof, F.; Backes, C.; Muller, M.; Plocke, T.; Maultzsch, J.; Thomsen, C.; Hauke, F.; Hirsch, A. Selective polycarboxylation of semiconducting single-walled carbon nanotubes by reductive sidewall functionalization. *J. Am. Chem. Soc.* **2011**, *133*, 19459–19473. [CrossRef]
33. Ryu, Y.; Yu, C.H. The influence of incorporating organic molecules or inorganic nanoparticles on the optical and electrical properties of carbon nanotube films. *Sol. State Commun.* **2011**, *151*, 1932–1935. [CrossRef]
34. Dillon, E.P.; Crouse, C.A.; Barron, A.R. Synthesis, characterization, and carbon dioxide adsorption of covalently attached polyethyleneimine-functionalized single-wall carbon nanotubes. *ACS Nano* **2008**, *2*, 156–164. [CrossRef]
35. Ludvigsson, M.; Tegenfeldt, J. Crystallinity in cast nafion. *J. Electrochem. Soc.* **2000**, *147*, 1303–1305. [CrossRef]
36. Devrim, Y. Preparation and testing of nafion/titanium dioxide nanocomposite membrane electrode assembly by ultrasonic coating technique. *J. Appl. Polym. Sci.* **2014**, *131*. [CrossRef]
37. Jia, Y.; Liu, C.; Liu, J.; Liu, C.; Xu, J.; Li, X.; Shen, L.; Jiang, Q.; Wang, X.; Yang, J.; et al. Efficient enhancement of the thermoelectric performance of vapor phase polymerized poly(3,4-ethylenedioxythiophene) films with poly(ethyleneimine). *J. Polym. Sci. Pol. Phys.* **2019**, *57*, 257–265. [CrossRef]
38. Piao, M.; Na, J.; Choi, J.; Kim, J.; Kennedy, G.P.; Kim, G.; Roth, S.; Dettlaff-Weglikowska, U. Increasing the thermoelectric power generated by composite films using chemically functionalized single-walled carbon nanotubes. *Carbon* **2013**, *62*, 430–437. [CrossRef]
39. Minnich, A.J.; Dresselhaus, M.S.; Ren, Z.F.; Chen, G. Bulk nanostructured thermoelectric materials: Current research and future prospects. *Energ. Environ. Sci.* **2009**, *2*, 466–479. [CrossRef]
40. Liu, S.Y.; Deng, H.; Zhao, Y.; Ren, S.J.; Fu, Q. The optimization of thermoelectric properties in a PEDOT: PSS thin film through post-treatment. *RSC Adv.* **2015**, *5*, 1910–1917. [CrossRef]
41. Zhang, Z.; Qiu, J.; Wang, S. Roll-to-roll printing of flexible thin-film organic thermoelectric devices. *Manuf. Lett.* **2016**, *8*, 6–10. [CrossRef]
42. Hong, C.T.; Kang, Y.H.; Ryu, J.; Cho, S.Y.; Jang, K.S. Spray-printed CNT/P3HT organic thermoelectric films and power generators. *J. Mater. Chem. A* **2015**, *3*, 21428–21433. [CrossRef]
43. Kim, J.Y.; Lee, W.; Kang, Y.H.; Cho, S.Y.; Jang, K.S. Wet-spinning and post-treatment of CNT/PEDOT: PSS composites for use in organic fiber-based thermoelectric generators. *Carbon* **2018**, *133*, 293–299. [CrossRef]
44. Jia, Y.H.; Shen, L.L.; Liu, J.; Zhou, W.Q.; Du, Y.K.; Xu, J.K.; Liu, C.C.; Zhang, G.; Zhang, Z.S.; Jiang, F.X. An efficient PEDOT coated textile for wearable thermoelectric generators and strain sensors. *J. Mater. Chem. C* **2019**, *7*, 3496–3502. [CrossRef]
45. Zhao, X.; Han, W.; Zhao, C.; Wang, S.; Kong, F.; Ji, X.; Li, Z.; Shen, X. Fabrication of transparent paper-based flexible thermoelectric generator for wearable energy harvester using modified distributor printing technology. *ACS Appl. Mater. Inter.* **2019**, *11*, 10301–10309. [CrossRef]
46. Pires, A.L.; Cruz, I.F.; Silva, J.; Oliveira, G.N.P.; Ferreira-Teixeira, S.; Lopes, A.M.L.; Araujo, J.P.; Fonseca, J.; Pereira, C.; Pereira, A.M. Printed flexible mu-thermoelectric device based on hybrid Bi_2Te_3/PVA composites. *ACS Appl. Mater. Inter.* **2019**, *11*, 8969–8981. [CrossRef]
47. Bharti, M.; Jha, P.; Singh, A.; Chauhan, A.K.; Misra, S.; Yamazoe, M.; Debnath, A.K.; Marumoto, K.; Muthe, K.P.; Aswal, D.K. Scalable free-standing polypyrrole films for wrist-band type flexible thermoelectric power generator. *Energy* **2019**, *176*, 853–860. [CrossRef]

48. Ni, D.; Song, H.; Chen, Y.; Cai, K. Free-standing highly conducting pedot films for flexible thermoelectric generator. *Energy* **2019**, *170*, 53–61. [CrossRef]
49. Park, K.T.; Choi, J.; Lee, B.; Ko, Y.; Jo, K.; Lee, Y.M.; Lim, J.A.; Park, C.R.; Kim, H. High-performance thermoelectric bracelet based on carbon nanotube ink printed directly onto a flexible cable. *J. Mater. Chem. A* **2018**, *6*, 19727–19734. [CrossRef]

Article

Wet-Spinning Assembly of Continuous, Highly Stable Hyaluronic/Multiwalled Carbon Nanotube Hybrid Microfibers

Ting Zheng [1,2,†], Nuo Xu [1,†], Qi Kan [3,†], Hongbin Li [4], Chunrui Lu [1], Peng Zhang [1], Xiaodan Li [1,5,*], Dongxing Zhang [1,*] and Xiaodong Wang [2,*]

1 School of Materials Science and Engineering, Harbin Institute of Technology, Harbin 150001, China; zthappy1127@gmail.com (T.Z.); xunuo_hit@163.com (N.X.); luchunrui06@126.com (C.L.); ZP1249558527@126.com (P.Z.)

2 College of Materials Science and Chemical Engineering, Harbin Engineering University, Harbin 150001, China

3 AVIC Aero Polytechnology Establishment, Beijing 100028, China; 15945698715@163.com

4 College of Light Industry and Textile, Qiqihar University, Qiqihar 161000, China; lhb987258@163.com

5 School of Chemistry and Chemical Engineering, Jinggangshan University, Ji'an 343009, China

* Correspondence: lixiaodanlixiaodan@126.com (X.L.); zhangdongxing@hit.edu.cn (D.Z.); wangxiaodong@hrbeu.edu.cn (X.W.); Tel.: +86-451-8628-2455 (X.L. & D.Z.); +86-451-8256-8337 (X.W.)

† These authors are equally contribution to this work.

Received: 22 March 2019; Accepted: 5 May 2019; Published: 13 May 2019

Abstract: Effective multiwalled carbon nanotube (MWCNT) fiber manufacturing methods have received a substantial amount of attention due to the low cost and excellent properties of MWCNTs. Here, we fabricated hybrid microfibers composed of hyaluronic acid (HA) and multiwalled carbon nanotubes (MWCNTs) by a wet-spinning method. HA acts as a biosurfactant and an ionic crosslinker, which improves the dispersion of MWCNTs and helps MWCNT to assemble into microfibers. The effects of HA concentration, dispersion time, injection speed, and MWCNT concentration on the formation, mechanical behavior, and conductivity of the HA/MWCNT hybrid microfibers were comprehensively investigated through SEM, UV-Vis spectroscopy, tensile testing, and conductivity testing. The obtained HA/MWCNT hybrid microfibers presented excellent tensile properties in regard to Young's modulus (9.04 ± 1.13 GPa) and tensile strength (130.25 ± 10.78 MPa), and excellent flexibility and stability due to the superior mechanical and electrical properties of MWCNTs. This work presents an effective and easy-to-handle preparation method for high-performance MWCNT hybrid microfibers assembly, and the obtained HA/MWCNT hybrid microfibers have promising applications in the fields of energy storage, sensors, micro devices, intelligent materials, and high-performance fiber-reinforced composites.

Keywords: multiwalled carbon nanotube; hyaluronic acid; microfibers; wet-spinning; microstructures; tensile properties

1. Introduction

Carbon nanotubes (CNTs) are very promising materials for electronic and energy storage devices, sensors, biosensors, composites, and transparent conducting films, etc., due to their high specific surface areas, excellent mechanical properties, and good electrical conductivities [1–3]. In recent years, individual CNTs have often been assembled into macro/micro structures, such as CNT fibers, CNT films, and CNT arrays, to improve their application [4–6]. Among these CNT products, CNT fibers are microsized materials with suitable electrical conductivities, stable electrochemical properties, significant specific moduli, and specific strengths [7,8]. Furthermore, CNT fibers are very flexible and can easily be prepared in different shapes to meet the special requirements of an application, such as woven fabric or artificial muscle bundles [9,10]. For instance, Peining Chen et al. developed actuating fibers that can contract and rotate in response to solvents and vapors [11]. Huhu Cheng et al. wove CNT/G hybrid fibers into textile electrodes for the construction of flexible supercapacitors [12]. Recently, several approaches have been applied to assemble macroscopic fibers of either pure CNTs or CNT-polymer composites. These methods can be summarized as the direct growth of the CNT fibers, dry spinning from a nanotube mat, and coagulation spinning methods [13–16]. However, it cannot be ignored that the applications of CNT fibers prepared by the first two methods are limited by the high costs of the super-aligned carbon nanotube arrays, strict conditions, complex operational procedures, and low quantities from production [17]. In comparison, the latter process, reminiscent of the so-called wet-spinning method, is particularly simple and potentially scalable for large-scale production. In this method, homogeneous CNTs will be injected into a rotating coagulation medium by a syringe to form a fiber shape. The most common approach for getting stabilized and uniform CNTs dispersion involves the use of surfactants, which can be adsorbed on the out surface of each carbon nanotube to overcome the attractions of van der Waals. The common surfactants are sodium dodecyl sulfate (SDS), sodium dodecyl benzene sulfonate (SDBS), lithium dodecyl sulfate (LDS), and triton X-100. It is worth noting that biomolecules such as single-stranded DNA [10] and chitosan (CHI) [18] have also showed good dispersion property for CNTs in recent literature [19,20]. The typical coagulation medium is polyvinylalcohol (PVA), and other polymer coagulation mediums such as polyethylene-imine (PEI) and polylactic acid (PLA) are also reported. The diameter of CNT fibers varies from several microns to 100 microns, which is mainly affected by the processing conditions, such as the diameter of the syringe, the flow rate of the injection solution, and the condition of the polymer coagulant. Thus, wet-spinning preparation technology of CNT fibers can use any disordered carbon nanotube powder or array as raw material to process into fibers. The raw materials are low cost, equipment is simple, and the operation is easy. However, very few studies have been made on macroscopic finely assembled MWCNT fibers with a controllable, uniform, and large-scale synthetic method that can be used for potential applications until now. Because this method still has many shortcomings, deficiencies need to be further solved. For example, more surfactants need to be added when the concentration of carbon nanotubes is high, large amounts of surfactants can easily form micelles and influence the structure and properties of CNT fibers. Moreover, a large number of studies have found that the obtained performances of these CNT fibers are far less than those of single CNT [21,22]. The main reasons for this are the uneven dispersion of the CNTs and the large number defects in the CNT fibers that will lead to the concentration of stress during the tensile process. The stress cannot be transferred effectively between CNTs, which affects the strength of the CNT fibers. In addition, the defects in the fibers increase the contact resistance between CNTs and affect the electronic transmission in the fibers.

To address the problems above, hyaluronic acid, a biomolecule, was utilized as a surfactant to obtain a stable and homogeneous dispersion of MWCNTs. In contrast with molecular surfactants (SDS or LDS), proportionally equivalent or lower amounts of biomolecules to CNTs were employed to generate a homogenous dispersion, whereby ratios of molecular surfactants of at least 2:1, and in some cases 3:1, were required [18,23]. Secondly, a calcium chloride ($CaCl_2$) solution in ethanol was chosen to replace cement polymer coagulant medium to reduce the effect of the polymer on the electrical conductivity of the MWCNT fibers. The fiber structure was obtained through the formation of calcium bridges between the D-glucuronic acid residues on adjacent chains of HA. The effects of different spinning parameters and different concentrations of MWCNTs on the formation, and electrical and mechanical properties of HA/MWCNT hybrid microfibers were discussed, the optimization of the spinning process was completed, and the structure and flexibility of the obtained HA/MWCNT hybrid microfibers were investigated. Based on the results, our wet-spinning system is an effective and easy-to-handle method to assemble MWCNT hybrid microfibers with high strength, conductivity, stability, and flexibility. The obtained high performance HA/MWCNT hybrid microfibers have promising applications in the fields of energy storage, sensors, micro devices, intelligent materials, and high-performance fiber-reinforced composites.

2. Materials and Methods

2.1. Materials

MWCNTs purchased from TimesNano Company (Chengdu, China) were used in this study. HA with a molecular weight (MW) of 41–65 kDa, was purchased from Lifecore Biomedical LLC (Chaska, MN, USA). $CaCl_2$ was purchased from Sigma–Aldrich (St. Louis, MO, USA).

2.2. Fabrication of the HA/MWCNT Hybrid Microfibers

The HA/MWCNT hybrid microfibers were fabricated by the wet-spinning method as below. Firstly, an HA solution was prepared by dissolving HA in deionized (DI) water at room temperature. Then, MWCNTs were added and dispersed in the HA solution using a probe sonicator (YM-1000Y, Shanghai Yuming Instrument Ltd., Shanghai, China) to form the spinning solution. The spinning solution was injected into a rotating coagulation bath (20 rpm) using a syringe pump (LSP02-1B, Dichuang Electronic Technology Ltd., Baoding, China) with a detachable needle (diameter of 0.2 mm) to control the flow rate of injection. The coagulation bath was filled with 5 wt% $CaCl_2$ in 70% ethanol. Next, the wet-spun hybrid microfibers were removed from the coagulation bath and rinsed using ethanol and DI water to remove residual coagulating agents. Finally, the prepared hybrid microfibers were dried in air under tension to obtain several meters of the hybrid microfibers.

2.3. Characterization and Measurements

The morphologies and microstructures of the HA/MWCNT hybrid microfibers were examined by a scanning electron microscope (SEM) (Merlin Compact, Carl Zeiss AG, Jena, Germany) and 3D digital microscopy (DSX-CB, Olympus Ltd., Tokyo, Japan). UV–Vis scanning spectrophotometry was used to study the dispersion of CNTs suspension. Fourier transform infrared spectroscopy (FTIR, Spectrum One, PerkinElmer, Boston, MA, USA) measurements were recorded at room temperature to confirm the presence of HA. The thermal properties of the HA/MWCNT hybrid microfibers were investigated by differential scanning calorimetry and thermogravimetry analysis (DSC/DTA-TG, STA449F3, Netzsch, Selb, Germany) in a nitrogen environment. The samples were heated from room temperature to 800 °C with a heating rate of 20 °C/min under a nitrogen flow of 50 mL/min. The percentages of the HA and MWCNTs in the HA/MWCNT hybrid microfibers were quantified by the thermogravimetric analysis (TGA) results. The mechanical properties of the HA/MWCNT hybrid microfibers were determined using a uniaxial tensile tester machine (T150 UTM, Agilent, Santa Clara, CA, USA) with a cell load capacity of 10 N at a 0.5 mm/min rate. The electrical conductivities of the HA/MWCNT hybrid microfibers were measured by the four-point probe method using a SourceMeter (Keithley 2400, Tektronix INC., Beaverton, OR, USA). Silver paste was used at the contact points between the hybrid microfibers and the electrode probes to eliminate the contact resistance.

3. Results

3.1. Dispersion of the MWCNTs in Different Surfactant Suspensions

The dispersion of the CNTs is one of the important primary parameters, which is used as the starting point for further processing into fibers, films, and composites. In general, ionic surfactants (SDS) and nonionic surfactants (TritonX-100) are usually used as dispersion systems for CNT suspensions [24,25]. Therefore, we first compared the bio-surfactant (HA) dispersion stability of the MWCNTs with those of SDS and TritonX-100. The dispersal of MWCNTs in different dispersion systems were studied by 3D digital microscopy and UV–Vis scanning spectrophotometry, as shown in Figure 1. The MWCNT dispersions were too dark to clearly discern the amount of sediment, as shown in the inset of Figure 1c.

Thus, the MWCNT dispersions with different surfactants were further observed under a microscope. The results and schematic diagrams of the dispersal states are shown in Figure 1a,b. In the case of a low degree of dispersion by SDS, several agglomerates are observed in Figure 1a. As the degree of dispersion increases, agglomerates become looser, and HA shows the best dispersion behavior. These results were also confirmed by UV–Vis spectroscopy. Figure 1c shows the UV–Vis spectra and photos of the MWCNTs dispersion with different surfactants on day 1 and day 180. The MWCNT dispersions exhibit a characteristic peak at approximately 300 nm, and the absorbance gradually decreases from the UV to near-IR region due to scattering. Increasing the amount of dispersed MWCNTs will result in an increase in the area below the spectral lines that represent the absorbance [26,27]. In addition, HA and TritonX-100 show high absorbance intensities and good dispersion stabilities over a long timeframe. Hence, HA has the highest dispersing power of these three surfactants according to the above experimental results.

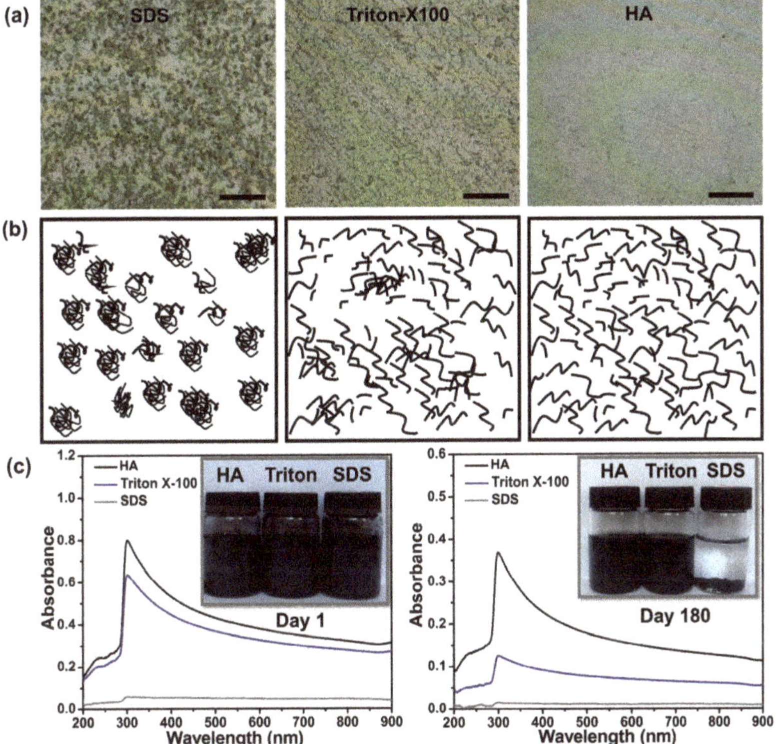

Figure 1. Dispersion of the MWCNTs in different surfactant suspensions. (**a**) Optical images of the MWCNT dispersions (scale bar: 500 μm); (**b**) schematic diagrams of the MWCNT dispersions and (c) photos and UV-Vis spectral curves of the MWCNT dispersions.

3.2. The Morphologies and Microstructures of the HA/MWCNT Hybrid Microfibers

Controlled injections of the HA/MWCNT dispersions into the $CaCl_2$ in ethanol coagulating medium afforded a continuous fiber structure due to the formation of calcium bridges between the D-glucuronic acid residues on adjacent chains of HA. A rotating coagulation bath was used to produce a continuous meter-long spinning of the HA/MWCNT hybrid hybrid microfibers; this process is shown in Figure 2a. The long wet HA/MWCNT microfiber was then dried by stretching. Figure 2b shows a microscope photograph of the long fibers collected from the $CaCl_2$ solution by wet spinning of a

HA/MWCNT dispersion. The fibers were uniformly circular in both the wet and dry states and were swollen and flexible when wet but became brittle and less flexible upon drying. The dry HA/MWCNT hybrid microfibers do not swell when submerged into solution. SEM images were then used to characterize the morphologies of the HA/MWCNT hybrid microfibers. Figure 2c clearly demonstrates that the HA/MWCNT hybrid microfibers are rigid and have a uniformly cylindrical shape. The diameter of the dry HA/MWCNT hybrid microfibers is approximately 70 μm. The enlarged SEM image of the HA/MWCNT microfiber suggests that the microfiber is composed of a number of carbon nanotubes, and the outer surface is very rough, with numerous wrinkles on the surface, which are likely attributable to the roughness of the inner surface of the needle orifice and the shrinkage of the fiber under stretching during drying.

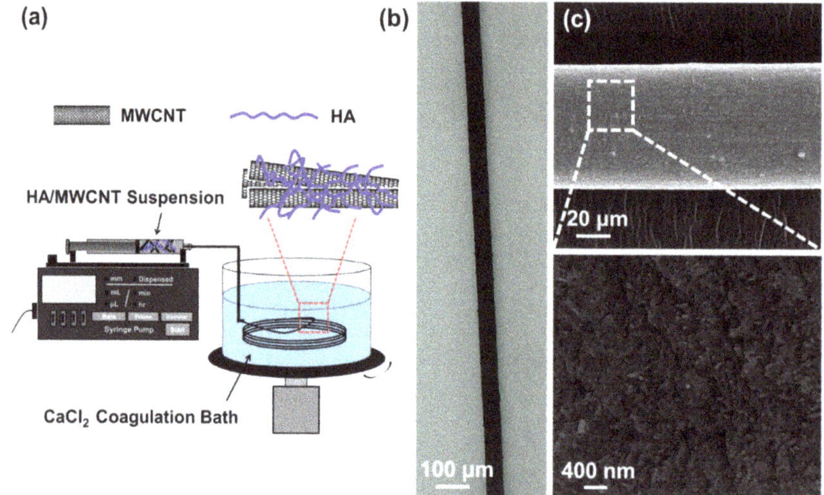

Figure 2. (**a**) A schematic of the experimental design of the wet-spinning method. (**b**) 3D microscopy image of the HA/MWCNT hybrid microfibers drawn from the $CaCl_2$ coagulation bath. (**c**) SEM image of the HA/MWCNT hybrid microfibers and an enlarged image of the fibers.

3.3. *Effects of HA/MWCNT Ratio*

HA not only influences the dispersion of the MWCNTs but also influences the content ratio of HA to MWCNTs in the hybrid microfibers and further determines the composition and properties of the obtained HA/MWCNT hybrid microfibers. To optimize the amount of HA, the mass ratios of MWCNTs to HA were studied with 1.2% MWCNTs under a 60 mL/h injection speed, and the results are displayed in Figure 3. Figure 3a shows the stress–strain curves of the HA/MWCNT hybrid microfibers during tensile testing, and two stages can be identified from the curve. First, HA/MWCNT hybrid microfibers show a linear relationship between stress and strain, suggesting an initially elastic-like deformation process during stretching, which is reversible. During the elastic-like deformation, the van der Waals forces and the friction between MWCNTs carry little stress. Next, the stress of the HA/MWCNT hybrid microfibers increased nonlinearly with strain before breaking, as fibers were further strained. MWCNTs will bear more of the load, leading to stretching and sliding of the MWCNTs inside the fiber. This is a plastic behavior that is irreversible [18]. Finally, once the tensile stress exceeds the critical value, local stresses cannot be further transferred to neighboring CNTs, and the fibers break. The tensile strengths and Young's moduli of the HA/MWCNT hybrid microfibers were similar when the mass ratios of MWCNTs to HA were 1:1 and 2:1 and then decreased significantly after reducing the contents of HA according to the results shown in Figure 3b. It is worth noting that the mass ratio of 4:1 produced weak and brittle hybrid microfibers that were difficult to handle during dry processing.

This is because the continuous fiber structure was induced by the formation of calcium bridges between D-glucuronic acid residues on adjacent chains of HA, low concentrations of HA will influence the dispersion of the MWCNTs and the formation of the MWCNT fibers. The uneven dispersion of the MWCNTs and the large number of defects in the MWCNT fibers will further lead to the concentration of stress during the tensile process. The stress cannot be transferred effectively between MWCNTs, which affects the strength of the MWCNT fibers. Meanwhile, the defects in the fibers increase the contact resistance between MWCNTs and affect the electronic transmission in the fibers. Therefore, the resistivity of HA/MWCNT hybrid microfibers increased after an initial decline, and the lowest resistivity was 0.92 ± 0.10 Ω·mm when the mass ratio of MWCNTs to HA was 2:1. The electrons could not be transferred effectively if there were a number of defects in the fibers. Considering the balance between properties, a 2:1 mass ratio of MWCNTs:HA is determined to be the optimal conditions for combining the results of the mechanical and conductivity properties.

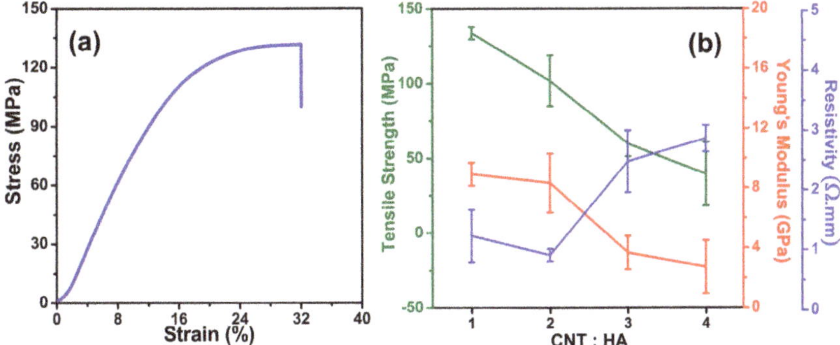

Figure 3. The mechanical and conductivity properties of the HA/MWCNT hybrid microfibers with different ratios of HA to CNTs. (**a**) The stress-strain curve of a HA/MWCNT hybrid microfiber; (**b**) The tensile strengths, Young's moduli and conductivities of HA/MWCNT hybrid microfibers prepared from different HA:MWCNT ratios.

3.4. *Effects of Sonication Duration*

Ultrasonication is an external mechanical energy that can help MWCNT bundles overcome the attractive van der Waals forces to disentangle [28]. Therefore, an effective way to disperse CNT solutions is by controlling the sonication time and supplied energy. Additionally, the absorbance of the CNT suspension at a specific wavelength can be related to the degree of debundling of the CNTs in solution [29]; thus, the peak intensity of the obtained UV–Vis spectra is an effective tool to monitor the sonication dispersion process. In this case, the ultrasonic power was set to a fixed value, and HA/MWCNT solutions treated by different ultrasonication times were investigated. Figure 4a illustrates the UV–Vis spectra of HA/MWCNT solutions with different ultrasonication times. The UV–Vis spectra exhibited a characteristic peak at approximately 300 nm, as discussed above. Additionally, the longer the ultrasonication duration, the greater the intensity of the characteristic peak. This means that extending the ultrasonication time can help to well disperse the MWCNTs in the HA solution. However, the ultrasonication process contains two mutually antagonistic effects, one being the deagglomeration of the MWCNTs and the other being the fragmentation of individual CNTs [30]. In other words, individual CNTs will be broken, if the ultrasonication duration is too long. In fact, the dispersion of the MWCNTs is an important factor that can not only affect the length and diameter of the MWCNTs but also further influence the formation and properties of the HA/MWCNT hybrid microfibers. The mechanical performance is one of the crucial parameters of MWCNTs for their practical application. Therefore, the mechanical properties of the HA/MWCNT hybrid microfibers prepared by suspensions with different ultrasonication times were investigated, and the results are displayed in Figure 4b. The tensile strength first increased as the ultrasonication times were increased due to the efficient dispersion of MWCNTs and then tended to decrease slightly for longer sonication times due to the destruction of the aspect ratio of individual CNTs by the sonication energy. Hence, 40 min was chosen as the optimal ultrasonication time for later experiments by combining the results of the UV–Vis spectra and tensile properties.

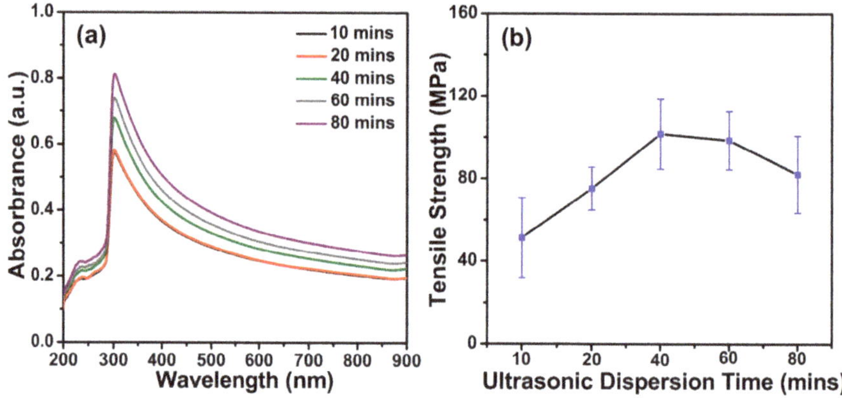

Figure 4. The UV–Vis spectra and tensile strength of the HA/MWCNT hybrid microfibers prepared by HA/MWCNT dispersions with different ultrasonication times. (**a**) UV-Vis spectra; (**b**) Tensile strength results.

3.5. Effects of Injection Speed

After sonication, HA/MWCNT dispersions were injected into a rotating coagulation bath to produce continuous meter-long fibers as required. In a previous report, the ratio of the injection speed to the rotating speed of the rotator acted as an important parameter for the formation and performance of wet-spun CNT fibers [18,31]. In our case, a 1.2% optimal HA/MWCNT suspension was injected into a 20 r/min rotating coagulation bath at different speeds (30 mL/h, 40 mL/h, 50 mL/h, 60 mL/h, 70 mL/h, and 80 mL/h). The HA/MWCNT hybrid microfibers were difficult to handle for subsequent processing and characterization under injection rates of 30 mL/h and 80 mL/h and the graphs of HA/MWCNTs hybrid microfibers prepared with other injection speeds (40 mL/h~70 mL/h) were shown in Figure S1. Only short fibers could be formed and removed from the coagulation bath when a 40 mL/h injection speed was applied. As shown in Figure 5, the diameter of the HA/MWCNT hybrid microfibers prepared by 40 mL/h injection speed was 55.42 ± 5.84 μm, and increased to 65.8 ± 8.28 μm and 88.12 ± 6.72 μm for the microfibers fabricated under 50 mL/h and 60 mL/h injection speed, respectively. Moreover, ribbon-like fibers were obtained and then became hollow and tubular after drying when higher injection rates were used to prepare the samples, as shown in Figure 5d. This morphology is likely caused by the shaking of the needle under faster injection rates. The tensile strength of the HA/MWCNT hybrid microfibers prepared at different injection speeds was also calculated. The results are shown in Figure 5e. The HA/MWCNT hybrid microfibers made with a 50 mL/h injection speed formed well in the coagulation bath (Figure 5b, inset) and possessed the highest mechanical properties of the different samples. As discussed above, the HA/MWCNT hybrid microfibers were easily broken during drying and a low fiber-forming rate was obtained when the injection speed was too low. When the injection speed was too fast, the HA/MWCNT hybrid microfibers became ribbons with poor mechanical properties that tended to curl easily. In this case, an injection speed of 50 mL/h was chosen as the suitable parameter for the subsequent MWCNT fiber preparations.

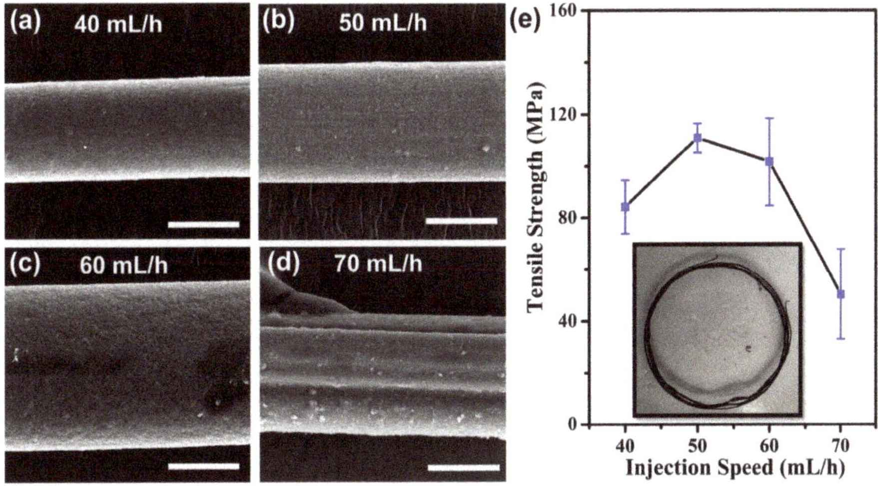

Figure 5. SEM images and tensile strength of the HA/MWCNT hybrid microfibers prepared with different injection speeds (scale bars: 40 μm). (**a**) Injection speed is 40 mL/h; (**b**) Injection speed is 50 mL/h; (**c**) Injection speed is 60 mL/h; (**d**) Injection speed is 70 mL/h; (**e**) Tensile strength results.

3.6. Effects of MWCNT Concentration

Figure 6 shows the comparison of the mechanical properties and resistivities of HA/MWCNT hybrid microfibers prepared with different MWCNT concentrations. As expected, the mechanical properties of the HA/MWCNT hybrid microfibers exhibited an increasing tendency with increasing MWCNT concentration due to the interfacial interactions, the efficient transfer of load and energy, reducing concentrated stress and dissipating the energy induced by stretching the fibers [32]. The Young's modulus of the samples varies from 4.24 ± 1.11 GPa for the 0.8% MWCNT hybrid microfibers to 9.04 ± 1.13 GPa for the 1.4% MWCNT microfiber, while the tensile strength increases from 50.27 ± 10.37 MPa to 130.25 ± 10.78 MPa, respectively, as shown in Figure 6c,d. However, when the MWCNT concentration increased to 1.6%, both the Young's modulus and tensile strength exhibited a decreasing trend, declining to 8.44 ± 1.10 GPa and 117.70 ± 12.68 MPa, respectively. The fracture mechanism model of the HA/MWCNT hybrid microfibers under tensile stress is shown in Figure 6a. During the tensile test, the stress is first applied to both ends of the microfiber in the axial direction, and subsequently, the fibrous elements endured a pulling force to straighten the MWCNTs. Next, fibers were further strained under force, and the MWCNTs bore a greater load, which lead to the stretching and sliding of the MWCNTs inside the hybrid microfibers. As the force further increased, the stress could no longer be transferred effectively [21]. Hence, MWCNTs in the fibers separated from one another, and the fiber broke. During this process, the interfacial binding and friction force between the MWCNTs may play an important role, and the existence of HA may improve the connections between individual MWCNTs. In all probability, the slight decrease in mechanical properties for high MWCNT concentration samples was caused by the poor distribution and weak interfacial contact between MWCNTs or MWCNT bundles.

The variation in the MWCNT concentrations could also control the electrical properties of the HA/MWCNT hybrid microfibers, which is another important parameter for their future application in fields such as electrical and electrochemical. Accordingly, the resistivities of the HA/MWCNT hybrid microfibers made with different CNT concentrations were investigated, and the results are displayed in Figure 6d. It is expected that the resistivity of the HA/MWCNT hybrid microfibers would decrease with increasing CNT concentrations due to the formation of efficient electrical pathways. However, the resistivity of HA/MWCNT hybrid microfibers with different CNT concentrations shows similar results to those of the mechanical properties. The resistivity decreased from 2.35 ± 0.40 Ω·mm to 0.77 ± 0.55 Ω·mm as the CNT concentration was varied from 0.8% to 1.4% and then increased to 0.91 ± 0.37 Ω·mm when the CNT concentration was further increased. This change can be explained by a number of reasons [5,33,34]. First, some of the MWCNTs are wrapped with a thin coating around a single tube, which may be an impediment to forming a conductive path. Importantly, the higher the concentration of MWCNT hybrid microfibers, the more HA hydrogel is in the fibers. It is generally known that materials made up of nonconducting materials and conducting carbon nanotubes exhibit a significant increase in their conductivities due to higher concentrations being needed to form electrical pathways through the material, i.e., the so-called percolation threshold [16]. For high-concentration HA/MWCNT hybrid microfibers, numerous overlapping carbon nanotubes form a well-connected electrical pathway with increasing concentrations of MWCNTs. However, the resistance of the fibers is controlled by the resistance of the junctions between overlapping MWCNTs. Therefore, it is most likely that HA may cause an increase in the junction resistance of the MWCNT network, which would produce similar results. The combination of the trends observed in the SEM images, mechanical properties and resistivities indicates that HA/MWCNT hybrid microfibers with the optimal properties were prepared by injecting a homogeneous HA/MWCNT dispersion by a 0.2 mm diameter syringe at a rate of 50 mL/h into a $CaCl_2$ coagulation bath, which was simultaneously rotated at 20 r/min. The homogeneous HA/MWCNT dispersion with a ratio of 1:2 was ultrasonicated in a water bath using a high-powered sonic tip (500 W, 36% amplitude) under pulse mode (1 s on, 1 s off) for 40 min.

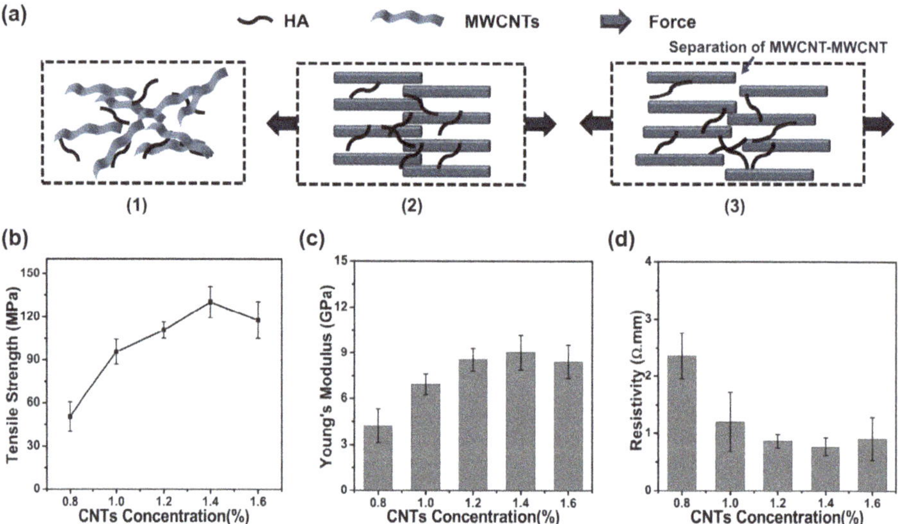

Figure 6. Mechanical properties and resistivities of the HA/MWCNT hybrid microfibers produced with different MWCNT concentrations. (**a**) The fracture mechanism model of the HA/MWCNT hybrid microfibers under tensile stress; (**b**) Tensile strength; (**c**) Young's Modulus; (**d**) Resistivity.

3.7. Characterization of the HA/MWCNT Hybrid Microfibers

Further testing was carried out to characterize the functional groups in the fibers by FTIR. The FTIR spectra of the MWCNTs, HA, and HA/MWCNT hybrid microfibers are shown in Figure 7a. The spectrum of the CNTs shows weak absorption bands and could not be detected by FTIR. For the spectrum of the HA/MWCNT hybrid microfibers, the absorption band between 3600 and 2800 cm^{-1} is assigned to the stretching vibrations of O–H and C–H, which are typical characteristics of polysaccharides. The typical absorption peak at 1567 cm^{-1} corresponds to overlapping vibrations of the acetamide and carboxylate groups [35]. These two obvious absorption bands can be utilized to verify the inclusion of HA in the

MWCNT hybrid microfibers. However, many of the HA absorption peaks are weak due to the low content of HA in the hybrid microfibers. For instance, the absorption peak at 1033 cm^{-1} is attributed to the skeletal vibrations of the saccharide structure, which involve C–O stretching. The peaks at 1413 cm^{-1} (–C–H bending) and 597 cm^{-1} (N–H vibration) are also present in the spectrum of HA [36].

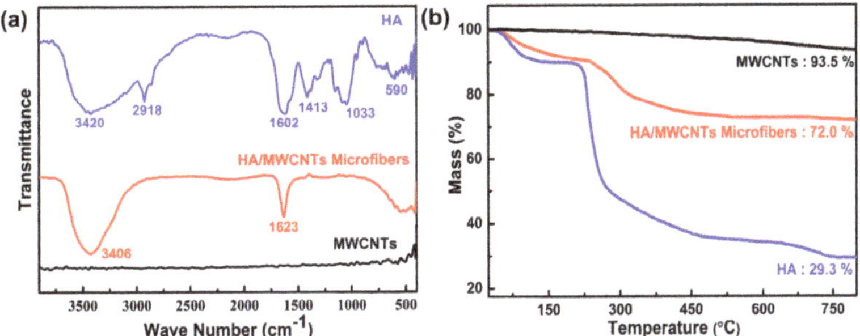

Figure 7. FTIR spectra and TGA curves of HA, MWCNTs, and HA/MWCNT hybrid microfibers. (**a**) FTIR spectra; (**b**) TGA curves.

To demonstrate the thermal stability of the HA/MWCNT hybrid microfibers and quantify the MWCNT and HA contents in the hybrid microfibers, TG analysis of the MWCNTs, HA, and HA/MWCNT hybrid microfibers was obtained and the results are shown in Figure 7b. The HA/MWCNT hybrid microfibers followed a similar decomposition trend as that of HA. Namely, before 100 °C, there is a gradual rate of weight loss for both HA and the HA/MWCNT hybrid microfibers, which is primarily attributed to the expulsion of absorbed water. Then, both samples start to rapidly degrade at approximately 210 °C due to the random chain scission of the HA polymer chains, and the maximum rate of weight loss temperature occurs at approximately 230 °C. More importantly, it should be noted that the microfiber peak positions of differential thermal gravity (DTG) move towards higher temperatures compared with those of the HA samples, as shown in Figure S2, suggesting an excellent thermal stability of the HA/MWCNT composite after assembling into hybrid microfibers. This is a consequence of the high thermal stability and low weight loss of the carbon nanotubes, which allows for very efficient heat transport in the samples. Moreover, this enhancement could reveal the strong interactions between the MWCNTs and HA, which may noticeably decrease the segmentation motions of the molecular chain, hence slowing down the decomposition process. Additionally, TGA was also performed to study the quantity of CNTs within the fibers according to the residual weight at 800 °C. The residual mass of MWCNTs, HA, and HA/MWCNTs hybrid microfibers was approximately 93.5%, 29.3% and 72.0%, respectively. Therefore, the exact amounts of HA and MWCNTs in the fibers were 33.5% and 66.5 wt%, respectively, which is consistent with the intended contents of these hybrid microfibers.

3.8. The Flexibility and Stability of the HA/MWCNT Hybrid Microfibers

In practical applications, fiber-shaped materials should possess excellent flexibility and mechanical stability. Here, the flexibility and stability of the HA/MWCNT hybrid microfibers were characterized by a resistance test, as shown in Figure 8. For convenient and accurate measurements, a fiber was placed on a thin paper substrate, and its two ends were connected to copper wires with silver conducting paste, as shown in the inset of Figure 8a. Figure 8a shows the curve and photographs illustrating the effect of bending angles on the resistance of the HA/MWCNT hybrid microfibers. The results show that the largest relative change of the resistance is 1.00 ± 0.06, revealing a high stability of the HA/MWCNT hybrid microfibers. In addition, the stability of the electrical conductance during cyclical mechanical deformation was examined by folding-unfolding cycles applied to the microfiber using a mechanical tester and the simultaneous measurement of the resistivity (Figure 8b insets).

As Figure 8b depicts, more than 1200 cycles were applied to the microfiber, and the resistance was recorded every 100 cycles. The relative change of the resistance was 1.15 ± 0.03 after 1200 cycles, which further confirmed the high stability of the HA/MWCNT hybrid microfibers. Furthermore, the HA/MWCNT hybrid microfibers were twisted into four circles (Figure 8c) to demonstrate their flexibility. As shown in Figure 8c, the relative change in the resistance shows a negligible deviation after twisting into two circles and increases to 1.11 ± 0.02 after being tied in four circles, confirming the flexibility of these fibers, which could be used in micromachine applications. The conductivity stability of MWCNT hybrid microfibers depends not only on electrical properties but also on the mechanical properties of the fibers. In our case, HA provides good mechanical support, while the MWCNTs provide a more stable conductive path. Overall, the HA/MWCNT hybrid microfibers exhibit stabilized electrical conductivity and excellent mechanical properties and could be used in microsized materials, conductive materials, electrode materials, intelligent materials and high-performance fiber reinforced composite materials applications

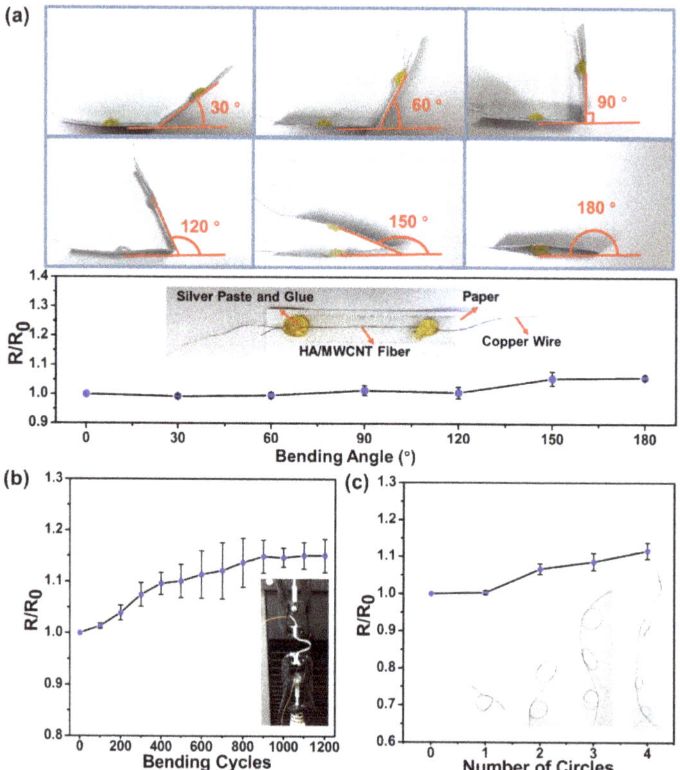

Figure 8. The stability properties of the HA/MWCNT hybrid microfibers. (**a**) Effects of bending angles on the resistance of HA/MWCNT microfibers on a paper substrate; inset: A graphic illustration of the test method for the HA/MWCNT microfibers; (**b**) The resistance stability results of the HA/MWCNT microfibers over 1200 folding-unfolding cycles; inset: A photograph showing the fiber being operated on by the tensile machine; and (**c**) The dependence of resistance on different numbers of tied circles.

4. Conclusions

In conclusion, continuous, conducting, high strength, and flexible HA/MWCNT hybrid microfibers have been successfully produced by a wet-spinning method using HA as a surfactant and ion-conducting binder in the spinning solution. The effects of HA concentration, dispersion sonication time, injection speed, and MWCNT concentration on the formation, conductivity, and mechanical behavior of the HA/MWCNT hybrid microfibers were comprehensively investigated. An effective and easy-to-handle manufacturing method for MWCNT hybrid microfibers is presented, and show that the obtained HA/MWCNT hybrid microfibers with excellent electrical conductivity, mechanical properties, and stable behavior are a promising material for microsized materials, conductive materials, electrode materials, intelligent materials, and high-performance fiber reinforced composite materials.

Supplementary Materials: The following are available online at http://www.mdpi.com/2073-4360/11/5/867/s1, Figure S1: The graphs of HA/MWCNTs microfibers prepared with different injection speed, Figure S2: DTG curves of HA and HA/MWCNTs microfibers.

Author Contributions: Conceptualization, T.Z. and X.W.; methodology, T.Z. and Q.K.; formal analysis, N.X., Q.K. and P.Z.; data curation, T.Z. and H.L.; writing—original draft preparation, T.Z.; writing—review and editing, N.X., C.L., and X.W.; supervision, X.L. and D.Z.; project administration, X.W. and D.Z.; funding acquisition, X.W.

Acknowledgments: The authors would like to express thanks to the financial support from Central Universities Grant [grant numberHEUCFG201816].

Conflicts of Interest: The authors declare no conflict of interest.

References

1. Zhang, J.; Zheng, T.; Alarçin, E.; Byambaa, B.; Guan, X.; Ding, J. Porous Electrospun Fibers with Self-Sealing Functionality: An Enabling Strategy for Trapping Biomacromolecules. *Small* **2017**, *1701949*, 1–15. [CrossRef] [PubMed]
2. Senokos, E.; Reguero, V.; Palma, J.; Vilatela, J.J.; Marcilla, R. Macroscopic fibres of CNTs as electrodes for multifunctional electric double layer capacitors: From quantum capacitance to device performance. *Nanoscale* **2016**, *8*, 3620–3628. [CrossRef] [PubMed]
3. Di, J.; Zhang, X.; Yong, Z.; Zhang, Y.; Li, D.; Li, R.; Li, Q. Carbon-Nanotube Fibers for Wearable Devices and Smart Textiles. *Adv. Mater.* **2016**, *28*, 10529–10538. [CrossRef]
4. Zhang, L.; Wang, J.; Fuentes, C.A.; Zhang, D.; Van Vuure, A.W.; Seo, J.W.; Seveno, D. Wettability of carbon nanotube fibers. *Carbon N. Y.* **2017**, *122*, 128–140. [CrossRef]
5. Mei, H.; Xu, Y.; Sun, Y.; Bai, Q.; Cheng, L. Carbon nanotube buckypaper-reinforced SiCN ceramic matrix composites of superior electrical conductivity. *J. Eur. Ceram. Soc.* **2016**, *36*, 1893–1898. [CrossRef]
6. Zheng, T.; Wang, G.; Xu, N.; Lu, C.; Qiao, Y.; Zhang, D.; Wang, X. Preparation and Properties of Highly Electroconductive and Heat-Resistant CMC / Buckypaper / Epoxy Nanocomposites. *Nanomaterials* **2018**, *8*, 969. [CrossRef]
7. Pu, X.; Li, L.; Liu, M.; Jiang, C.; Du, C.; Zhao, Z.; Hu, W.; Wang, Z.L. Wearable Self-Charging Power Textile Based on Flexible Yarn Supercapacitors and Fabric Nanogenerators. *Adv. Mater.* **2016**, *28*, 98–105. [CrossRef]
8. Michardie, A.; Mateo-mateo, C.; Derre, A.; Correa-duarte, M.A.; Mano, N.; Poulin, P. Carbon Nanotube Micro fi ber Actuators with Reduced Stress Relaxation. *J. Phys. Chem. C* **2016**, *120*, 6851–6858. [CrossRef]
9. Zheng, W.; Razal, J.M.; Whitten, P.G.; Ovalle-Robles, R.; Wallace, G.G.; Baughman, R.H.; Spinks, G.M. Artificial muscles based on polypyrrole/carbon nanotube laminates. *Adv. Mater.* **2011**, *23*, 2966–2970. [CrossRef] [PubMed]
10. Shin, S.R.; Lee, C.K.; Eom, T.W.; Lee, S.H.; Kwon, C.H.; So, I.; Kim, S.J. DNA-coated MWNT microfibers for electrochemical actuator. *Sensors Actuators, B Chem.* **2012**, *162*, 173–177. [CrossRef]
11. Chen, P.; Xu, Y.; He, S.; Sun, X.; Pan, S.; Deng, J.; Chen, D.; Peng, H. Hierarchically arranged helical fibre actuators driven by solvents and vapours. *Nat. Nanotechnol.* **2015**, *10*, 1077–1083. [CrossRef]
12. Cheng, H.; Dong, Z.; Hu, C.; Zhao, Y.; Hu, Y.; Qu, L.; Chen, N.; Dai, L. Textile electrodes woven by carbon nanotube-graphene hybrid fibers for flexible electrochemical capacitors. *Nanoscale* **2013**, *5*, 3428–3434. [CrossRef]
13. Behabtu, N.; Behabtu, N.; Young, C.C.; Tsentalovich, D.E.; Kleinerman, O.; Wang, X.; Ma, A.W.K.; Bengio, E.A.; Waarbeek, R.F.; De Jong, J.J.; et al. Fibers of Carbon Nanotubes with Ultrahigh Conductivity. *Science* **2013**, *182*, 182–187. [CrossRef]

14. Sun, G.; Zhou, J.; Yu, F.; Zhang, Y.; Pang, J.H.L.; Zheng, L. Electrochemical capacitive properties of CNT fibers spun from vertically aligned CNT arrays. *J. Solid State Electrochem.* **2012**, *16*, 1775–1780. [CrossRef]
15. Huang, T.; Zheng, B.; Kou, L.; Gopalsamy, K.; Xu, Z.; Gao, C.; Meng, Y.; Wei, Z. Flexible high performance wet-spun graphene fiber supercapacitors. *RSC Adv.* **2013**, *3*, 23957–23962. [CrossRef]
16. Chou, T.W.; Gao, L.; Thostenson, E.T.; Zhang, Z.; Byun, J.H. An assessment of the science and technology of carbon nanotube-based fibers and composites. *Compos. Sci. Technol.* **2010**, *70*, 1–19. [CrossRef]
17. Vigolo, B.; Poulin, P.; Lucas, M.; Launois, P.; Bernier, P. Improved structure and properties of single-wall carbon nanotube spun fibers. *Appl. Phys. Lett.* **2002**, *81*, 1210–1212. [CrossRef]
18. Lynam, C.; Moulton, S.E.; Wallace, G.G. Carbon-nanotube biofibers. *Adv. Mater.* **2007**, *19*, 1244–1248. [CrossRef]
19. Yeol, H.; Koo, H.; Young, K.; Chan, I.; Choi, K.; Hyung, J.; Kim, K. Biomaterials Photo-crosslinked hyaluronic acid nanoparticles with improved stability for in vivo tumor-targeted drug delivery. *Biomaterials* **2013**, *34*, 5273–5280. [CrossRef]
20. Filip, J.; Šefčovičová, J.; Tomčík, P.; Gemeiner, P.; Tkac, J. A hyaluronic acid dispersed carbon nanotube electrode used for a mediatorless NADH sensing and biosensing. *Talanta* **2011**, *84*, 355–361. [CrossRef]
21. Wang, P.; Zhang, X.; Varghese, R.; Sun, G.; Zhang, H.; Zheng, L.; Yu, T.X.; Lu, G.; Yang, J. Strengthening and failure mechanisms of individual carbon nanotube fi bers under dynamic tensile loading. *Carbon N. Y.* **2016**, *102*, 18–31. [CrossRef]
22. Zheng, L.; Sun, G.; Zhan, Z. Tuning Array Morphology for High-Strength Carbon-Nanotube Fibers. *Small* **2010**, *6*, 132–137. [CrossRef]
23. Zheng, T.; Pour Shahid Saeed Abadi, P.; Seo, J.; Cha, B.-H.; Miccoli, B.; Li, Y.-C.; Park, K.; Park, S.; Choi, S.-J.; Bayaniahangar, R.; et al. Biocompatible Carbon Nanotube-based Hybrid Microfiber for Implantable Electrochemical Actuator and Flexible Electronic Applications. *ACS Appl. Mater. Interfaces* **2019**. [CrossRef] [PubMed]
24. Chen, J. Kinetics and Mechanism Studies on Dispersion of CNT in SDS Aqueous Solutions. *J. Chin. Chem. Soc.* **2014**, *61*, 481–489. [CrossRef]
25. Alafogianni, P.; Dassios, K.; Farmaki, S.; Antiohos, S.K.; Matikas, T.E.; Barkoula, N. Colloids and Surfaces A: Physicochemical and Engineering Aspects On the efficiency of UV—Vis spectroscopy in assessing the dispersion quality in sonicated aqueous suspensions of carbon nanotubes. *Colloids Surfaces A Physicochem. Eng. Asp.* **2016**, *495*, 118–124. [CrossRef]
26. Vaisman, B.L.; Marom, G.; Wagner, H.D. Dispersions of Surface-Modified Carbon Nanotubes in Water-Soluble and Water-Insoluble Polymers **. *Adv. Funct. Mater.* **2006**, *16*, 357–363. [CrossRef]
27. Lee, S.; Lee, W.; Yi, J.; Jeong, B.; Um, M.; Materials, C.; Materials, N. Relationship between Dispersion and UV-Visible Transmittance in Nanocarbons Reinforced Composites. In Proceedings of the 18th International Conference on Composite Materials, Jeju Island, Korea, 21–26 August 2011; pp. 1–4.
28. Yu, J.; Grossiord, N.; Koning, C.E.; Loos, J. Controlling the dispersion of multi-wall carbon nanotubes in aqueous surfactant solution. *Carbon N. Y.* **2007**, *45*, 618–623. [CrossRef]
29. Zhang, S.; Liu, W.B.; Hao, L.F.; Jiao, W.C.; Yang, F.; Wang, R.G. Preparation of carbon nanotube/carbon fiber hybrid fiber by combining electrophoretic deposition and sizing process for enhancing interfacial strength in carbon fiber composites. *Compos. Sci. Technol.* **2013**, *88*, 120–125. [CrossRef]
30. Kim, S.; Lee, Y.; Kim, D.; Lee, K.; Kim, B.; Hussain, M.; Choa, Y. Estimation of dispersion stability of UV/ozone treated multi-walled carbon nanotubes and their electrical properties. *Carbon N. Y.* **2012**, *51*, 346–354. [CrossRef]
31. Razal, J.M.; Gilmore, K.J.; Wallace, G.G. Carbon nanotube biofiber formation in a polymer-free coagulation bath. *Adv. Funct. Mater.* **2008**, *18*, 61–66. [CrossRef]
32. Wang, P.; Yang, J.; Sun, G.; Zhang, X.; Zhang, H. Twist induced plasticity and failure mechanism of helical carbon nanotube fi bers under di ff erent strain rates. *Int. J. Plast.* **2018**, *110*, 74–94. [CrossRef]
33. Lin, H.; Li, L.; Ren, J.; Cai, Z.; Qiu, L.; Yang, Z.; Peng, H. Conducting polymer composite film incorporated with aligned carbon nanotubes for transparent, flexible and efficient supercapacitor. *Sci. Rep.* **2013**, *3*, 1–6. [CrossRef]
34. Granero, A.J.; Razal, J.M.; Wallace, G.G.; In Het Panhuis, M. Spinning carbon nanotube-gel fibers using polyelectrolyte complexation. *Adv. Funct. Mater.* **2008**, *18*, 3759–3764. [CrossRef]

35. Umerska, A.; Corrigan, O.I.; Tajber, L. Intermolecular interactions between salmon calcitonin, hyaluronate, and chitosan and their impact on the process of formation and properties of peptide-loaded nanoparticles. *Int. J. Pharm.* **2014**, *477*, 102–112. [CrossRef]
36. Quiñones, J.P.; Brüggemann, O.; Covas, C.P.; Ossipov, D.A. Self-assembled hyaluronic acid nanoparticles for controlled release of agrochemicals and diosgenin. *Carbohydr. Polym.* **2017**, *173*, 157–169. [CrossRef]

Article

A Combined In-Mold Decoration and Microcellular Injection Molding Method for Preparing Foamed Products with Improved Surface Appearance

Wei Guo [1,2,3], Qing Yang [1,2,3], Huajie Mao [2,3,4,*], Zhenghua Meng [1,2,3,*], Lin Hua [1,2,3] and Bo He [2,3,4]

1 School of Automotive Engineering, Wuhan University of Technology, Wuhan 430070, China; whutgw@126.com (W.G.); yqwhut@163.com (Q.Y.); hualin@whut.edu.cn (L.H.)
2 Hubei Key Laboratory of Advanced Technology for Automotive Components, Wuhan University of Technology, Wuhan 430070, China
3 Hubei Collaborative Innovation Center for Automotive Components Technology, Wuhan 430070, China
4 School of Materials Science and Engineering, Wuhan University of Technology, Wuhan 430070, China; hbwhut@163.com
* Correspondence: maohj_whut@163.com (H.M.); meng@whut.edu.cn (Z.M.); Tel.: +86-13807171614 (H.M.); +86-13277972099 (Z.M.)

Received: 10 April 2019; Accepted: 19 April 2019; Published: 1 May 2019

Abstract: A combined in-mold decoration and microcellular injection molding (IMD/MIM) method by integrating in-mold decoration injection molding (IMD) with microcellular injection molding (MIM) was proposed in this paper. To verify the effectiveness of the IMD/MIM method, comparisons of in-mold decoration injection molding (IMD), conventional injection molding (CIM), IMD/MIM and microcellular injection molding (MIM) simulations and experiments were performed. The results show that compared with MIM, the film flattens the bubbles that have not been cooled and turned to the surface, thus improving the surface quality of the parts. The existence of the film results in an asymmetrical temperature distribution along the thickness of the sample, and the higher temperature on the film side leads the cell to move toward it, thus obtaining a cell-offset part. However, the mechanical properties of the IMD/MIM splines are degraded due to the presence of cells, while specific mechanical properties similar to their solid counterparts are maintained. Besides, the existence of the film reduces the heat transfer coefficient of the film side so that the sides of the part are cooled asymmetrically, causing warpage.

Keywords: in-mold decoration injection molding; microcellular injection molding; surface quality; mechanical properties; warpage

1. Introduction

Microcellular injection molding (MIM) technology originated from the idea of Nam Suh [1]. Compared with conventional injection molding (CIM), it has some advantages including weight reduction, cost saving, and excellent dimensional stability [2–5]. Despite these advantages, however, surface defects-such as gas flow marks and lack of smoothness-still remain one of the main drawbacks of the MIM parts, which limit MIM's application to a large extent [6].

In MIM, the polymer melt and supercritical fluid (SCF) are mixed to form the polymer/SCF single-phase solution in the barrel; the polymer/SCF single-phase solution is then injected into the mold cavity under the action of the screw. The large pressure drop causes thermodynamic instability, which provides conditions for bubble nucleation and growth. As the bubble grows, it is stretched, compressed, and even broken under the action of the shear flow. Due to the fountain flow behavior at the front of the melt, the bubble turns over to the cavity surface and solidifies under the cooling of the mold, leaving a so-called bubble mark on the surface of the part, which greatly affects the surface quality [7]. There have been many attempts to eliminate bubble marks to improve the surface quality. For example, rapid thermal cycle molding (RTCM) technology was developed to eliminate surface defects of the microcellular injection molded parts [8–12]. By increasing the temperature of the mold cavity surface, the fluidity of the melt can be increased, thereby increasing the filling ability of the melt to dissolve the bubble marks back into the melt [13]. Besides, the high-temperature mold flattens the bubble marks which have not been cooled and formed to be turned over to the mold cavity surface, and the microcellular foamed parts with a good surface quality are obtained.

Integrating co-injection with microcellular injection molding can also solve the surface appearance issue [14–17]. Microcellular co-injection molding injects the surface layer material before injecting the core material to form a sandwich structure in which the surface layer is not foamed, and the core layer is foamed. Since the surface layer is solid, the parts with good surface appearance are obtained.

It is also possible to use a combination of gas counter pressure (GCP) and microcellular injection molding to eliminate flow marks to improve surface quality [18–20]. GCP is the way to apply high pressure to the mold cavity before melt injection and to maintain a constant back pressure during the melt injection. Constant back pressure can suppress the nucleation and growth of the bubbles, and the bubbles will start to grow after the back pressure is released. At this time, the surface layer has solidified, thus improving the surface quality by reducing the bubble marks on the surface.

There have also been efforts to improve surface quality by gas-assisted microcellular injection molding [21]. Hou at el. found that the cells generated during the melt filling process could be dissolved back into the polymer melt through the high-pressure assisted gas from the GAMIM, thus improving the part's surface appearance by eliminating the sliver marks. Although these methods do have improved surface quality, they all require additional equipment, which means improvements of the experimental equipment, thus greatly increasing the cost of the experiment.

In addition, since the conventional injection molded parts need to be decorated with some patterns and logos after the injection to improve the aesthetics and surface properties, the conventional surface decoration techniques such as spraying and plating have many drawbacks, such as high defect rate, wear resistance, long production cycle, and severe environmental pollution. To solve these problems, the in-mold decoration (IMD) technique has been proposed. The IMD method integrates injection molding and decoration, thus eliminating the secondary processing of conventional injection molding. Compared with conventional injection molding, it not only reduces labor costs, but also improves work efficiency, and the injection molded parts have better stability and durability [22–25].

To solve the shortcomings of microcellular injection molding, a combined in-mold decoration and microcellular injection molding (IMD/MIM) method was proposed in this paper. The IMD, CIM, IMD/MIM and MIM simulations and experiments were conducted to verify the effectiveness of the IMD/MIM method. The mechanical properties, forming defects and cellular structure of the samples were characterized to compare the IMD, CIM, IMD/MIM and MIM.

2. Combined In-Mold Decoration and Microcellular Injection Molding

The schematic diagram of the combined in-mold decoration and microcellular injection molding (IMD/MIM) method is illustrated in Figure 1. The PET film is attached to the mold cavity before injection, the polymer melt and supercritical fluid (SCF) are mixed under the stirring of the screw to form the polymer/SCF single-phase solution in the barrel for injection, as can be seen in Figure 1a. In Figure 1b, the polymer/SCF single-phase solution is injected into the cavity under the action of the screw. The large pressure drop causes thermodynamic instability, which provides conditions for bubble nucleation and growth. Due to the presence of the film, the heat transfer coefficient on the film side is reduced, resulting in an asymmetrical temperature distribution along the thickness of the sample, thus the flow is asymmetrical. In Figure 1c, the melt filling is completed, the bubbles continue to grow until the melt solidifies. At the end of the filling, the melt temperature on the film side is higher than that on the non-film side, and the viscosity is lower than that of the non-film side.

Finally, the mold is opened, and the flexural sample is taken out as depicted in Figure 1d. The bubble is shifted toward the film side due to the higher temperature on the film side, and some cells undergo significant deformation under shear, extension, and compression of the shear flow, which is caused by the fountain flow behavior at the melt front.

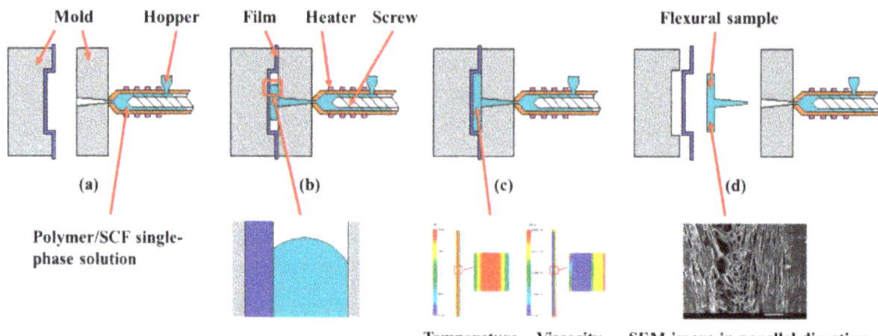

Figure 1. Schematic diagram of the IMD/MIM: (**a**) pasting film before injection; (**b**) injecting polymer/SCF single-phase solution; (**c**) cooling molding; (**d**) ejecting flexural sample.

3. Experimental Setup

3.1. Numerical Simulation

3.1.1. Finite Element Model

To correspond to the experiment, a 4-cavity was used, and the flexural sample (80 × 10 × 4 mm) was selected as the research object. The decorative film (PET) with the thickness of 0.2 mm was attached to the mold cavity surface, and the cooling system and runner system were established as required in Moldflow as shown in Figure 2. The main processing parameters were set as follows: melt temperature 220 °C, mold temperature 50 °C, coolant temperature 25 °C, foaming agent's content 0.5%; the bubble nucleation model used the fitted classical nucleation model due to the asymmetric mold cavity temperature distribution which affects bubble nucleation.

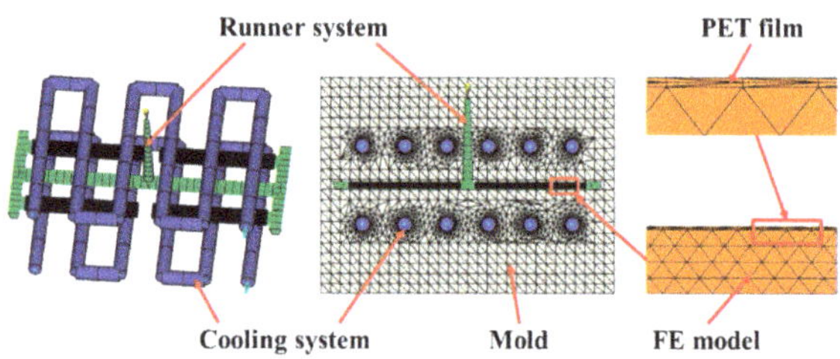

Figure 2. Finite element model.

3.1.2. Experimental Apparatus

A PVT testing machine (PVT-6000, Gotech, Beijing, China) was used to measure the PVT performance; the method adopted was the isostatic cooling method, the pressure was set to 30, 60, 90 and 120 MPa, the temperature was heated to 200 °C, and the cooling rate was 5 °C/min.

Viscosity was measured using the capillary rheometer (CR-6000, Gotech, Beijing, China); the experimental method used was to keep the temperature constant, pressurize, and then viscosity was measured at different shear rates, and the experiment was repeated at different temperatures (180, 200, 220 and 240 °C).

Cellular structure and surface topography were observed using a JSM-IT300 (JEOL Ltd., Tokyo, Japan) scanning electron microscope (SEM). The samples were ruptured after immersion in liquid nitrogen for 3 h before SEM observation. Then, platinum was sprayed on the fracture surface to conduct electricity.

A 3D optical profiler (ST400, NANOVEA Corporation, Irvine, CA, USA) was used to analyze the surface profile of the samples. Warpage was measured with a 3D laser scan (Handy scan 700, Creaform, Wuxi, China).

The tensile, flexural and impact tests were performed to characterize the mechanical properties of the samples. Tensile and flexural tests were performed using an electromechanical universal testing machine (CMT6104, MTS Systems Corp, Eden Prairie, MN, USA). The impact test was performed using an impact tester (XJUD-5.5, Chengde Jinjian Testing Instrument Co., Ltd., Chengde, China). The tensile test method was ISO 527-1:1993 with a crosshead speed of 50 mm/min. The flexural test method was ISO 178:2001 with a speed of 2 mm/min. The impact strength was measured according to ISO 180:2000. To avoid errors caused by the contingency of the test, five samples were tested under each test condition, and their average values were taken as the results.

3.1.3. Materials

The polymer material and physical foaming agent were modified PP, grade AIP-1927 (supplied by Kingfa Sci & Tech Co., Ltd., Guangzhou, China) and N_2 (supplied by Wuhan Xiangyun Industry and Trade Co., Ltd., Wuhan, China), respectively, and PET film with the thickness of 0.2 mm was used as the decorative film. Their properties were depicted in Table 1.

Table 1. Physical properties of the materials.

Family Abbreviation	Modified PP	PET
Brand	AIP-1927	543C
Solid density (g/cm^3)	1.0278	1.4050
Melt density (g/cm^3)	0.8682	1.1696
Elastic modulus (MPa)	1300	3450
Sheer modulus (MPa)	780	1225
Poisson's ratio	0.350	0.408

To perform simulation analysis more accurately, the PVT performance and viscosity property of the modified PP were tested, thus obtaining the PVT and viscosity curves as shown in Figures 3a and 4a, respectively. Then, based on the Cross-WLF model and the modified dual-domain Tait equation of state, the PVT and viscosity curves were fitted using the combination of the Mc Quaid method and the general global optimization method (LM-UGO); the fitted PVT and viscosity curves were obtained as illustrated in Figures 3b and 4b, respectively. The fitted PVT and viscosity characteristic parameters were imported into Moldflow for simulation [26].

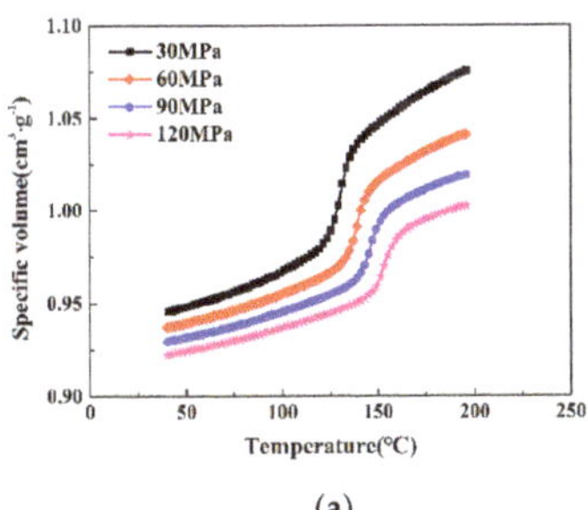

(**a**)

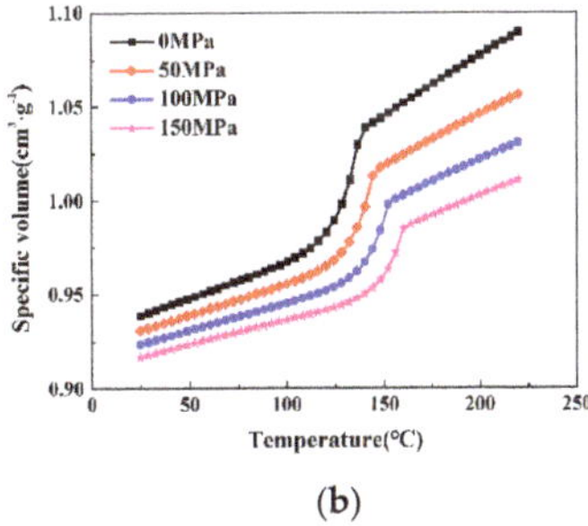

(**b**)

Figure 3. PVT performance of modified PP: (**a**) PVT curve of modified PP; (**b**) fitted PVT curve of modified PP.

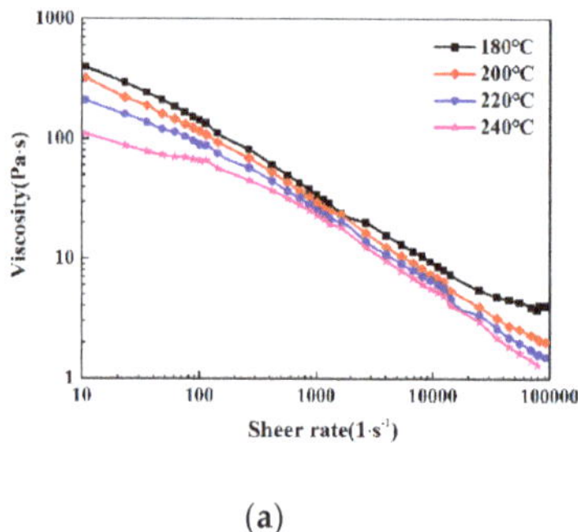

(a)

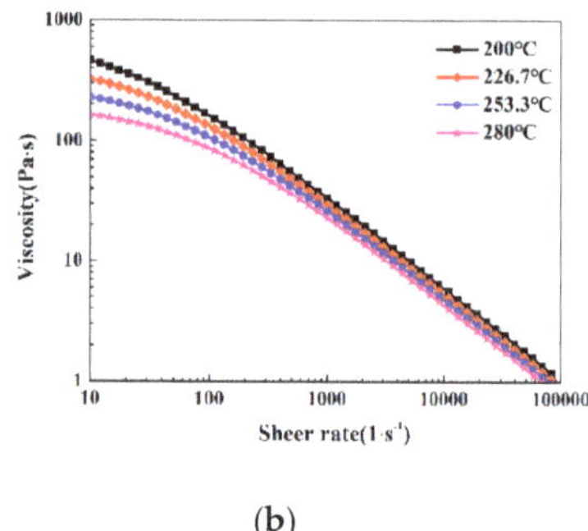

(b)

Figure 4. Viscosity property of modified PP: (**a**) viscosity curve of modified PP; (**b**) fitted viscosity curve of modified PP.

3.2. Experimental Design

The IMD, CIM, IMD/MIM and MIM experiments were carried out using an injection molding machine (HDX50, Ningbo Haida Plastic Machinery Co., Ltd., Ningbo, China) and high-pressure compressor (GBL-200/350, Beijing Zhongtuo Machinery Group Co., Ltd., Beijing, China). The processing parameters used in all experiments are as follows: injection speed 60 g/s, injection pressure 70 MPa, melt temperature 220 °C, mold temperature 50 °C, coolant temperature 25 °C, back pressure 10 MPa, cooling time 25 s, foaming agent's content 0.5%, gas dosing pressure 17.5 MPa, gas dosing time 3 s. A holding pressure of 35 MPa was applied in the CIM and IMD experiments for a holding time of 12 s.

3.3. Characterizations

Three flexural samples on the same cavity were selected as the observation object, as shown in Figure 5. The vertical section (10 × 4 mm), which represents the cross-section of the sample perpendicular to the melt flow direction, was observed to assess the cell structure and distribution. The parallel section (10 × 4 mm), which represents the cross-section of the sample parallel to the melt flow direction, was examined to obtain the orientation and deformation of the cells. The cell diameter and cell number were measured by Image Pro Plus 6.0 based on SEM micrographs. The cell average radius (R) was calculated using Equation (1):

$$R = \frac{\sum_{i=1}^{n} d_i}{2n} \quad (1)$$

where d_i indicates the diameter of the i_{th} cell within the designated area; n represents the number of cells in the given SEM micrograph. The cell density (N) was calculated by using Equation (2):

$$N = \left(\frac{nM^2}{A}\right)^{\frac{3}{2}} \quad (2)$$

where M represents the magnification of the SEM micrograph; A represents the area of the SEM micrograph.

The ratio of the length to diameter (c) was used to characterize the cell deformation and the tilt angle (θ) to characterize the cell orientation. Similarly, the ratio of the length and tilt angle were obtained by Image Pro Plus 6.0. As can be seen from Figure 5, the ratio of length to diameter (c) could be calculated as Equation (3):

$$c = \frac{L}{B} \quad (3)$$

where L represents the major axis length of the cell, and B represents the minor axis length of the cell. The tilt angle (θ) was defined as the angle between the major axis of the cell and the melt flow direction (sharp angle), which is shown in Figure 6 [27].

The rectangular sample (10 mm × 10 mm) which is located at the center of the flexural sample along the melt flow direction was used for observing surface appearance; the observation surface is depicted in Figure 5. The arithmetic mean roughness (Ra) and root mean square roughness (Rq) were used to characterize the surface roughness. Both Ra and Rq were calculated from the 3D optical profiler with a precision of 0.01 μm. Five samples were analyzed for each condition and the average surface roughness obtained were taken as the results.

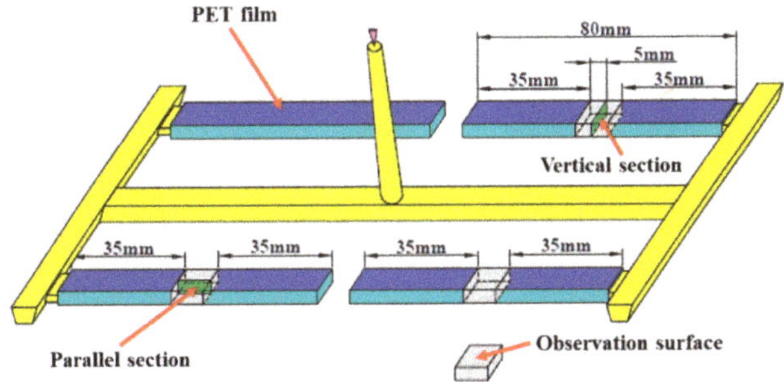

Figure 5. Preparation of samples for SEM observation from flexural samples.

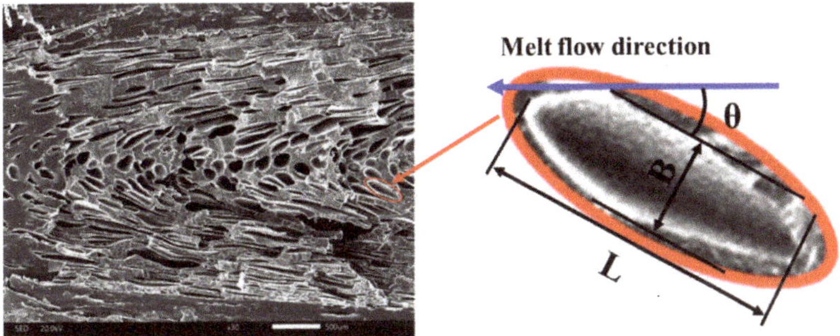

Figure 6. Schematic diagram of ratio of length to diameter and tilt angle.

4. Results and Discussion

4.1. Mechanical Properties

4.1.1. Tensile Properties

The tensile stress-strain curves of the IMD, CIM, IMD/MIM and MIM samples are given in Figure 7a. As can be seen from Figure 7a, the tensile strength of foamed samples is lower than that of their solid counterparts. This is attributed to the presence of cells, which reduces the bearing area, and some larger cells also cause stress concentration, thus reducing the tensile strength of the spline. The decrease in tensile strength of CIM sample is approximately 0.9%, which is almost negligible when compared with that of IMD sample. It is no exception that the tensile strength of the IMD/MIM sample is only a little higher than that of MIM sample. It can be inferred that the presence of the film has no significant effect on the tensile strength. The specific tensile stress-strain curves of the IMD, CIM, IMD/MIM and MIM samples are obtained by taking the weight reduction into account, as shown in Figure 7b. It can be observed that the specific tensile strength of the foamed samples is a little lower than that of the solid ones, which shows that the foamed samples maintain specific tensile strength similar to that of their solid counterparts. The quantitative comparison of the specific tensile properties such as specific tensile strength, strain-at-break, specific Young's modulus, and specific toughness of the four kinds of samples are given in Figure 8. The density of the foamed specimens of IMD/MIM and MIM is 13.75% and 15.47% less than that of their solid counterparts, respectively. It is found that the specific tensile strength and specific Young's modulus of the foamed samples are a litter lower than those of their solid counterparts, which means foamed samples can maintain the same specific tensile properties as those of the solid ones.

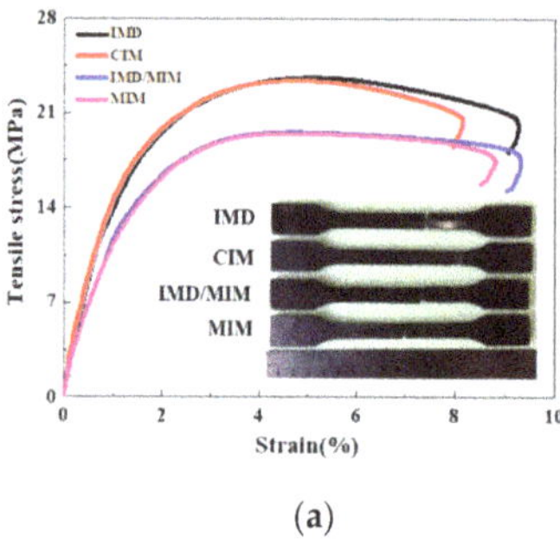

(a)

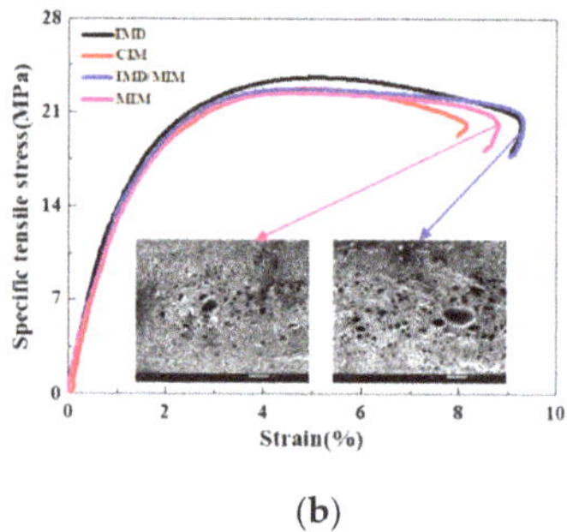

(b)

Figure 7. The tensile properties of the IMD, CIM, IMD/MIM and MIM samples: (**a**) tensile stress-strain curves; (**b**) specific tensile stress-strain curves.

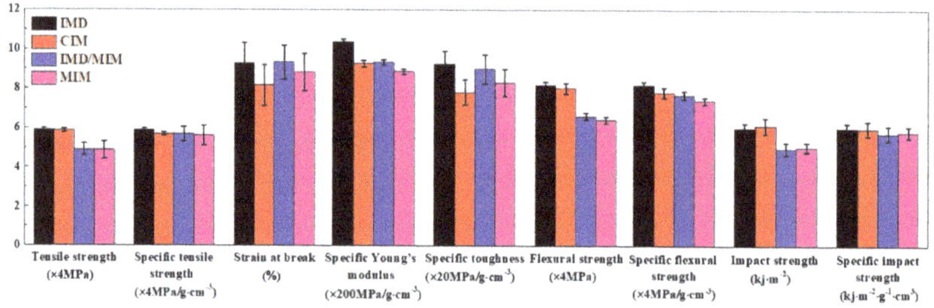

Figure 8. Comparison of mechanical properties among the IMD, CIM, IMD/MIM and MIM samples (the error bars in the figure represent the standard deviation).

4.1.2. Flexural Properties

Figure 9a gives the flexural stress-strain curves of the IMD, CIM, IMD/MIM and MIM samples. As shown in Figure 9a, the foamed samples have a higher flexural strength and elastic modulus than those of their solid counterparts, which indicates that the flexural strength and stiffness of foamed samples are reduced. This is attributed to the presence of cells, which reduces the material on the forced cross-section, thereby decreasing the rigidity and resulting in a reduction in the flexural strength of the spline. It can be seen from Figure 9a that the presence of the film has no significant effect on the flexural strength. The specific flexural stress-strain curves of the IMD, CIM, IMD/MIM and MIM samples are given in Figure 9b. It can be observed that the specific flexural strength of the foamed splines is a little lower than that of the solid ones, which shows that the foamed samples can maintain specific flexural strength similar to that of their solid counterparts. The quantitative comparison of the specific flexural properties (specific flexural strength) of the four kinds of samples is given in Figure 8. As shown in Figure 8, the specific flexural strength of the foamed samples is a little lower than that of the solid ones, which is almost negligible. This means that the specific flexural strength of the foamed samples does not decrease significantly due to the presence of the cells.

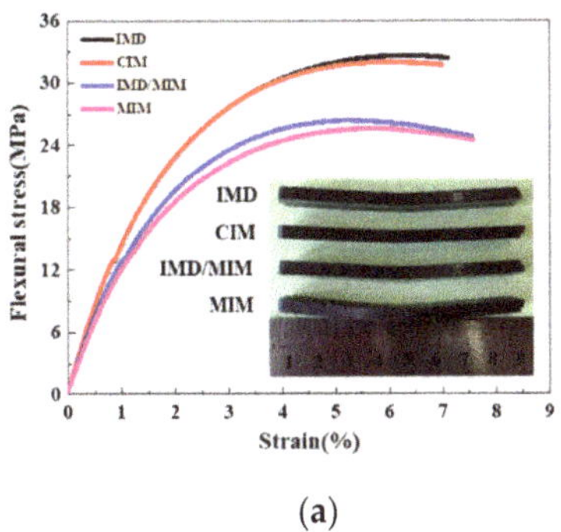

(a)

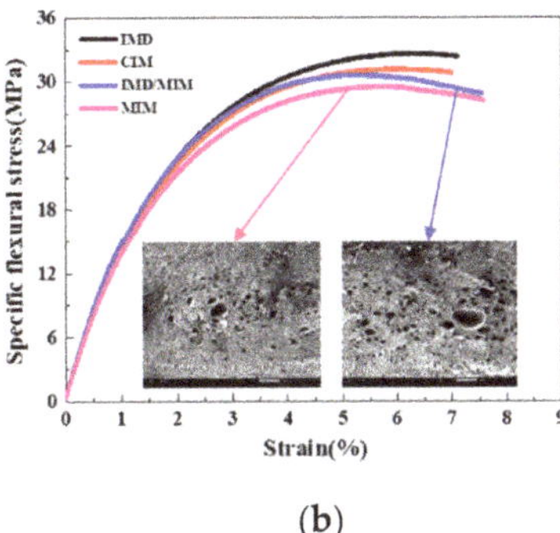

(b)

Figure 9. The flexural properties of the IMD, CIM, IMD/MIM and MIM samples: (**a**) flexural stress-strain curves; (**b**) specific flexural stress-strain curves.

4.1.3. Impact Properties

The quantitative comparison of the flexural properties (flexural strength and specific flexural strength) of the IMD, CIM, IMD/MIM and MIM samples is shown in Figure 8. It can be seen that the impact strength of the foamed samples is lower than that of their solid counterparts. The presence of cells helps to passivate the crack tip and effectively prevent further crack propagation. Besides, the deformation of the cells can absorb part of the energy, thereby improving the impact strength. However, uneven cell distribution causes local large cells to become stress concentration points, thereby reducing impact toughness. The combined effect of the two factors results in a decrease in the impact strength. However, it can be seen from Figure 8 that the specific impact strength of the foamed samples is only a little lower than that of their solid counterparts, which means that the foamed samples can maintain the same specific impact properties as those of the solid ones. Similarly, the presence of the film has no significant effect on the impact strength, as shown in Figure 8.

4.2. Formation Defects

4.2.1. Surface Topography

The surface topography, surface profile, experimental volume shrinkage and simulated volume shrinkage of the IMD, CIM, IMD/MIM and MIM samples are shown in Figure 10. It can be seen that the MIM sample has a rough surface with many silver streaks and cracked bubbles, while the surface of the IMD/MIM sample is as smooth as those of the IMD and CIM samples, with a small amount of burst bubbles. Due to the fountain flow behavior at the front of the melt, the grown bubbles turn to the sides, reach the mold surface and deform or even rupture under the shearing and stretching of the shear flow. Under the action of the mold, the bubbles are flattened on the surface to form bubble marks, which greatly reduces the surface quality of the part. However, the surface quality of IMD/MIM sample is almost as good as that of CIM sample, with few bubble marks on the surface. The presence of the film, which reduces the heat transfer coefficient on the film side, results in a higher melt temperature on the film side and a lower cooling rate. When the bubbles are turned over to the surface of the mold, they are flattened by the film before being completely cooled. From the surface profile in Figure 10, we can find that the solid samples have shrunk while the foamed splines have not, which is consistent with the results of simulation and experiment. Due to the growth of the cells, the voids created by the shrinkage of the melt are supplemented, allowing the melt to fill the entire cavity and reducing the shrinkage of the samples.

The surface roughness of the IMD, CIM, IMD/MIM and MIM samples is shown in Figure 11. It can be found that the surface roughness of IMD sample is smaller than that of CIM sample, and the roughness of IMD/MIM sample is smaller than that of MIM sample, which means that the presence of the film reduces the surface roughness, so that the sample has a better surface quality. Compared with the CIM sample, the MIM sample has much larger surface roughness, which is attributed to the presence of many bubble marks on the surface of MIM parts, resulting in the poor surface quality. However, the surface roughness of IMD/MIM is only a little smaller than that of IMD samples, which is almost negligible. The presence of the film compensates for the defect of the surface quality of the foamed samples. The result is consistent with that of the surface topography, so it can be concluded that the IMD/MIM method does improve surface quality of the foamed samples, which plays a significant role in the promotion and application of microcellular injection molding technology.

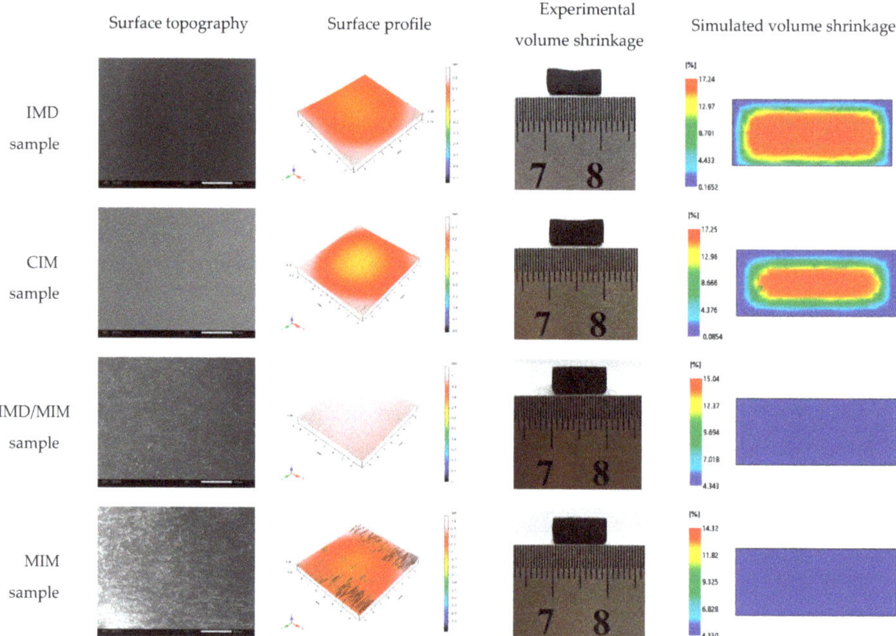

Figure 10. The comparison of the surface topography, surface profile, experimental volume shrinkage and simulated volume shrinkage among the IMD, CIM, IMD/MIM and MIM samples.

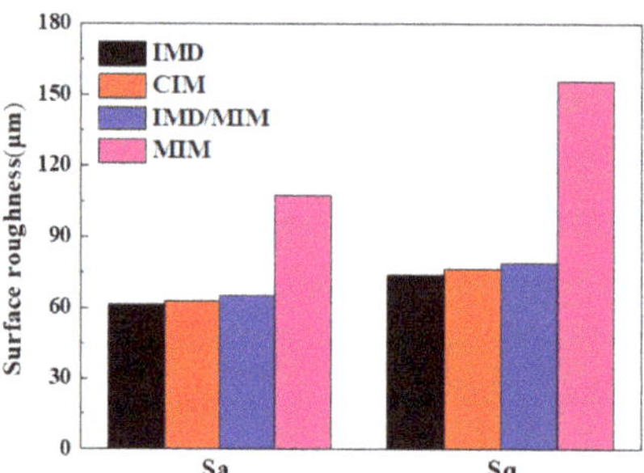

Figure 11. The comparison of the surface roughness indexes among the IMD, CIM, IMD/MIM and MIM samples.

4.2.2. Warpage

The warpage of the IMD, CIM, IMD/MIM, and MIM samples is shown in Figure 12. It is observed that the warpage value of the IMD sample is 2.961 mm, which is much greater than that (0.7144 mm) of CIM sample, which is consistent with the experimental result. The existence of the film reduces the heat transfer coefficient on the film side, so that the cooling rate of the melt on the film side is lower than that of the non-film side, resulting in an asymmetrical temperature on both sides of the spline, and the temperature difference between the two sides at the end of the melt filling (35.13 °C) is much greater than that (5.92 °C) of CIM sample, as shown in Figure 12. Therefore, the two sides are not uniformly shrunk, and the sample is bent toward the high-temperature side, that is, the film side. Besides, the temperature at each point on the part surface is also different, resulting in different shrinkage rates everywhere, which also increases the warpage. The warpage of the IMD/MIM sample (1.898 mm) is much larger than that of the MIM sample (0.5887 mm), but the warpage is reduced compared with that of the IMD sample, which is consistent with the experimental result. The growth of the cells facilitates the replenishment of the voids caused by the shrinkage of the melt, so that the melt fills the entire cavity, thereby reducing the warpage to some extent.

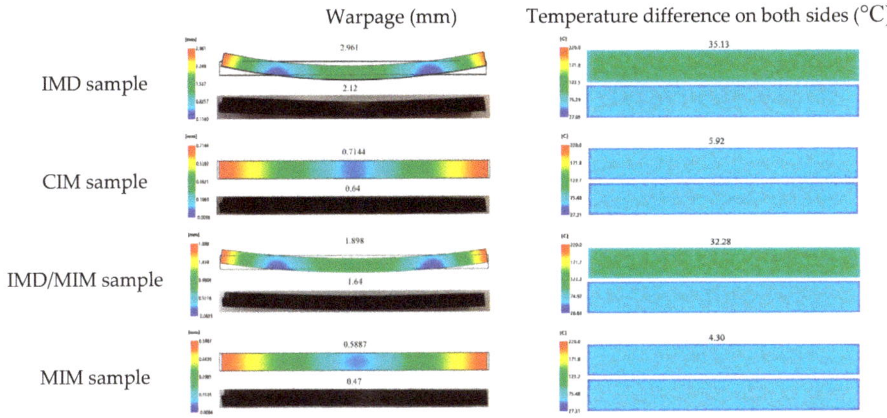

Figure 12. The comparison of the warpage and temperature difference on both sides among the IMD, CIM, IMD/MIM and MIM samples.

4.3. Cellular Structure

Since the cells in the vertical section are hardly deformed, we can divide the vertical section of the MIM and IMD/MIM samples into three layers, namely, the transition layer A, transition layer B and core layer C based on the cell size distribution, as shown in Figure 13. For the cells in the parallel section, they are stretched and compressed under the action of the shear flow, resulting in deformation. We no longer stratify according to the cell size distribution, but determine the boundary between the transition layer and the core layer according to the degree of cell deformation.

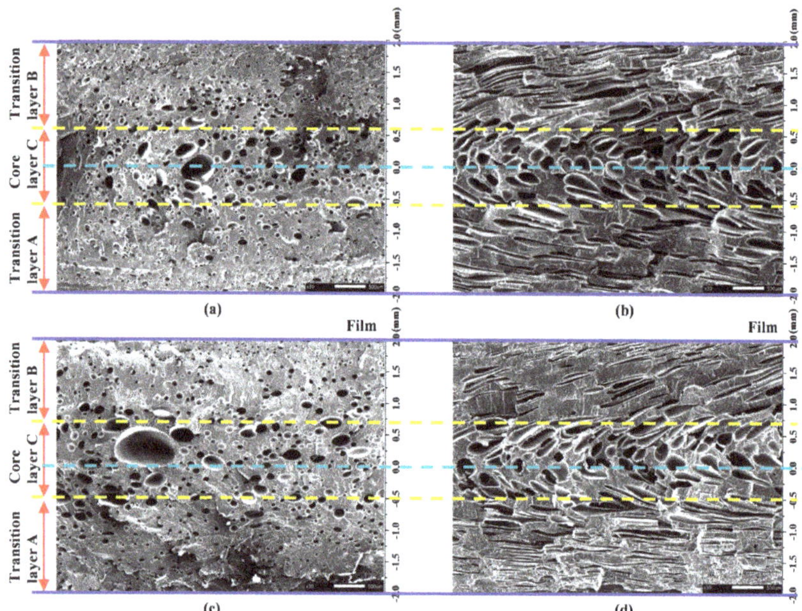

Figure 13. Layered schematic diagram: (**a**) and (**b**) layered schematic diagram of vertical section and parallel section of the MIM sample, respectively; (**c**) and (**d**) layered schematic diagram of vertical section and parallel section of the IMD/MIM sample, respectively.

4.3.1. Cellular Structure of Vertical Section

The simulation results and SEM micrographs of the vertical section of MIM and IMD/MIM samples are given in Figure 14. It can be seen that whether it is the MIM or IMD/MIM sample, the cell size of the core layer is larger than that of the transition layer, which is consistent with the cell size distribution in Figure 14a,b. Combined with the temperature curves in Figure 14e,f, it can be found that the temperature distribution along the thickness direction of the sample is not uniform, and the temperature of the core layer is higher than that of the transition layer, which provides a longer time for cell growth. Therefore, there is such a phenomenon that the core layer has large cells and the transition layer cells are small. As can be seen from Figure 14a,b, the cell radius and tensile modulus change in the opposite direction in the thickness direction, and the maximum cell radius corresponds to the minimum tensile modulus, which illustrates that the cell size affects the tensile modulus. The cell density of the core layer is less than that of the transition layer. In combination with the viscosity curves of Figure 14e,f, higher temperature of core layer provides longer growth times for cells, while lower viscosity results in lower melt strength, thus reducing cell growth resistance. Therefore, the cells are merged and collapsed more, resulting in the cell density of the core layer being lower than that of the transition layer.

As can be seen from Figure 15, it is not difficult to find that the cells of the two transition layers of the MIM sample are similar in size and density. However, the cell of transition layer B (with film side) of the IMD/MIM sample is larger than that of the transition layer A, while the cell density on both sides is not much different. The presence of the film reduces the heat transfer coefficient on the film side, so cells have a longer time to grow due to the higher temperature of the film side. At the same time, due to the higher temperature, the solubility of the gas in the melt is lowered, the thermodynamic instability of the gas is enhanced, thus increasing the nucleation rate of cells. However, the number of cell nucleation is primarily determined by the thermodynamic instability induced by the large pressure drop that occurs when the polymer/SCF single-phase solution is injected into the mold cavity, so that there is no significant difference in the cell density of the two transition layers.

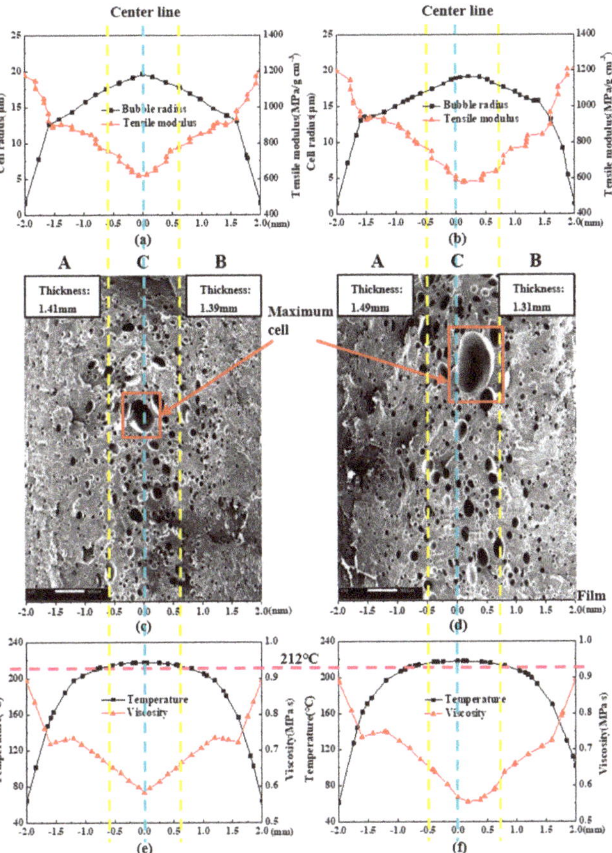

Figure 14. Simulation results and SEM micrographs of the vertical section of MIM and IMD/MIM samples: (**a**), (**b**) and (**e**), (**f**) the simulation results along the center line of vertical section of MIM and IMD/MIM samples, respectively; (**c**) and (**d**) the SEM micrograph of MIM and IMD/MIM samples, respectively.

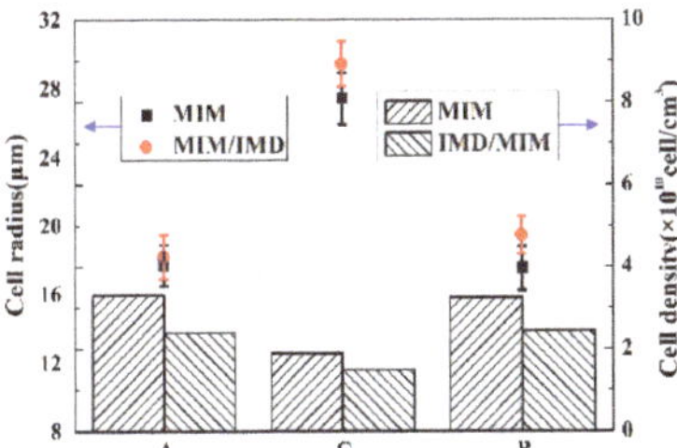

Figure 15. The cell radius and cell density of the transition layer A, core layer C and transition layer B of MIM and IMD/MIM samples.

From Figure 14c,d, we can see that the largest cell of the MIM sample is on the center line, while the largest cell of the IMD/MIM sample is offset from the center line toward the film side. Due to the presence of the film, the temperature on the film side is higher than that on the non-film side. The maximum temperature no longer appears on the center line of the sample thickness, but is shifted to a certain distance toward the film side, so the largest cell of the IMD/MIM is biased toward the film side. The distance from the left edge of the MIM core layer to the left surface is 1.41 mm, and the distance from the right edge to the right surface is 1.39 mm, while the distance from the left edge of the IMD/MIM core layer to the left surface is approximately 1.49 mm and the distance from the right edge to the right surface is approximately 1.31 mm. It can be clearly seen that the core layer of the MIM is symmetrical about the center line, while the core layer of the IMD/MIM is offset by 0.08 mm toward the film side.

The temperature field and viscosity field affect the boundary position of the core layer. As shown in Figure 14e,f, the highest point of the temperature curve and the lowest point of the viscosity curve of MIM are on the center line, while the highest point of the temperature curve and the lowest point of the viscosity curve of IMD/MIM are off the center line, approaching the film side, so the core layer of the IMD/MIM is offset toward the film side. The offset of the core layer affects the thickness of the transition layer. It can be seen from Figure 14c,d that the thicknesses of the two transition layers of MIM sample are similar. However, the thickness of transition layer B (with film side) of IMD/MIM sample is less than that of the transition layer A. The intersections of the four yellow dashed lines and the purple dashed line shown in Figure 14e,f represent the intersections of the above four boundary positions with the temperature profile at the end of the melt filling. It can be seen that the temperatures at these locations are almost all about 212 °C. Only the temperature at the left boundary of the IMD/MIM core layer in Figure 14f is slightly above 212 °C. The melt viscosity at this boundary is large, resulting in low melt strength, so cells easily collapse and merge, thus a higher temperature is required.

4.3.2. Cellular Structure of Parallel Section

SEM micrographs of the parallel section of MIM and IMD/MIM samples are given in Figure 16. It can be seen that whether it is the MIM or IMD/MIM sample, the core layer cells are spherical, while the transition layer cells are severely deformed and appear elliptical. When the transition layer cells grow during the melt filling stage, they are stretched and deformed by the strong shearing action of the melt. Higher temperature of the core layer results in lower melt viscosity, which means the shear stress subjected is smaller, so most of the core layer cells are still in a spherical state. The cell density of the core layer is lower than that of the transition layer. Higher core layer temperature provides longer growth time for cells, while lower melt strength results in reduced cell growth resistance. Therefore, cells are more likely to merge or collapse, resulting in the cell density of the core layer being lower than that of the transition layer.

Due to the severe deformation of the transition layer cells, we no longer study the cell size, but use the ratio of the length to diameter to characterize the cell deformation and tilt angle to characterize the cell orientation. As can be seen from Figure 17a, it is not difficult to find that the ratio of the length to diameter and the tilt angle of the two transition layer cells of MIM are both similar. The length to diameter ratio of the transition layer B (film side) cells of the IMD/MIM is smaller than that of the transition layer A, and the tilt angle is larger than that of the transition layer A. This is because the temperature of the film side is higher, and the melt viscosity is lower, so the cells are less sheared during the melt filling stage, thus resulting in a smaller aspect ratio and smaller angle of the cell from the vertical direction, so the tilt angle is larger. The cell density on the film side is similar to that on the non-film side, which is consistent with the cell distribution phenomenon in the vertical direction.

From Figure 16a,b, we can see that the distance from the left edge of the MIM core layer to the left surface is 1.41 mm, and the distance from the right edge to the right surface is 1.39 mm, while the distance from the left edge of the IMD/MIM core layer to the left surface is 1.49 mm and the distance from the right edge to the right surface is 1.31 mm. From these data, we can find that the thickness of

IMD/MIM core layer is the same as that of MIM, which is almost 1.2 mm, but the core layer of MIM is symmetrical about the center line, while the core layer of IMD/MIM is offset to the film side by 0.08 mm. Due to the reduced heat transfer coefficient on the film side, the highest temperature no longer appears on the center line of the thickness direction, but is offset toward the film side by a certain distance; therefore, the core layer of the IMD/MIM is biased toward the film side.

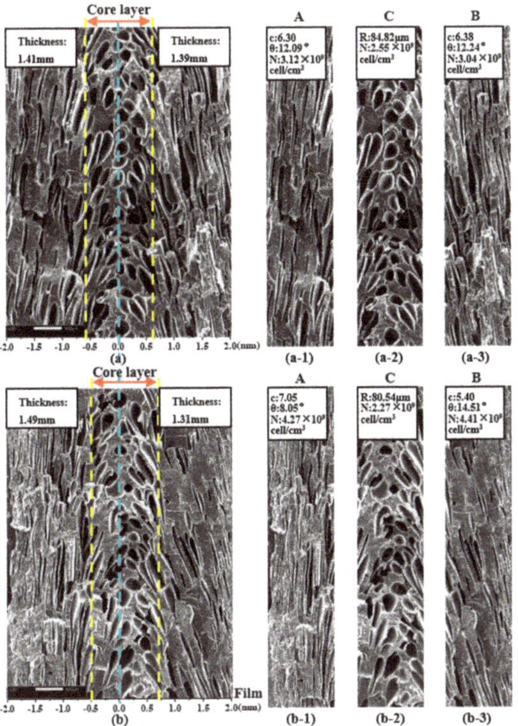

Figure 16. SEM micrographs of the parallel section of MIM and IMD/MIM samples: (**a**) and (**b**) the SEM micrograph of the parallel section of MIM and IMD/MIM samples, respectively; (**a-1**), (**a-2**), (**a-3**) and (**b-1**), (**b-2**), (**b-3**) the SEM micrographs of the transition layer A, core layer C and transition layer B of the parallel section of MIM and IMD/MIM samples, respectively.

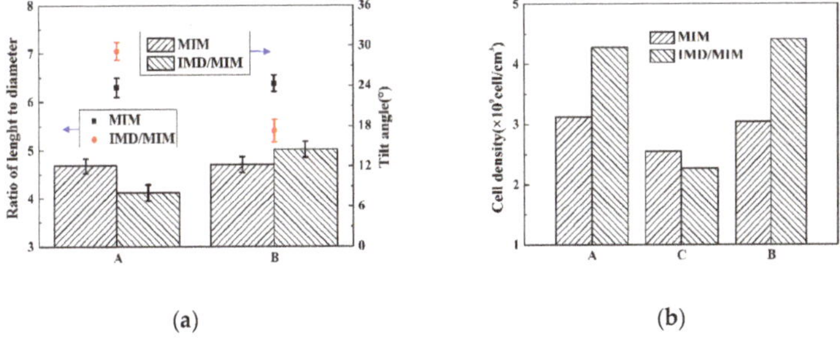

Figure 17. (**a**) The ratio of length to diameter and tilt angle of the transition layer A and transition layer B of MIM and IMD/MIM samples; (**b**) The cell density of the transition layer A, core layer C and transition layer B of MIM and IMD/MIM samples.

5. Conclusions

A combined in-mold decoration and microcellular injection molding (IMD/MIM) method was presented in this paper. The in-mold decoration injection molding (IMD), conventional injection molding (CIM), IMD/MIM and microcellular injection molding (MIM) simulations and experimental comparisons were performed to validate the effectiveness of the IMD/MIM method. The mechanical properties, forming defects and cellular structure of the samples were analyzed and compared.

The results show that the proposed IMD/MIM method can improve surface quality while maintaining almost no degraded mechanical properties and obtaining a cell-offset part. Compared with IMD, the mechanical properties of IMD/MIM samples are reduced due to the presence of cells. However, the specific mechanical properties are rarely reduced, and specific mechanical properties similar to their solid counterparts are maintained. Compared with MIM, the presence of the film flattens the bubbles that have not been cooled and turned to the surface, thus improving the surface quality by eliminating the bubble marks, and the IMD/MIM parts have almost the same good surface appearance as that of CIM. However, the existence of the film reduces the heat transfer coefficient on the film side, so that the two sides of the part are cooled asymmetrically, resulting in an asymmetry temperature, thus causing warpage. The presence of the film results in an asymmetrical temperature distribution along the thickness of the specimen. The higher temperature on the film side leads the cells to move toward it, obtaining a cell-offset part. Therefore, the method provides an effective way of producing foamed parts with improved surface appearance, which provides a broad development prospect for its applications in many industries such as furniture packaging, construction and automotive interior and exterior.

Author Contributions: Conceptualization, W.G. and Z.M.; Software, Q.Y.; Formal Analysis, Q.Y.; Investigation, Q.Y.; Resources, L.H.; Data Curation, Q.Y.; Writing-Original Draft Preparation, Q.Y.; Writing-Review & Editing, W.G. and B.H.; Supervision, W.G.; Funding Acquisition, H.M.

Funding: This research was funded by the National Natural Science Foundation of China Youth Fund (NO.51605356), the 111Project (B17034), Innovative Research Team Development Program of Ministry of Education of China (No. IRT_17R83) and the fundamental research funds for the central universities (WUT: 2017IVB035).

Acknowledgments: The authors would like to acknowledge the financial support from the National Natural Science Foundation of China Youth Fund (NO.51605356), the 111Project (B17034), Innovative Research Team Development Program of Ministry of Education of China (No. IRT_17R83) and the fundamental research funds for the central universities (WUT: 2017IVB035).

Conflicts of Interest: The authors declare no conflict of interest.

References

1. Kumar, V.; Suh, N.P. A process for making microcellular thermoplastic parts. *Eng. Sci.* **1990**, *30*, 1323–1329. [CrossRef]
2. Kramschuster, A.; Cavitt, R.; Ermer, D.; Shen, C.; Chen, Z.; Turng, L.-S. Quantitative Study of Shrinkage and Warpage Behavior for Microcellular and Conventional Injection Molding. *Tribology* **2005**, *45*, 535–543.
3. Cha, S.W.; Yoon, J.D. Label behavior property attached on microcellular foamed parts. *J. Appl. Sci.* **2005**, *98*, 289–293. [CrossRef]
4. Chen, S.-C.; Liao, W.-H.; Chien, R.-D. Structure and mechanical properties of polystyrene foams made through microcellular injection molding via control mechanisms of gas counter pressure and mold temperature. *Int. Commun. Heat Mass Transf.* **2012**, *39*, 1125–1131. [CrossRef]
5. Sun, X.; Kharbas, H.; Peng, J.; Turng, L.-S. A novel method of producing lightweight microcellular injection molded parts with improved ductility and toughness. *Polymer* **2015**, *56*, 102–110. [CrossRef]
6. Lee, J.; Turng, L.-S.; Dougherty, E.; Gorton, P. A novel method for improving the surface quality of microcellular injection molded parts. *Polymer* **2011**, *52*, 1436–1446. [CrossRef]
7. Dong, G.; Wang, G.; Zhang, L.; Zhao, G.; Li, S. Bubble morphological evolution and surface defect formation mechanism in the microcellular foam injection molding process. *RSC Adv.* **2015**, *5*, 70032–70050.
8. Chen, S.-C.; Lin, Y.-W.; Chien, R.-D.; Li, H.-M. Variable mold temperature to improve surface quality of microcellular injection molded parts using induction heating technology. *Adv. Technol.* **2008**, *27*, 224–232. [CrossRef]

9. Wang, G.L.; Zhao, G.Q.; Wang, J.C.; Zhang, L. Research on formation mechanisms and control of external and internal bubble morphology in microcellular injection molding. Polym. *Eng. Sci.* **2014**, *55*, 807–835.
10. Xiao, C.-L.; Huang, H.-X. Development of a rapid thermal cycling molding with electric heating and water impingement cooling for injection molding applications. *Appl. Eng.* **2014**, *73*, 712–722. [CrossRef]
11. Xiao, C.-L.; Huang, H.-X.; Yang, X. Development and application of rapid thermal cycling molding with electric heating for improving surface quality of microcellular injection molded parts. *Appl. Eng.* **2016**, *100*, 478–489. [CrossRef]
12. Zhao, G.Q.; Wang, G.L.; Guan, Y.J.; Li, H.P. Research and application of a new rapid heat cycle molding with electric heating and coolant cooling to improve the surface quality of large LCD TV panels. *Polym. Adv. Technol.* **2011**, *22*, 476–487. [CrossRef]
13. Lee, J.; Turng, L.-S. Improving surface quality of microcellular injection molded parts through mold surface temperature manipulation with thin film insulation. *Eng. Sci.* **2010**, *50*, 1281–1289. [CrossRef]
14. Zhang, K.Y.; Nagarajan, V.; Zarrinbakhsh, N.; Mohanty, A.K.; Misra, M. Co-injection molded new green composites from biodegradable polyesters and miscanthus fibers. Macromol. *Mater. Eng.* **2014**, *299*, 436–446.
15. Suhartono, E.; Chang, Y.-H.; Lee, K.H.; Chen, S.-C. Improvement on the surface quality of microcellular injection molded parts using microcellular co-injection molding with the material combinations of PP and PP-GF. *Int. J. Plast. Technol.* **2017**, *21*, 239–251. [CrossRef]
16. Turng, L.-S.; Kharbas, H. Development of a Hybrid Solid-Microcellular Co-injection Molding Process. *Int. Process.* **2004**, *19*, 77–86. [CrossRef]
17. Gong, S.; Yuan, M.; Chandra, A.; Kharbas, H.; Osorio, A.; Turng, L.S. Microcellular injection molding. *Int. Polym. Proc.* **2005**, *20*, 202–214. [CrossRef]
18. Chen, S.C.; Hsu, P.S.; Hwang, S.S. The effects of gas counter pressure and mold temperature variation on the surface quality and morphology of the microcellular polystyrene foams. *J. Appl. Polym. Sci.* **2013**, *127*, 4769–4776. [CrossRef]
19. Lee, J.W.; Lee, R.E.; Wang, J.; Jung, P.U.; Park, C.B. Study of the foaming mechanisms associated with gas counter pressure and mold opening using the pressure profiles. *Chem. Eng. Sci.* **2017**, *167*, 105–119. [CrossRef]
20. Li, S.; Zhao, G.; Wang, G.; Guan, Y.; Wang, X. Influence of relative low gas counter pressure on melt foaming behavior and surface quality of molded parts in microcellular injection molding process. *J. Cell. Plast.* **2014**, *50*, 415–435. [CrossRef]
21. Hou, J.J.; Zhao, G.Q.; Wang, G.L.; Dong, G.W.; Xu, J.J. A novel gas-assisted microcellular injection molding method for preparing lightweight foams with superior surface appearance and enhanced mechanical performance. *Mater. Des.* **2017**, *127*, 115–125. [CrossRef]
22. Phillips, C.; Claypole, T.; Gethin, D.; Phillips, C. Mechanical properties of polymer films used in in-mould decoration. *J. Mater. Process. Technol.* **2008**, *200*, 221–231. [CrossRef]
23. Chen, H.-L.; Chen, S.-C.; Liao, W.-H.; Chien, R.-D.; Lin, Y.-T. Effects of insert film on asymmetric mold temperature and associated part warpage during in-mold decoration injection molding of PP parts. *Int. Commun. Heat Mass Transf.* **2013**, *41*, 34–40. [CrossRef]
24. Chen, S.-C.; Huang, S.-T.; Lin, M.-C.; Chien, R.-D. Study on the thermoforming of PC films used for in-mold decoration. *Int. Commun. Heat Mass Transf.* **2008**, *35*, 967–973. [CrossRef]
25. Lee, D.; Chen, W.-A.; Huang, T.-W.; Liu, S.-J. Factors Influencing the Warpage in In-Mold Decoration Injection Molded Composites. *Int. Process.* **2013**, *28*, 221–227. [CrossRef]
26. Mao, H.J.; He, B.; Guo, W.; Zhang, M.Y.; Yuan, T.Q. Fitting and experimental verification of viscosity model and PVT model of modified PP material based on LM-UGO algorithm. *Plast. Sci. Technol.* **2018**, *46*, 37–42.
27. Mao, H.; He, B.; Guo, W.; Hua, L.; Yang, Q. Effects of Nano-CaCO3 Content on the Crystallization, Mechanical Properties, and Cell Structure of PP Nanocomposites in Microcellular Injection Molding. *Polymers* **2018**, *10*, 1160. [CrossRef]

Article

Electrically Self-Healing Thermoset MWCNTs Composites Based on Diels-Alder and Hydrogen Bonds

Guilherme Macedo R. Lima [1], Felipe Orozco [1], Francesco Picchioni [1], Ignacio Moreno-Villoslada [2], Andrea Pucci [3], Ranjita K. Bose [1,*] and Rodrigo Araya-Hermosilla [4,*]

1 Department of Chemical Product Engineering, ENTEG, University of Groningen, Nijenborgh 4, 9747AG Groningen, The Netherlands; g.de.macedo.rooweder.lima@rug.nl (G.M.R.L.); f.orozco.gutierrez@rug.nl (F.O.); f.picchioni@rug.nl (F.P.)

2 Laboratorio de Polímeros, Instituto de Ciencias Químicas, Facultad de Ciencias, Universidad Austral de Chile, Valdivia 5090000, Chile; imorenovilloslada@uach.cl

3 Department of Chemistry and Industrial Chemistry, University of Pisa, Via Moruzzi 13, 56124 Pisa, Italy; andrea.pucci@unipi.it

4 Programa Institucional de Fomento a la Investigación, Desarrollo e Innovación, Universidad Tecnológica Metropolitana, Ignacio Valdivieso 2409, P.O. Box 8940577, San Joaquín, Santiago 8940000, Chile

* Correspondence: r.k.bose@rug.nl (R.K.B.); rodrigo.araya@utem.cl (R.A.-H.); Tel.: +31-50-36-34486 (R.K.B.); +56-2-27877911 (R.A.-H.)

Received: 11 October 2019; Accepted: 11 November 2019; Published: 14 November 2019

Abstract: In this work, we prepared electrically conductive self-healing nanocomposites. The material consists of multi-walled carbon nanotubes (MWCNT) that are dispersed into thermally reversible crosslinked polyketones. The reversible nature is based on both covalent (Diels-Alder) and non-covalent (hydrogen bonding) interactions. The design allowed for us to tune the thermomechanical properties of the system by changing the fractions of filler, and diene-dienophile and hydroxyl groups. The nanocomposites show up to 1×10^4 S/m electrical conductivity, reaching temperatures between 120 and 150 °C under 20–50 V. The self-healing effect, induced by electricity was qualitatively demonstrated as microcracks were repaired. As pointed out by electron microscopy, samples that were already healed by electricity showed a better dispersion of MWCNT within the polymer. These features point toward prolonging the service life of polymer nanocomposites, improving the product performance, making it effectively stronger and more reliable.

Keywords: Paal-Knorr reaction; polyketone; carbon nanotubes; nanocomposite; Diels-Alder; click-chemistry; hydrogen bonding; self-healing; re-workability; recycling; Joule heating

1. Introduction

Self-healing thermoset polymer materials represent an outstanding approach and cost-effective solution for non-recyclable thermoset applications [1]. These materials possess the remarkable ability of being mended, which thus prolongs their service life [2]. Particularly, the self-healing approach in thermoset polymer materials relies on the ability of the linking moieties to cleave—and—reform upon exposure to a certain stimulus. Thus, microcracks on the material can be repaired based on such reversible features [3,4]. For instance, such a healing effect can be triggered by heat [5–8] or light [9–13].

Intensive research efforts are also dedicated to the synthesis of self-healing thermoset polymers by including organic/inorganic nanofillers during in-situ polymerization to improve the nanofiller distribution, strength, modulus, and toughness of the final thermoset nanocomposite material to achieve further improvements [14–17]. As a normal procedure, the weight content of the nanofiller (e.g., carbon nanotubes) in the matrix is tailored to gain the maximal reinforcement from the filler while considering its aspect ratio and loading. Accordingly, the chemical functionalization of nanofillers (via covalent attachment of compatibilizing functional groups or polymers) has been reported to increase the homogeneous distribution of the filler and the mechanical performance of the nanocomposites, even while using relatively high filler loadings [16,18–20]. Although nanocomposites with exceptional mechanical and self-healing properties have already been produced [4,21,22], the healing process for most of the applications still relies on direct heating as external stimulus for repairing. The latter hinders macro-scale applications of these materials due to the high-energy source used for in-situ (i.e., in-service) damage repairing [23].

Recently, many authors have reported approaches for in-situ damage repairing with remarkable achievements [24–27]. Among those, a newly emerging system is represented by electrically self-healing nanocomposite. Electrically-induced self-healing is a relatively new concept that is currently used in the design of light long-lasting nanocomposites for electronic applications and actuators [28–33]. The healing process occurs via heat generation when an electrical current passes through a conductive matrix, such as a percolated multi-walled carbon nanotubes (MWCNT) network. The so-called Joule-effect activates the intrinsic self-healing ability of thermally self-mendable matrices to heal damage in local areas [34,35]. Among the several chemical routes used for thermal healing, the Diels–Alder (DA) reversible cycloaddition is one of the most effective alternatives [36]. It allows for the formation of three-dimensional and reversible network structures by means of the DA and retro-Diels-Alder (r-DA) reactions. Besides, thermally-reversible polymers based on DA chemistry and combined with MWCNTs have proven to be feasible systems for generating electrically self-healing nanocomposites based on the DA reaction [37]. In this particular example, the activation of the Joule heating, using electricity in the presence of macroscale damages (i.e., cuts), generates local changes in electrical resistivity at the crack tip, which leads to a local temperature increase, within the range of the r-DA reaction temperatures. The elevated temperature generates sufficient chain mobility to close and seal the cut upon cooling by means of the DA reaction. Remarkably, this approach introduced a new concept for polymer healing while using Joule heating and set the basis for the development of self-healing electrically conductive nanocomposites. It is, in fact, a fundamental point, since the higher interfacial interaction between the components via DA chemistry, the better the recovery of mechanical properties. For instance, it has been already reported that MWCNTs promotes a catalytic effect upon DA reaction for damage recovery in thermally self-healing matrices that are based on reversible DA chemistry [38].

Owing to the need to produce scalable self-healing thermoset nanocomposite systems, the chemical modification of polyketones (PK) with amine compounds via the Paal–Knorr reaction represents a feasible alternative for the industrial production of these materials [39]. This is related to the tolerance of this reaction towards many functional groups, particularly sterically hindered amines. This synthetic pathway offers several advantages, such as high yield under relatively mild conditions, high reaction kinetics, even in the absence of any catalyst, and water as the only by-product. In 2009, Picchioni and his co-workers published the first attempt for the chemical modification of PK with furan amino-substituted compounds aimed at preparing thermoset polymer networks that are able to undergo reversible reactions via DA chemistry [40]. The furan groups that were directly grafted on the PK backbone chain allowed for the formation of a three-dimensional network structure after being cross-linked with aromatic bismaleimide (B-Ma). The material indeed formed a thermally reversible and self-healing thermoset polymer network by means of the DA and r-DA sequence. In another approach, polyketones bearing DA and hydrogen bonding (HB) active groups were prepared by the same Paal–Knorr chemical route. The resulting materials displayed self-healing properties while using heat as external stimulus [41]. The novelty of this approach allowed for the preparation of a series of compounds displaying the same backbone structure but a systematic variation in the amount of hydrogen bonding and Diels–Alder pendant groups. This allowed for precisely pinpointing the influence of both kinds of interactions on the thermal and mechanical properties and thermal self-healing of the systems.

In a previously reported work, we have demonstrated the electrical-conductivity of a composite material consisting of a polyketone functionalized with furan groups, cross-linked with bismaleimide, and reinforced with MWCNTs [42]. Herein, we report on electrically self-healing nanocomposites that display tunable thermomechanical properties and re-workability by means of tuning the ratio between DA and HB functional groups and the wt.% loading of MWCNTs. These electrically conductive nanocomposites stem from the chemical modification of an alternating aliphatic polyketone grafted with furan (PK-Fu) and propyl alcohol groups (PK-Fu-A2P) via the Paal–Knorr reaction. The polymers are cross-linked with B-Ma and reinforced with MWCNTs via reversible Diels–Alder cycloaddition and hydrogen bonding. Spectroscopic techniques were used to study the chemical modification of the polyketone backbone polymer and crosslinking of the resulting polymer nanocomposites via DA reaction. Thermomechanical tests allowed for evaluating the mechanical performance, recyclability, and re-workability of the composites via reversible DA/r-DA cycloaddition, hydrogen bonding, and MWCNTs loading. The modulus and electrical performance of the material were measured. Additionally, the effect of Joule heating was followed by IR thermography. The morphology of the systems was studied by optical microscopy. The dispersion of MWCNTs in the polymeric matrices was analyzed by electronic microscopy before and after self-healing by Joule heating.

2. Materials and Methods

The alternating aliphatic polyketone (PK30) was synthesized according to previously reported works [39–41]. The resulting co- and terpolymers of carbon monoxide present a total olefin content of 30% of ethylene and 70% of propylene (PK30, MW 2687 Da). Furfurylamine (FU, Sigma-Aldrich, ≥99%, Zwijndrecht, The Netherlands), and amino-2-propanol (A2P, Sigma Aldrich, 99%, Darmstadt, Germany) were freshly distilled before used. Multi-walled carbon nanotubes (MWCNTs, O.D. 6–9 nm, average length 5 μm, Sigma-Aldrich 95% carbon, St. Louis, MO, USA), dimethyl sulfoxide-d6 (DMSO-d6, Sigma-Aldrich 99.5%, St. Louis, MO, USA), (1,1-(methylenedi-4,1-phenylene)bismaleimide (B-Ma, Sigma-Aldrich 95%, St. Louis, MO, USA), tetrahydrofuran (THF, Boom BV ≥95%, Meppel, The Netherlands), chloroform ($CHCl_3$, Avantor 99.5%, Gliwice, Poland), and deuterated chloroform ($CDCl_3$, Sigma-Aldrich 99.8 atom% D, St. Louis, MO, USA) were purchased and used as received.

2.1. Functionalization of Polyketone with Furan and Propyl Alcohol Groups

The reaction between PK30, FU, and A2P (Figure 1) was carried out as described by Araya-Hermosilla et al. [41] at different ratios between the 1,4-dicarbonyl groups of PK30 and the primary amine groups of FU and A2P via the Paal–Knorr reaction. The molar ratio between the reactants in the feed and methodology are described in the Table S1. The chemical modifications of PK30 yielded different polymers bearing pendant furan and hydroxyl groups ($PK30_xFU_yA2P_z$).

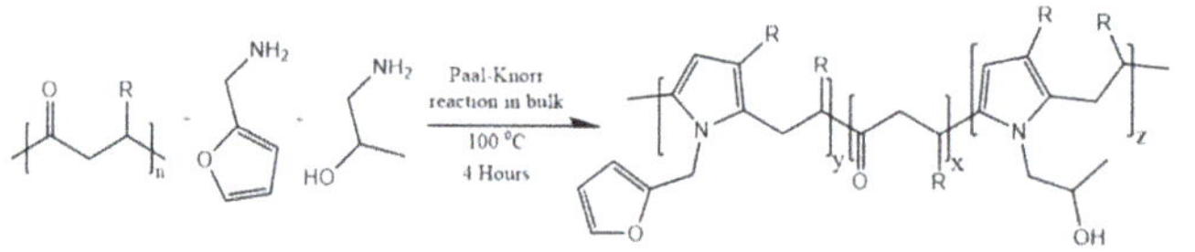

Figure 1. Schematic representation of Paal–Knorr functionalization of Polyketone with furfurylamine and amino-2-propanol (PK-FU-A2P).

2.2. Preparation of $PK30_xFU_yA2P_z$/B-Ma/MWCNT Composite

A three-dimensional and reversible network structure was produced by means of the DA reaction of the furan-derived PK30 and the aromatic bismaleimide cross-linker agent. MWCNTs may also undergo DA with furan and maleimide groups by means of the diene/dienophile character of the graphitic surface of the MWCNTs [43]. First, the MWCNTs were suspended in $CHCl_3$ (0.75 wt.%), a solvent reported as an effective dispersant for MWCNTs [44,45], and sonicated for 30 min. Polymers that were grafted with FU and A2P groups at different ratios, bismaleimide at equimolar amounts with the furan groups, and already sonicated MWCNTs at specific percentages from 0.1 to 5 wt.% (Table S2) were mixed in chloroform (comprised roughly 90% of the total volume) in a round-bottom flask with a condenser. The reaction was set under vigorous stirring at 50 °C for 24 h while using an oil bath that was equipped with a temperature controller. The resulting mixtures were put on Teflon plates to evaporate the solvent in a vacuum oven at 60 °C for 48 h. Around 500 mg of the powder acquired from ground materials after DA reaction was set in stainless steel molds that were lined with Teflon paper. The powders were then pressed using a heated press (Schwabentan Polystat 100T) at 150 °C and 40 bar pressure for 30 min. to form rectangular bars with dimensions of 35 mm × 6 mm × 1 mm. The bars were cooled down to room temperature and stored for further testing.

2.3. Characterization

The elemental composition of the polymers was analyzed while using a Euro EA elemental analyzer (Langenselbold, Germany) for nitrogen, carbon, and hydrogen. ^{1}H-NMR spectra were recorded on a Varian Mercury Plus 400 MHz apparatus (Agilent, Santa Clara, CA, USA) while using $CDCl_3$ or DMSO-d6 as solvent. FT-IR spectra were collected using a Perkin-Elmer Spectrum 2000 (San Francisco, CA, USA), transmission measurements were recorded at the range of 4000 cm^{-1} to 500 cm^{-1} at a resolution of 4 cm^{-1} averaged over 64 scans. Differential scanning calorimetry (DSC) analysis was performed on a Perkin Elmer Pyris Diamond under a nitrogen atmosphere (Shelton, CT, USA). The samples were weighed (5–12 mg) in an aluminum pan, which was then sealed. Subsequently, the samples were heated from 0 °C to 150 °C and then cooled to 0 °C. Four heating-cooling cycles were performed at a rate of 10 °C/min. Gel Permeation Chromatography (GPC) measurements were performed with an HP1100 Hewlett-Packard (Wilmington, Philadelphia, PA, USA). The equipment consists of three 300 × 7.5 mm PLgel 3 m MIXED-E columns in series and a GBC LC 1240 RI detector (Dandenong, Victoria, Australia). The samples were dissolved in THF (1 mg/mL) and eluted at a flow rate of 1 mL/min. and a pressure of 100–140 bar. The calibration curve was made using polystyrene as standard and the data were interpolated using the PSS WinGPC software. Thermomechanical analyses were conducted on a Perkin Elmer Dynamic Mechanical Analyzer DMA 8000 (Waltham, MA, USA) using single cantilever mode at an oscillation frequency of 1 Hz and heating rate of 3 °C/min. The samples for DMA analysis were prepared by compression molding of 500 mg of the composite into rectangular bars (6 mm wide, 1 mm thick, 35 mm long) at 150 °C for 30 min. under a pressure of 40 bar to ensure full homogeneity and then annealed in an oven at 50 °C for 24 h. Electrical measurements were performed on the rectangular bars used for DMA analysis. The setup consisted of a Velleman Power supply single output DC switching bench 60V, 5A, and a multimeter (Gossen Metrawatt Metrahit 18S) (Figure S1). Electrical parameters were measured on samples that were connected to a conventional circuit using copper clamps holders. In addition, silver paste was used at both ends of the bars and cover into aluminum foil to improve the contact area between the copper clamps holder and the sample. The sample resistivity (ρ) was calculated, as follows

$$\boldsymbol{\rho} = \mathbf{R}\frac{\mathbf{A}}{\mathbf{l}} \tag{1}$$

where **ρ** is given in Ω·m, **R** (Ω) is the electrical resistance, **A** (m^2) is the cross-sectional area, and **l** (m) is the length between copper clamps holders. The electrical resistance is calculated, as follows

$$\mathbf{R} = \frac{\mathbf{V}}{\mathbf{I}} \tag{2}$$

where **V** stands for voltage (V) supplied in the electrical circuit and **I** is the current (A) measured in amperes passing through the sample. Thermal images of the samples that were subjected to electrical current were obtained while using a Fluke IR thermometer camera (VT02 Everett, WA, USA) (Figure S1).

For safety and practical reasons, when the temperature of the different samples reached 150 °C, the stimulus that triggered the rise in temperature was switched off, such as heat exchange in DSC and DMA, and voltage in conductivity analyses.

3. Results and Discussion

3.1. PK Functionalized with Furan and Propyl Alcohol Groups Via Paal-Knorr Reaction

The Paal–Knorr reaction on the polyketone with FU and A2P resulted in dicarbonyl conversion of higher than 65%, as seen in Table 1. This is a good result, given that the maximum conversion expected has been demonstrated to be statistically 80% [41], due to the occurrence of reactions leaving the monocarbonyl segments in the chains. The total conversion efficiency ($y + z$) could be calculated by EA through the relative content of N in the samples. The relative weight of y and z could be obtained by ^{1}H-NMR.

Table 1. Elemental Analysis and Gel Permeation Chromatography (GPC) measurements of PK30 alone and modified with Furfurylamine (FU) and A2P at different ratio.

Sample	x (%)	y (%)	z (%)	CO (%)	Mn (×10³ M)	Mw (×10³ M)	PDI
PK30	100	-	-	-	2.4	5.4	2.2
$PK30_x$-FU_y	35	65	-	65	2.1	5.4	2.6
$PK30_x$-FU_y-$A2P_z$	28	55	17	72	1.8	5.0	2.7
$PK30_x$-FU_y-$A2P_z$	27	47	26	73	2.0	5.4	2.7
$PK30_x$-FU_y-$A2P_z$	28	36	36	72	1.9	5.3	2.7
$PK30_x$-$A2P_z$	28	-	72	72	1.7	4.1	2.4

Figure 2 shows the ^{1}H-NMR spectra of the polymers before and after the different functionalizations. In all cases, a successful grafting process was indicated by the signals that were attributed to the formed pyrrole rings around 5.7 ppm (H6). Peaks at 4.9 (H1), 5.9 (H2), 6.2 (H3), and 7.3 ppm (H4) confirm the presence of furan groups, while 3.9 ppm (H5) signal stands for the confirmation for the hydroxyl moieties. The spectra clearly show how the signals attributed to each group increase or decrease with the respective FU/A2P ratio used in the formulation. From the relative intensity of the resonance peaks, the values of x and y shown in Table 1 were obtained.

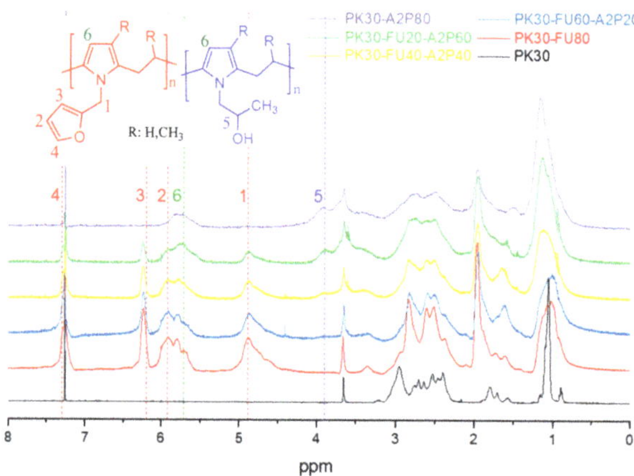

Figure 2. ^{1}H-NMR spectra of PK30 functionalized FU and A2P at different ratios.

DSC experiments of the polymers show differences in their respective T_g, as can be seen in Figure 3. The experimental plots can be seen in Figure S2, where the first curves were neglected to remove the thermal history. As expected, polymers with stronger intermolecular HB showed higher T_g. The T_g increased as a function of the A2P/FU ratio for all of the different copolymers synthesized.

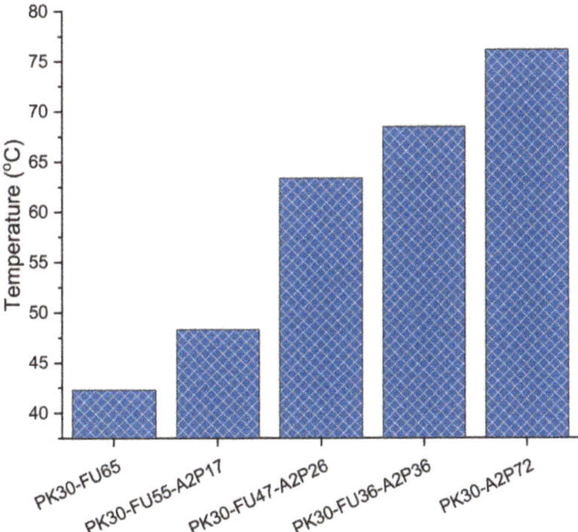

Figure 3. Glass transition of PK30 grafted with FU, A2P, and their respective ratios according to DSC (see Figure S2 for thermal history of all polymer series).

3.2. PK-Fu-A2P / B-Ma/ MWCNT Composite

The polymers were mixed with bismaleimide and MWCNT while using $CHCl_3$ as solvent at 50 °C for 24 h to produce the reinforced composite materials through the DA cycloaddition reaction. Figure 4 and Figure S3 show the FT-IR spectra of the composites before and after the cross-linking reaction in the absence and presence of 5 wt.% MWCNT [43,46,47]. There are emerging bands centred at approximately 1180 cm^{-1} attributed to the C-O-C ether moiety from the DA adduct. The band at 1009 cm^{-1} assigned to C-O-C of unreacted furans, and a band at approximately 1378 cm^{-1}, representing C-N stretching in the maleimide ring disappear over time.

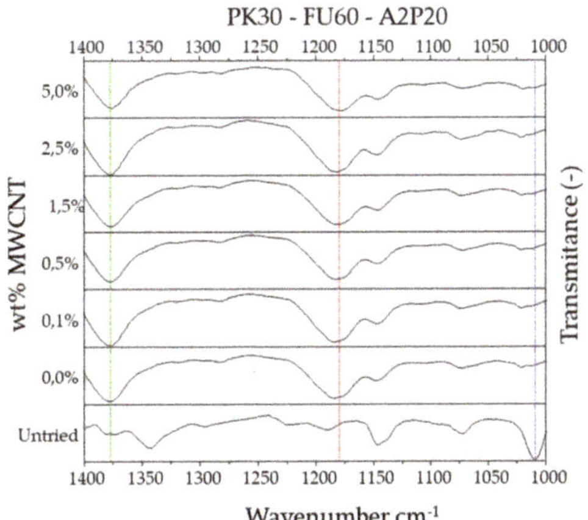

Figure 4. FT-IR of polyketones (PK) grafted with the ratios of FU and A2P Untried (not cross-linked) and cross-linked with different wt.% of multi-walled carbon nanotubes (MWCNTs).

The DSC experiments were then performed with the cross-linked nanocomposites to follow the reversibility process related to the DA and r-DA sequence of the materials. For brevity, only one DSC study is shown in Figure 5 (see Figure S4 for a full review of all crosslinking series of nanocomposites). The resulting graph displays a broad endothermic transition in the range of temperature 100–140 °C for each consecutive thermal cycle. The resemblance between the cycle curves in the DSC graphs confirm the reversible cross-linked character of the samples, in other words, the resilience of the material upon thermal cycling.

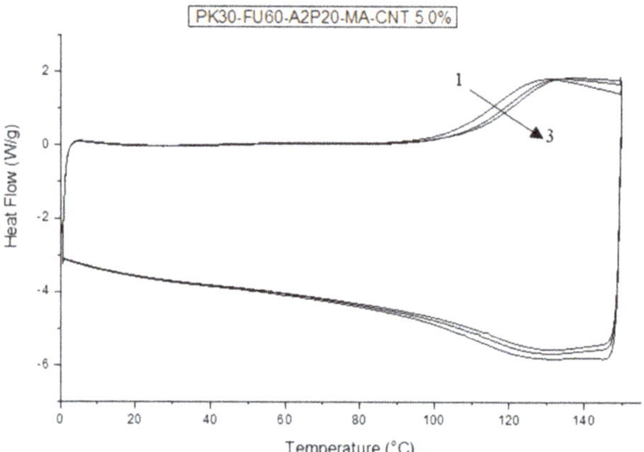

Figure 5. Differential scanning calorimetry (DSC) thermal cycles of PK-Fu60-A2P20 cross-linked with b-Ma and reinforced with 5.0 wt.% of MWCNTs.

In addition, from DSC analysis it was possible to evaluate the de-cross-linked samples. We can assign many parameters to compare the samples since the endothermic transition corresponds to the r-DA process. Thus, the peak of the curves in Figure 5 is the indication of the moment at which the DA cleaved is at its maximum. The area under the curve associated to this peak is therefore related to the energy that is absorbed during the cleavage of the DA adducts [48]. That is to say, the results (Figure 6A) present no relevant changes or trends relating temperature peaks to the ratio of amines or the weight percentage of MWCNT in each sample. This result can be interpreted as the intermolecular bonds playing an important role in the system. Hence, a greater amount of FU results in a higher crosslinking density. The latter means that, even though the DA cross-linking density diminished, physical interactions via hydrogen bonding can compensate the thermal stability of the systems. This is supported by the DSC results shown in Figure 3, where the higher amount of OH groups results in higher glass transition temperatures of the different polymers [8,42].

Likewise, the area under the curve, also being addressed as the endothermic integral, does not present the trend of relation in the manner of different weight percentage of MWCNT (Figure 6B). Nonetheless, there is a pertinent difference of area if we take the different amine ratios into consideration, and clearly, the percentage of FU that is grafted onto the polymer dictates the energy absorbed during the cleavage of the DA adducts. This is in agreement with the cross-linking density of the material as already mentioned [46].

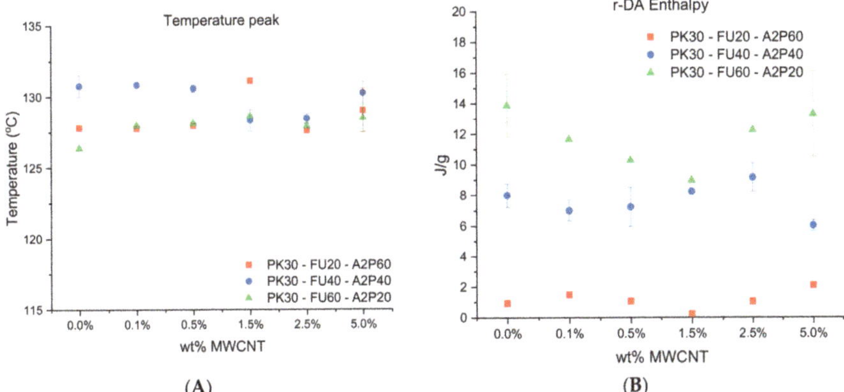

Figure 6. (**A**) Temperature peak where most of the Diels–Alder (DA) adducts are cleaved for all the series of cross-linked nanocomposites with bismaleimide and different wt.% of MWCNTs. (**B**) Enthalpy as determined by DSC analysis of cross-linked samples according to wt.% of MWCNT and ratios of amines compounds.

3.3. Mechanical Properties

Successful bar specimens for DMA analysis were obtained after hot compression molding of ground cross-linked samples using 150 °C and 40 bar during 30 min. This process favors the r-DA mechanism that leads to the de-crosslinking of the thermoset nanocomposites (Figure S5 grinded/molded samples). This probes the reversibility and recyclability character of these materials. The variation in storage modulus (E′-ability of the material to store energy), loss modulus (E″-ability of the material to dissipate energy), and ratio of the loss modulus to the storage modulus (tan δ) of all the samples were measured while using the single cantilever method in the DMA analyses. The peak of tan δ is where the viscous behavior of the network is at its maximum. For simplicity, we refer to this as the softening point in this discussion. Figure 7A shows that the filler considerably enhances the loss/elastic moduli and softening point (tan(δ) peak), with respect to the neat cross-linked thermoset.

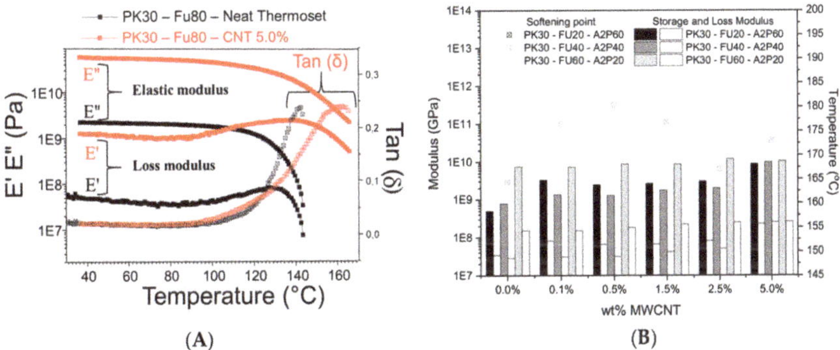

Figure 7. (**A**) Dynamic mechanical analysis of PK30-FU80 (reference sample) cross-linked with bismaleimide and reinforced with 5 wt.% of MWCNTs. Black line represents the sample without filler and the red one crosslinked with 5 wt.% of MWCNT (**B**) E′ E″(bars–left axis), tan (δ) delta (points–right axis) vs Temperature for formulations that vary on the Fu/A2P ratio and different wt.% of MWCNTs (see Figure S6 for the whole series of DMA analysis).

Remarkably, the storage and loss moduli increased more than one order of magnitude when comparing the neat thermoset system to the reinforced one containing 5 wt.% of MWCNTs (Figure 7A). Likewise, the softening point (tan(δ)) increased from ~140 °C (neat thermoset) to ~160 °C for the reinforced nanocomposite. This can be explained by the MWCNTs reinforced structure of the thermoset matrix showing increased rigidity. This effect might also be a consequence of strong interfacial interactions between the polymer matrix and MWCNTs. As furan and maleimide moieties both have been reported to undergo DA cycloaddition with CNT surface [38,42]. In Figure 7B, the ratio of DA and HB active groups grafted on the neat thermosets (without MWCNTs) shows increasing E′, E″, and tan δ with increasing crosslinking density. However, substantial differences are not found in E′, E″ and tan δ between systems with the same Fu/A2P ratio in the presence of increasing wt.% of MWCNTs. This suggests the lack of interaction between the nanofiller and the cross-linked thermoset matrices. We also observed that systems with different Fu/A2P ratios and same 5 wt.% of MWCNTs show similar E′, E″, but the tan(δ) is 10 °C higher for the highest cross-linked PK30–Fu60–A2P20 system. As compared to the PK30–Fu80 reference system with 5 wt.% MWCNTs, the nanocomposites PK30–Fu60–A2P20 showed a 22 °C higher softening point (tan(δ)). This reveals the greater capacity for dissipating energy most likely due to the better distribution of MWCNTs and the contribution of HB in the matrix.

3.4. Electrical Conductivity Properties

The electrical conductivity measurements were performed to evaluate the minimum amount of MWCNTs needed to achieve an efficient percolation pathway to heat the sample by Joule heating. The sample bar was placed between two copper clamps (Figure S1). Silver paste and aluminum foil were used to improve the contact between the copper clamps and the samples. An infrared camera is used in addition to a digital thermometer with a thermocouple to monitor the temperature change during the conductivity tests (Figure S1).

The minimum amount of MWCNTs needed to obtain electrical conductivity is 1.5 wt.% for all three different ratios between Fu/A2P functional groups, as can be seen in Figure 8A. Conductivity increases as the amount of MWCNTs increases by two orders of magnitude higher for either composite system containing 5 wt.% MWCNTs, as compared to systems only containing 1.5 wt.%. The systems containing FU60/A2P20 ratio conduct more electricity than systems containing lower FU/A2P ratios, even though the amount of MWCNTs is the same in all instances. This characteristic has been already been reported previously and attributed to the effective primary (covalent) and secondary interactions between MWCNT and FU. Both types of interactions cooperatively favor the homogeneous distribution of the filler along the sample [42,49].

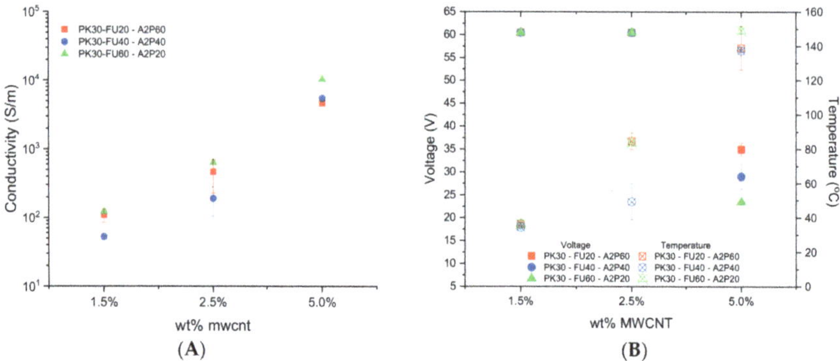

Figure 8. (**A**) Electrical conductivity of sample bars and (**B**) Temperature reached with each sample bars according to voltage.

Importantly, the temperature that the bars achieve by means of the Joule heating under different voltages was measured with the same set up using the thermal camera. Temperatures above 120 °C are needed to activate the r-DA process for self-healing with samples containing 5 wt.% of MWCNTs. This required using voltages between 25 to 35 V (Figure 8B). For lower MWCNTs content, the temperature did not reach such a high value, even upon applying voltage of 60 V. This is clearly related to the ratio between Fu/A2P compounds, the higher amount of FU grafted, the higher the achieved temperature with lower voltage.

Notably, the conductivity and Joule heating are reproducible during heating and cooling cycles (e.g., turning on/off the circuit), thus testifying that the percolation pathway does not undergo relevant alteration in the course of temperature changes. Moreover, according to the conductivity tests, the higher the FU content in the sample, the greater was the overall bar conductivity, which resulted in improved resistive heating being needed to perform electrical self-healing. That is to say that the MWCNTs distribution was superior in the sample PK30-FU60-A2P20 than in other FU/A2P ratios.

3.5. Thermal and Electrical Self-Healing

Two qualitative experiments were done in order to demonstrate the self-healing features of the nanocomposites. The first test attempted to validate the self-healing process by breaking bar samples. For all the wt.% of MWCNT, the broken samples were placed back in the molds where they were first made. The initial process conditions of 150 °C and 40 bar for 30 min. were applied. The bars recover their full integrity, as can be seen in Figure 9. The effectiveness of this method is expected, since applied pressure enhances the contact between the damaged faces after the r-DA has taken place, allowing for the DA reaction to proceed upon cooling. This method not only proves self-healing ability, but also the reworkability and recyclability of the material.

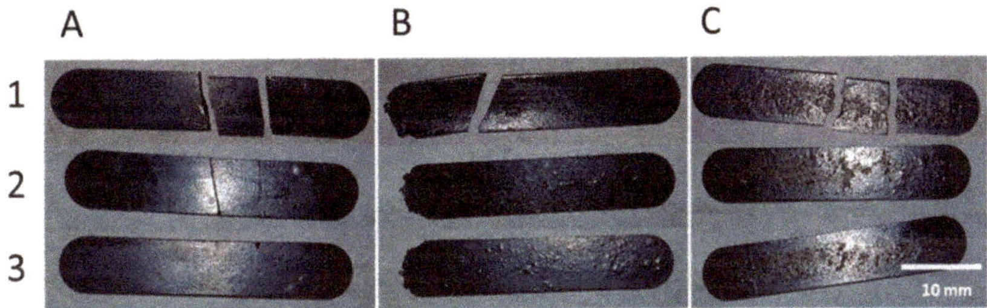

Figure 9. Thermally self-healing of sample during the re-process of the bars with press using 150 °C and 40 bar for 30 min. Sample (**A**) PK30-FU20-A2P60-CNT 0.0% (**B**) PK30-FU20-A2P60-CNT 1.5% (**C**) PK30-FU60-A2P20-CNT 0.1% at the moment (1) broken, (2) healed front face (3) healed back face.

In the second test, samples with 5.0% MWCNTs were set in the same frame used to evaluate the conductivity of the bars, under a microscope. Figure 10A,B show the samples before and after damaging, respectively. The scratches on the surface of the samples were done by hand with a sharp blade. Subsequently, an electrical current is applied to produce the Joule heating in the materials under a voltage ranging from 25 to 35 V. After the sample reached a temperature of about 145 °C, healing was verified, as can be seen in Figure 10C, showing nearly full recovery from the damage. It is important to mention that all of the samples with different ratios of amines successfully healed after 10 min.

The SEM micrographs were obtained from the bar sample PK30-FU60-A2P20-CNT 5% subjected to different treatments. Figure 11A shows the freshly broken surfaces of the composite after molding, where delamination and poor dispersion of the MWCNTs were observed. Figure 11B displays the same sample after being molded and annealed by conventional heating at 150 °C for 24 h. It is worth noticing that a better MWCNTs dispersion was achieved and delamination was diminished, suggesting the better interaction between the components [42]. Notably, Figure 11C showed that the sample subjected to electrical current rendered a composite with well dispersed MWCNTs. As compared to Figure 11,B, the Joule heating increased the polymer mobility that favors the redistribution of macromolecular chains, thus increasing the MWCNT dispersion aided by the better interfacial adhesion between the components.

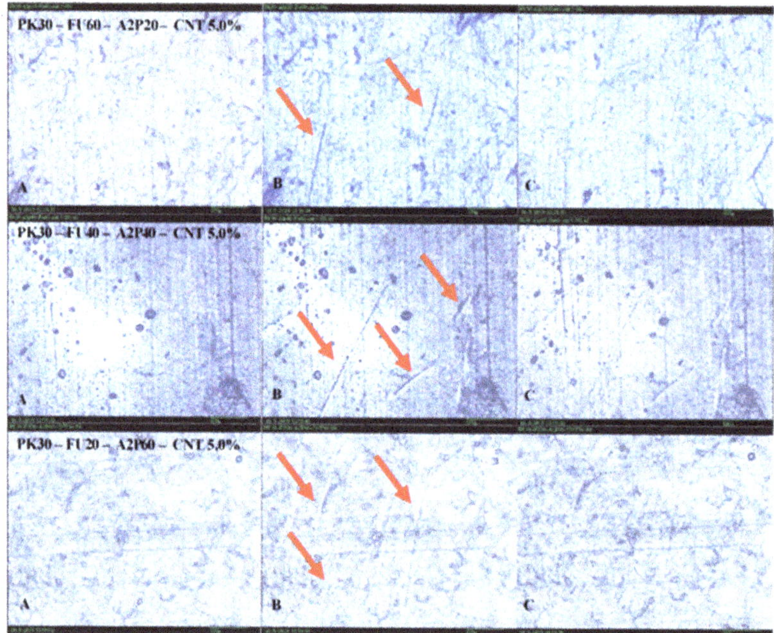

Figure 10. Microscope images of polyketone grafted with different ratio of amines (Fu/A2P) and cross-linked during the healing process. Same area with (**A**) no damage, (**B**) damaged, and (**C**) healed.

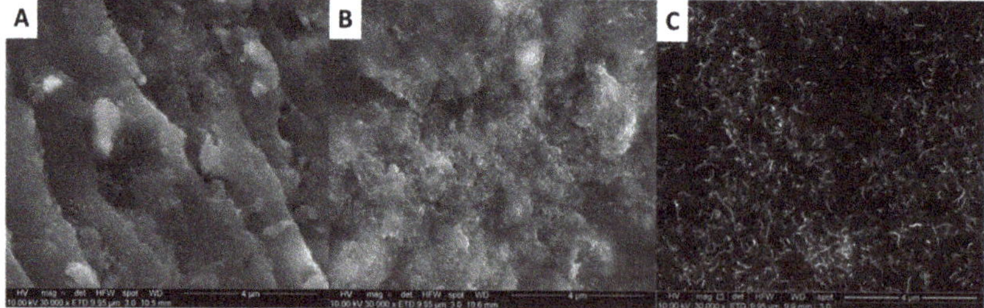

Figure 11. SEM micrographs of PK30-FU60-A2P20-CNT 5% cross-linked composite after molding (**A**), heating in an oven at 150 °C for 24 h (**B**) and after Joule heating at 145 °C using 25 Volt during 24h (**C**).

4. Conclusions

In this project, we have demonstrated the design and development of thermoset polymer nanocomposites showing tunable mechanical, electrical, and self-healing properties. The materials are capable of recovering structural damage due to thermally reversible DA bonds that are assisted by HB through Joule heating. The effect the amine content and the fraction of the fillers on the self-healing, electrical and mechanical properties of the final products were systematically investigated. The self-healing nanocomposite materials were prepared via the Paal–Knorr modifications of an alternating aliphatic PK with the association of DA reversible reaction (Furfurylamine) and HB

(Amino-2-propanol) to assist the healing process. In addition, MWCNT was included to provide reinforcement as well as an electrical pathway. The reactions were successfully carried out in a one-pot method and FT-IR, ^{1}H-NMR analysis, elemental analysis, and GPC verified its high-efficiency yield. The thermal properties of the polymer evaluated by DSC before crosslinking showed the relevant contribution of HB groups to the glass transition.

Likewise, FT-IR analysis confirmed the presence of the DA adduct proving successful cross-linking reaction. DSC confirmed the thermal reversibility of the nanocomposites where consecutive cycles approximately overlap each other in all of the samples. The electrical measurements showed the correlation of weight content of MWCNT and resistivity. It proved that a moderate amount of MWCNT filler, more precisely 1.5 wt.%, was enough to build an electrical network inside the nanocomposite; however, this amount of filler was not enough to heat the sample with a voltage of 60V. However, samples with 5.0 wt.% of MWCNT could reach temperatures high enough to undergo r-DA reaction. Interestingly, the electrical measurements also established that samples with a higher ratio of FU conduct electricity better if compared to lower ratios due to the better interfacial adhesion between the components. This suggests the ability of FU groups to react with MWCNTs and possibly assist the distribution of MWCNT during the dispersion. With regard to self-healing and reworkability, the nanocomposite presented excellent results for all of the methods applied. The conventional way in the oven and the resistive heating could both achieve almost a complete disappearance of the damage in the samples.

This research might contribute towards a cost-effective industrial production of electrically-conductive, recyclable thermoset nanocomposites featuring re-connection that is triggered by Joule heating. As compared to published works on electrically self-healing systems, the present materials are crosslinked and recyclable nanocomposite systems that can be used as an additive or matrix for the manufacture of materials. Particularly, alternating aliphatic polyketones containing furan pendant groups act as diene polymer materials that are of interest to academia and are industrially scalable. In line with that, the article also shows a robust method for obtaining nanocomposites that can be used as thermosetting, thermo-reversible, thermo-adhesive, thermo-conductive, electrically-conductive, and self-repairing systems.

Supplementary Materials: The following are available online at http://www.mdpi.com/2073-4360/11/11/1885/s1. Table S1: Composition of polyketones functionalized with FU and A2P groups, Table S2: Experimental conditions of grafted polymers mixed with different wt.% of MWCNTs, Figure S1: Set up for electrical conductive easements of a bar (A) with clamps and (B) Infrared camera, Figure S2: DSC first cycle after thermal history erase of PK30 grafted with FU, A2P and their respective co-polymers, Figure S3: FT-IR of PK grafted with the ratios of FU and A2P Untried (non-cross-linked) and cross-linked with different wt.% of MWCNTs, Figure S4: DSC thermal cycles of PK-Fu-A2P series cross-linked with b-Ma and reinforced with different wt.% of MWCNTs, Figure S5: Three hot compression molded bars made with a grinded cross-linked sample (see powder in the middle of the picture) using 150 °C and 40 bar during 30 min., Figure S6 Dynamic mechanical analysis of bar samples made with PK30-FU60-A2P20, PK30-FU40-A2P40, PK30-FU20-A2P60 cross-linked with bismaleimide and reinforced with different wt.% of MWCNTs.

Author Contributions: G.M.R.L.: Article writing, experimental design, laboratory work, results processing, F.O.: Laboratory guidance, discussion of results, writing corrections. F.P.: Polymer design, idea developer, writing corrections. I.M.-V.: writing corrections, literature recommendation. A.P.: Idea developer, writing corrections, literature recommendation. R.K.B.: Polymer design, discussion of results, laboratory guidance, writing corrections. R.A.-H.: Idea developer, polymer and experimental design, results processing, discussion of results, SEM analysis, article writing.

Funding: R.A.-H. grafetully acknowledges Fondecyt Postdoctorado-Chile grant number 3170352. And The APC was funded by the University of Groningen.

Acknowledgments: We gratefully acknowledge the support of Esteban Araya-Hermosilla, Patrizio Raffa, Nicola Migliore, Léon Rohrbach, Mario E. Flores, Franck Quero, Marcel H. de Vries, and Marcos dos Santos Lima.

Conflicts of Interest: The authors declare no conflict of interest.

References

1. Binder, W.H. *Self-Healing Polymers: From Principles to Applications*; Wiley-VCH: Weinheim, Germany, 2013; ISBN 9783527670185.
2. Zhang, M.Q.; Rong, M.Z. Intrinsic self-healing of covalent polymers through bond reconnection towards strength restoration. *Polym. Chem.* **2013**, *4*, 4878–4884. [CrossRef]
3. Billiet, S.; Hillewaere, X.K.D.; Teixeira, R.F.A.; Du-Prez, F.E. Chemistry of Crosslinking Processes for Self-Healing Polymers. *Macromol. Rapid Commun.* **2013**, *34*, 290–309. [CrossRef] [PubMed]
4. Thakur, V.K.; Kessler, M.R. Self-healing polymer nanocomposite materials: A review. *Polymer (Guildf.)* **2015**, *69*, 369–383. [CrossRef]
5. Long, K.N. The mechanics of network polymers with thermally reversible linkages. *J. Mech. Phys. Solids* **2014**, *63*, 386–411. [CrossRef]
6. Nielsen, C.; Weizman, H.; Nemat-Nasser, S. Thermally reversible cross-links in a healable polymer: Estimating the quantity, rate of formation, and effect on viscosity. *Polymer (Guildf.)* **2014**, *55*, 632–641. [CrossRef]
7. Zhang, R.; Yu, S.; Chen, S.; Wu, Q.; Chen, T.; Sun, P.; Li, B.; Ding, D. Reversible Cross-Linking, Microdomain Structure, and Heterogeneous Dynamics in Thermally Reversible Cross-Linked Polyurethane as Revealed by Solid-State NMR. *J. Phys. Chem. B* **2014**, *118*, 1126–1137. [CrossRef]
8. Araya-Hermosilla, R.; Broekhuis, A.A.; Picchioni, F. Reversible polymer networks containing covalent and hydrogen bonding interactions. *Eur. Polym. J.* **2014**, *50*, 127–134. [CrossRef]
9. Song, Y.-K.; Chung, C.-M. Repeatable self-healing of a microcapsule-type protective coating. *Polym. Chem.* **2013**, *4*, 4940. [CrossRef]
10. Wang, Z.; Urban, M.W. Facile UV-healable polyethylenimine–copper (C_2H_5N–Cu) supramolecular polymer networks. *Polym. Chem.* **2013**, *4*, 4897–4901. [CrossRef]
11. Chang, J.-M.; Aklonis, J.J. A photothermal reversibly crosslinkable polymer system. *J. Polym. Sci. Polym. Lett. Ed.* **1983**, *21*, 999–1004. [CrossRef]
12. Zhang, Q.; Qu, D.-H.; Wu, J.; Ma, X.; Wang, Q.; Tian, H. A Dual-Modality Photoswitchable Supramolecular Polymer. *Langmuir* **2013**, *29*, 5345–5350. [CrossRef]
13. Burnworth, M.; Tang, L.; Kumpfer, J.R.; Duncan, A.J.; Beyer, F.L.; Fiore, G.L.; Rowan, S.J.; Weder, C. Optically healable supramolecular polymers. *Nature* **2011**, *472*, 334–337. [CrossRef] [PubMed]
14. Wu, G.; Liu, D.; Liu, G.; Chen, J.; Huo, S.; Kong, Z. Thermoset nanocomposites from waterborne bio-based epoxy resin and cellulose nanowhiskers. *Carbohydr. Polym.* **2015**, *127*, 229–235. [CrossRef] [PubMed]
15. Feldman, D. Polyblend Nanocomposites. *J. Macromol. Sci. Part A* **2015**, *52*, 648–658. [CrossRef]
16. McNally, T.; Pötschke, P. *Polymer-Carbon Nanotube Composites: Preparation, Properties and Applications*; Woodhead Publisher: Cambridge, UK, 2011; ISBN 9781845697617.
17. Zhang, X. *Multifunctional Polymer Nano-Composites*, 1st ed.; Taylor and Francis Group/CRC Press: Boca Raton, FL, USA, 2011; ISBN 9781138111806.
18. Calisi, N.; Giuliani, A.; Alderighi, M.; Schnorr, J.M.; Swager, T.M.; Di Francesco, F.; Pucci, A. Factors affecting the dispersion of MWCNTs in electrically conducting SEBS nanocomposites. *Eur. Polym. J.* **2013**, *49*, 1471–1478. [CrossRef]
19. Koval'chuk, A.A.; Shevchenko, V.G.; Shchegolikhin, A.N.; Nedorezova, P.M.; Klyamkina, A.N.; Aladyshev, A.M. Effect of Carbon Nanotube Functionalization on the Structural and Mechanical Properties of Polypropylene/MWCNT Composites. *Macromolecules* **2008**, *40*, 7536–7542. [CrossRef]
20. Zhang, Y.; Broekhuis, A.A.; Stuart, M.C.A.; Landaluce, T.F.; Fausti, D.; Rudolf, P.; Picchioni, F. Cross-linking of multiwalled carbon nanotubes with polymeric amines. *Macromolecules* **2008**, *41*, 6141–6146. [CrossRef]
21. Yuan, Y.C.; Yin, T.; Rong, M.Z.; Zhang, M.Q. Self healing in polymers and polymer composites. Concepts, realization and outlook: A review. *Express Polym. Lett.* **2008**, *2*, 238–250. [CrossRef]
22. Zhong, N.; Post, W. Self-repair of structural and functional composites with intrinsically self-healing polymer matrices: A review. *Compos. Part A Appl. Sci. Manuf.* **2015**, *69*, 226–239. [CrossRef]
23. Li, G. *Self-Healing Composites: Shape Memory Polymer-Based Structures*; John Wiley & Sons: Oxford, UK, 2014; ISBN 9781118452424.
24. Bailey, B.M.; Leterrier, Y.; Garcia, S.J.; Zwaag, S. Van Der Progress in Organic Coatings Electrically conductive self-healing polymer composite coatings. *Prog. Org. Coat.* **2015**, *85*, 189–198. [CrossRef]

25. Ren, D.; Chen, Y.; Li, H.; Ur, H.; Cai, Y.; Liu, H. High-e ffi ciency dual-responsive shape memory assisted self-healing of carbon nanotubes enhanced polycaprolactone/thermoplastic polyurethane composites. *Colloids Surf. A* **2019**, *580*, 123731. [CrossRef]
26. Lin, C.; Sheng, D.; Liu, X.; Xu, S.; Ji, F.; Dong, L.; Zhou, Y.; Yang, Y. NIR induced self-healing electrical conductivity polyurethane / graphene nanocomposites based on Diels À Alder reaction. *Polymer (Guildf.)* **2018**, *140*, 150–157. [CrossRef]
27. Wu, X.; Wang, J.; Huang, J.; Yang, S. Room temperature readily self-healing polymer via rationally designing molecular chain and crosslinking bond for flexible electrical sensor. *J. Colloid Interface Sci.* **2020**, *559*, 152–161, in press. [CrossRef] [PubMed]
28. Hohlbein, N.; Shaaban, A.; Schmidt, A.M. Remote-controlled activation of self-healing behavior in magneto-responsive ionomeric composites. *Polymer (Guildf.)* **2015**, *69*, 301–309. [CrossRef]
29. Li, H.; Li, H.; Li, Z.; Lin, F.; Liu, D.; Wang, W.; Wang, B.; Xu, Z. T pattern fuse construction in segment metallized film capacitors based on self-healing characteristics. *Microelectron. Reliab.* **2015**, *55*, 945–951. [CrossRef]
30. Chuo, T.-W.; Wei, T.-C.; Liu, Y.-L. Electrically driven self-healing polymers based on reversible guest-host complexation of β-cyclodextrin and ferrocene. *J. Polym. Sci. Part A Polym. Chem.* **2013**, *51*, 3395–3403. [CrossRef]
31. Swait, T.J.; Rauf, A.; Grainger, R.; Bailey, P.B.S.; Lafferty, A.D.; Fleet, E.J.; Hand, R.J.; Hayes, S.A. Smart composite materials for self-sensing and self-healing. *Plast. Rubber Compos.* **2012**, *41*, 215–224. [CrossRef]
32. Wang, C.C.; Ding, Z.; Purnawali, H.; Huang, W.M.; Fan, H.; Sun, L. Repeated Instant Self-healing Shape Memory Composites. *J. Mater. Eng. Perform.* **2012**, *21*, 2663–2669. [CrossRef]
33. Cui, H.P.; Song, C.L.; Huang, W.M.; Wang, C.C.; Zhao, Y. Rubber-like electrically conductive polymeric materials with shape memory. *Smart Mater. Struct.* **2013**, *22*, 055024. [CrossRef]
34. Kim, J.-W.; Sauti, G.; Siochi, E.J.; Smith, J.G.; Wincheski, R.A.; Cano, R.J.; Connell, J.W.; Wise, K.E. Toward High Performance Thermoset/Carbon Nanotube Sheet Nanocomposites via Resistive Heating Assisted Infiltration and Cure. *ACS Appl. Mater. Interfaces* **2014**, *6*, 18832–18843. [CrossRef]
35. Lee, J.; Stein, I.Y.; Kessler, S.S.; Wardle, B.L. Aligned carbon nanotube film enables thermally induced state transformations in layered polymeric materials. *ACS Appl. Mater. Interfaces* **2015**, *7*, 8900–8905. [CrossRef] [PubMed]
36. Khan, N.I.; Halder, S.; Gunjan, S.B.; Prasad, T. A review on Diels-Alder based self-healing polymer composites. *IOP Conf. Ser. Mater. Sci. Eng.* **2018**, *377*, 012007. [CrossRef]
37. Willocq, B.; Bose, R.K.; Khelifa, F.; Garcia, S.J.; Dubois, P.; Raquez, J.-M. Healing by the Joule effect of electrically conductive poly(ester-urethane)/carbon nanotube nanocomposites. *J. Mater. Chem. A* **2016**, *4*, 4089–4097. [CrossRef]
38. Kuang, X.; Liu, G.; Dong, X.; Wang, D. Enhancement of Mechanical and Self-Healing Performance in Multiwall Carbon Nanotube/Rubber Composites via Diels-Alder Bonding. *Macromol. Mater. Eng.* **2016**, *301*, 535–541. [CrossRef]
39. Zhang, Y.; Broekhuis, A.A.; Stuart, M.C.A.; Picchioni, F. Polymeric amines by chemical modifications of alternating aliphatic polyketones. *J. Appl. Polym. Sci.* **2008**, *107*, 262–271. [CrossRef]
40. Zhang, Y.; Broekhuis, A.A.; Picchioni, F. Thermally Self-Healing Polymeric Materials: The Next Step to Recycling Thermoset Polymers? *Macromolecules* **2009**, *42*, 1906–1912. [CrossRef]
41. Araya-Hermosilla, R.; Lima, G.M.R.; Raffa, P.; Fortunato, G.; Pucci, A.; Flores, M.E.; Moreno-Villoslada, I.; Broekhuis, A.A.; Picchioni, F. Intrinsic self-healing thermoset through covalent and hydrogen bonding interactions. *Eur. Polym. J.* **2016**, *81*, 186–197. [CrossRef]
42. Araya-Hermosilla, R.; Pucci, A.; Raffa, P.; Santosa, D.; Pescarmona, P.P.; Gengler, R.Y.N.; Rudolf, P.; Moreno-Villoslada, I.; Picchioni, F. Electrically-responsive reversible Polyketone/MWCNT network through Diels-Alder chemistry. *Polymers* **2018**, *10*, 1076. [CrossRef]
43. Eremin, Y.S.; Kolesnikova, A.A.; Grekhov, A.M. Agglomeration and Sedimentation of MWCNTS in Chloroform. *Phys. Procedia* **2015**, *72*, 56–61. [CrossRef]
44. Kong, H.; Gao, C.; Yan, D. Controlled Functionalization of Multiwalled Carbon Nanotubes by in Situ Atom Transfer Radical Polymerization. *J. Am. Chem. Soc.* **2004**, *126*, 412–413. [CrossRef]

45. Toncelli, C.; De Reus, D.C.; Picchioni, F.; Broekhuis, A.A. Properties of Reversible Diels–Alder Furan/Maleimide Polymer Networks as Function of Crosslink Density. *Macromol. Chem. Phys.* **2012**, *213*, 157–165. [CrossRef]
46. Gaina, C.; Ursache, O.; Gaina, V.; Varganici, C.D. Thermally reversible cross-linked poly(ether-urethane)s. *Express Polym. Lett.* **2013**, *7*, 636–650. [CrossRef]
47. Lubomír, R.; Šebek, P.; Havlas, Z.; Hrabal, R.; Čapek, P.; Svatoš, A.; Svatoš, A. An Experimental and Theoretical Study of Stereoselectivity of Furan−Maleic Anhydride and Furan−Maleimide Diels−Alder Reactions. *J. Org. Chem.* **2005**, 6295–6302.
48. Araya-Hermosilla, R.; Pucci, A.; Araya-Hermosilla, E.; Pescarmona, P.P.; Raffa, P.; Polgar, L.M.; Moreno-Villoslada, I.; Flores, M.; Fortunato, G.; Broekhuis, A.A.; et al. An easy synthetic way to exfoliate and stabilize MWCNTs in a thermoplastic pyrrole-containing matrix assisted by hydrogen bonds. *RSC Adv.* **2016**, *6*, 85829–855837. [CrossRef]
49. Araya Hermosilla, R.A.; Picchioni, F. *Thermally Reversible Thermoset Materials Based on the Chemical Modification of Alternating Aliphatic Polyketones*; University of Groningen: Groningen, The Netherlands, 2016; ISBN 9789036792479.

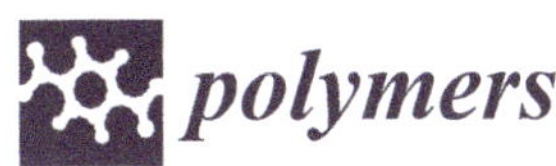

Article

Melt-Mixed Thermoplastic Nanocomposite Containing Carbon Nanotubes and Titanium Dioxide for Flame Retardancy Applications

C. Cabello-Alvarado [1,2], P. Reyes-Rodríguez [2], M. Andrade-Guel [2], G. Cadenas-Pliego [2,*], M. Pérez-Alvarez [3], V.J. Cruz-Delgado [4], L. Melo-López [1,2], Z.V. Quiñones-Jurado [5] and C.A. Ávila-Orta [2,*]

1 CONACYT-Consorcio de Investigación y de Innovación del Estado de Tlaxcala, C.P. 90000 Tlaxcala, Mexico
2 Centro de Investigación en Química Aplicada, Saltillo, 25315 Coahuila, Mexico
3 CONACYT-Instituto Mexicano del Petróleo, Eje Central Lázaro Cárdenas Norte 152, 07730 Ciudad de Mexico, Mexico
4 CONACYT-Unidad de Materiales, Centro de Investigación Científica de Yucatán, A.C., Mérida C.P. 97205, Yucatán, Mexico
5 Innovación y Desarrollo en Materiales Avanzados A.C., Grupo POLYnnova, Carr. San Luis Potosí-Guadalajara 1510, Nivel 3, Local 12, Lomas del Tecnológico, San Luis Potosí S.L.P. C.P. 78211 Mexico, Mexico
* Correspondence: gregorio.cadenas@ciqa.edu.mx (G.C.-P.); carlos.avila@ciqa.edu.mx (C.A.A.-O.)

Received: 20 June 2019; Accepted: 15 July 2019; Published: 19 July 2019

Abstract: The study of polymeric nanocomposites is a possible alternative to conventional flame retardants. The aim of the present work is to investigate the effects of carbon-nanotubes (CNT) and TiO_2 nanoparticles (NPs) on the thermo-mechanical, flammability, and electrical properties of polypropylene (PP). In this work, PP-TiO_2/CNT nanocomposites were obtained with TiO_2/CNT mixtures (ratio 1:2) through the melt extrusion process, with different weight percentage of nanoparticles (1, 5, and 10 wt %). The PP-TiO_2/CNT nanocomposites were characterized by DSC, TGA, MFI, FTIR, XRD, and SEM. It was possible to determine that the thermal stability of the PP increases when increasing the content of NPs. A contrary situation is observed in the degree of crystallinity and thermo-oxidative degradation, which decreased with respect to pure PP. The TiO_2 NPs undergo coalition and increase their size at a lower viscosity of the nanocomposite (1 and 5 wt %). The mechanical properties decreased slightly, however, the Young's modulus presented an improvement of 10% as well as electrical conductivity, this behavior was noted in nanocomposites of 10 wt % of NPs. Flammability properties were measured with a cone calorimeter, and a reduction in the peak heat release rate was observed in nanocomposites with contents of nanoparticles of 5 and 10 wt %

Keywords: nanocomposite; polypropylene; carbon nanotube; titanium dioxide

1. Introduction

Avoiding or preventing fires helps decrease injuries or human losses in automotive industry [1]. Different causes arise for these types of events, such as mechanical problems, electrical failures, or car accidents, since most automotive liquids are flammable. The automotive parts manufactured with different polymers are flammable and must be protected against fires [2]. For these materials, some flame retardants are applied only in the exterior, and some carbonaceous and ceramic films help to improve their flame-retardant effect [3–6].

One of the most common polymers used to obtain polymeric nanocomposites is polypropylene (PP), which is a semi-crystalline thermoplastic widely used in applications such as textile, films, bottles production, and piping, and is also the most used in the automotive industry [7–9]. However, PP has some limitations, such as being sensitive to heat and oxidative degradation, as well as its fragility. This polymer by itself is prone to crack or vulnerable to mechanical failures and, like most polymers, has low electrical and thermal conductivity; for these reasons, the addition of nanoparticles or nanocharges has been adopted with the purpose of changing properties of neat polymer [10,11].

The incorporation of nanoparticles to improve polymers properties depends in great extent on wt % of nanocharge used. A clear example of this effect was observed in ethylene vinyl acetate (EVA) nanocomposites with different nanoclay content (1–15 wt %). Thermal stabilization was achieved when the nanoclay content was between 2.5–10 wt %. Outside this range, the EVA thermal stability was not significant. In general, polymers properties are improved with low percentages in weight of nanocharge (<3 wt %), however, several researchers have studied high nanocharges percentages [12–14]. The flame-retardant effect of carbon nanotubes (CNT) has been researched in epoxy, polystyrene, polyaniline, polypropylene, and polyurethane polymeric matrixes, and the polymer/CNT nanocomposites are effective in producing continuous structured network [15,16]. A significant reduction in the peak heat release rate was observed in PP/CNT nanocomposites that contain 0.5–4.0 wt % of CNT [17].

Antunes et al. obtained PP and carbon nanofibers (CNF) nanocomposites by melting extrusion with electrical properties, and, taking into account the percolation threshold, formation of this nanocomposite showed similar conduction of Hall effect at CNF concentrations of 5%. At higher CNF contents, no significant improvements were achieved since tunnels conduction decreased in the extent polymer crystallinity increased [18]. The PP/CNT nanocomposites improve their electrical properties when increasing CNT content; the electrical percolation threshold was reached at CNT content of 1 and 2 wt % [19].

The addition of inorganic nanoparticles in polymeric matrix and the combination of different types of particles are used for obtaining multifunctional materials. The nanoparticles such as TiO_2, SiO_2, $CaCO_3$, ZnO, Ag, and nanoclays help to improve the physical and mechanical properties as well as flame retardant activity, thermal stability, oxidation, and permeability, among others [20,21]. The titanium dioxide (TiO_2) nanoparticles are ceramic materials widely used thanks to their properties as reinforcement for polymeric materials due to their long-term stability. Aydemir et al. [22] report the production of PP/TiO_2 nanocomposites, where mechanical properties as stress resistance and elasticity module were widely favored with the addition of TiO_2 in the polymeric matrix. Esthappan et al. [23] report the production of PP-TiO_2 nanocomposites for use like fibers, since addition of TiO_2 nanoparticles improve thermal stability of polypropylene, besides improving polymer crystallinity.

There are different methods for polymeric nanocomposites synthesis. Among them is found the melting extrusion method, which favors homogeneous dispersion of nanocharges, and is considered eco-friendly and economically viable for the industry, since it does not require the use of solvents and can yield high production volumes [24,25].

This research work studies the synthesis and evaluation of nanocomposites of polypropylene with TiO_2 and CNT nanoparticles blend. The nanocomposites were obtained by melt extrusion method. The purpose of using these additives was to improve the properties of thermal stability, electrical conductivity, and flame retardant of PP.

2. Materials and Methods

2.1. Material

MWNTC were provided by Cheap Tubes Inc. and have an average diameter from 20 to 40 nm, length of 10–30 μm, and purity of ≥ 90% wt. The TiO_2 nanoparticles were provided by DuPont, with a particle size from approximately 200 nm and semispherical morphology rutile phase. The polymeric matrix PP with fluid index of 0.76 g/10 min. supplied by Polímeros Nacionales (México, México) was used.

2.2. *Methods*

Synthesis of PP/CNT Nanocomposites

The preparation of PP nanocomposites with a mixture of TiO_2/CNT was performed by the melting extrusion process. The masterbatch preparation of PP/TiO_2/CNT was conducted in a twin-screw extruder lab size from Thermo Scientific model Prism TSE-24MC, screw diameter of 24 mm, L/D ratio 40:1, and temperature profile of 180 °C and rotational speed of 100 rpm. Low shear strengths were used to improve particles dispersion in polymeric matrix in the screw configuration, which is shown in Figure 1.

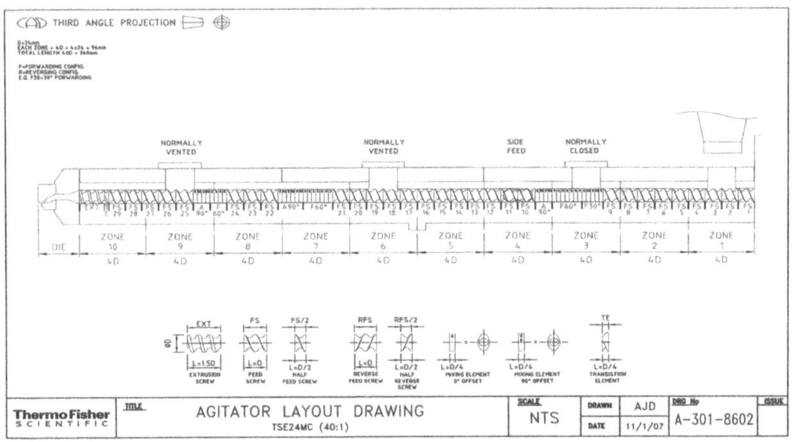

Figure 1. Configuration for the extrusion process.

There were prepared three nanocomposites of PP (1, 5, 10 wt %) with different weight percentages of CNT and titanium dioxide (TiO_2) nanoparticles. Table 1 lists the amounts used. The nanocomposites are identified as PP-TiO_2/CNT-X, where X means the weight percentage of nanocharges (CNT+ TiO_2) and PP without particles. It was decided to use these percentages due to some authors reporting improvements in the physical and chemical properties of polymers when adding CNT [26].

Table 1. Compounding formulations polypropylene-carbon-nanotubes (PP-TiO_2/CNT-X).

Sample	CNT + TiO_2 (wt %)	PP (g)	TiO_2/CNT (g)
PP	0	200	0
PP-TiO_2/CNT-1	1	198	2
PP-TiO_2/CNT-5	5	190	10
PP-TiO_2/CNT-10	10	180	20

2.3. *Characterization*

2.3.1. X-Ray Diffraction (XRD)

For nanocomposites structural analysis, wide angle X-ray diffraction technique (WAXD) was used, conducted in diffractometer from Siemens model D-5000, operating at current intensity of 25 mA and voltage of 35 kV, to obtain Cu Kα radiation with a wavelength equivalent to 1.54056 Å.

2.3.2. Fourier Transform Infrared Spectroscopy FTIR (ATR)

For FT-IR analysis a Thermo Nicolet infrared spectrometer, model MAGNA 550 (GMI, Minneapolis, Minnesota, USA), was used. The conditions at which these analyses were performed are the following: scanner 100, resolution of 16 cm^{-1}, and wave interval from 4000 to 500 cm^{-1} with ATR support.

2.3.3. Melt Flow Index (MFI)

The melt flow index was obtained using Dynisco plastometer, which consists of heating a barrel to melt material. Aw piston press was loaded with melted material to make it flow through die with a circular orifice of 2.1 mm of diameter and length of 8 mm. This test was performed under ASTM D1238-40 standard.

2.3.4. Thermal Stability (DSC and TGA)

For evaluating thermal properties, thermogravimetric analysis was used. The equipment used was Dupont Instruments model 951 (TA Instruments, New Castle, Pennsylvania, USA), operated at heating rate of 10 °C/min in nitrogen atmosphere with gas flow of 50 mL/min. The approximate weight of samples was of 10 mg and was analyzed in the interval of temperature from 25 to 800 °C. The Differential Scan Calorimetry was carried out under ASTM D3418 standard and thermal analyzer Q2000 from TA Instruments (New Castle, Pennsylvania, USA), with standard cell.

The degree of crystallinity (Xc) was calculated using with the following equation:

$$X_c(\%) = \left[\frac{\Delta H_f}{(1-\varnothing)\Delta H^*}\right] \times (100) \tag{1}$$

where ΔH_f is the fusion heat or formation of PP in the nanocomposites, ΔH^* is the formation heat of PP with crystallinity of 100% equivalent to 209 J/g and Ø is the fraction of weight of TiO_2 and CNT [27].

2.3.5. Scanning Electron Microscopy (SEM)

For the determination of size and morphology for each one of the components, a JOEL Field Emission Scanning Electron Microscope model JSM-7401F (JEOL, Peabody, MA, USA) was used. The microscope acceleration voltage was of 3.0 kV using the LEI secondary electrons detector.

2.3.6. Mechanical Properties Analysis

For measuring mechanical properties, tension tests were performed on model 4301 Instron universal machine (Instron corporation, Norwood, Massachusetts, USA), at 5 mm min^{-1}, with different lengthening percentages (0, 60, 400, and 700%).

2.3.7. Electrical Resistivity

These evaluations were made three times. For electrical resistivity, thickness plates were used, and they were covered in both sides with silver paint. The device used was Keysight LCR [inductance (L), capacitance (C), and resistance (R)] meter model E 4980 A, over 20 Hz to 2 MHz and LCR meter model ZM2372, from 0.001 Hz to 100 kHz. The electrical conductivity was calculated using the Equation (2) [28]:

$$\sigma = \frac{1}{\rho} \tag{2}$$

where σ is the electrical conductivity and ρ is the electrical resistivity.

2.3.8. Calorimetric Cone

For evaluation of combustion properties of nanocomposites, a calorimetric cone from Fire Testing Technology was used following the method described under ASTM E1354 standard. The evaluation of samples was made in horizontal position and the position of heat flow generator cone was also in horizontal position. The samples measurements were from 100 mm × 100 mm × 3 mm and were obtained by compression molding. The calorimetric cone was calibrated at 5 kW with methane flow, the flow in the extraction duct was of 24 L/s, and the analyzer was calibrated with 20.95% of oxygen. Heat flow for assessing samples was 35 kW/m^2. The sample was placed in aluminum paper tray with same dimension of sample and 1 cm height, leaving the surface to evaluate with free area of 100 mm^2. This was placed in the sample holder adjusting distance between cone and surface of sample to 25 mm.

3. Results and Discussion

3.1. X-ray Diffraction

In Figure 2, there are shown PP, and CNT and TiO_2 nanoparticles XRD diffractograms with the aim to compare diffraction patterns of base materials with nanocomposites of PP-TiO_2/CNT. The diffractograms of samples with PP showed signals located at 2θ angles of 14.1, 16.1, 16.8, 21.5 and 25.5°, which correspond to planes (110), (300), (040), (111), and (060) of PP crystalline phase, and the other signal located at 2θ angle of 21.1° can be correlated to phases α(111) and β(311) of PP. In the nanocomposites, signals characteristic of CNT [29.1° plane (100)] were not detected, since polypropylene signals are superimposed. The nanocomposites containing TiO_2 showed the characteristic signal located at 28.5° (110) [29,30].

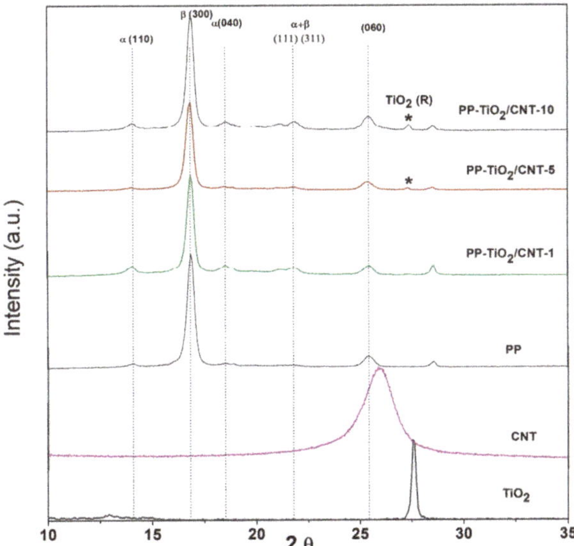

Figure 2. X-ray diffraction patterns of nanocomposite PP-TiO_2/CNT (1,5,10%), polypropylene (PP), carbon-nanotubes (CNT) and TiO_2.

The crystallinity can be more affected in the PP-TiO_2/CNT-10 nanocomposite where change related to polymer crystallinity is detected, increasing the nanocomposites signals. In 2013, Wang et al. studied PP nanocomposites with carboxylate nucleating agent (NTC) and the XRD diffractograms showed a signal at 2θ of 16.51°, which corresponds to crystalline plane (300) of β-hexagonal crystalline phase, showing that NTC has clear effect of nucleation in PP [31]. Similar results were obtained by Zohrevand et al., in 2014 when studying PP/TiO_2 nanocomposites with 1, 3, and 5 vol%. The intensity of this β phase is significant only in nanocomposites with 1 vol%, while with 3 and 5 vol%, the intensity of peak is not significant. Besides this, they report that presence of a peak in 2θ at 21.1° can be correlated to alpha and beta phase, and these results confirm that presence of TiO_2 nanoparticles induces β-form crystal formation in PP [32].

3.2. Fourier Transform Infrared FTIR (ATR)

The spectroscopy results of FT-IR are shown in Figure 3, where it can be noted that nanocomposites (Figure 3d–f) show transmittance signals characteristic of PP located in the range from 3000 to 2800 cm^{-1}, which correspond to asymmetric and symmetric C-H stretching vibration of methylene (CH_2) and methyl (CH_3) groups. Signals corresponding to flexions of CH_2 and CH_3 bonds are localized in 1448 cm^{-1} and 1373 cm^{-1}, respectively. The FT-IR spectrum of TiO_2 nanoparticles (Figure 3a) shows three signals: in 3438 cm^{-1} corresponding to hydroxyl group O-H, in 1624 cm^{-1} of Ti-OH, and at 735 cm^{-1} corresponding to Ti-O bond [33,34]. The FT-IR spectrum of CNT is shown in Figure 3b, and the signal of stretching of O-H bonds was detected at 3464 cm^{-1}, which existing in CNT. Also, in 2918 and 2841 cm^{-1}, signals of C-H bond were detected, and in 1650 cm^{-1} there were detected signals from stretching of C=C vibrations and at 1041 cm^{-1} of C-O bond [35,36].

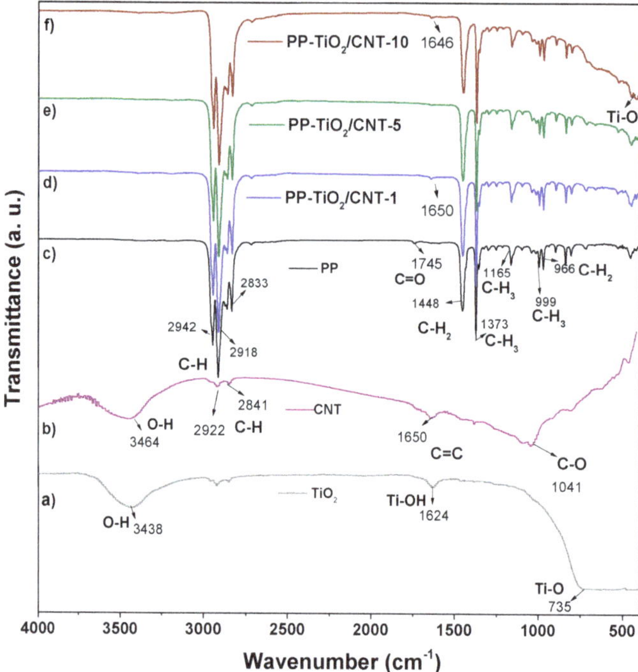

Figure 3. FT-IR spectra of nanocomposite PP-TiO_2/CNT (1, 5 and 10%), polypropylene (PP), carbon nanotubes (CNT) and TiO_2.

The nanocomposites of PP with contents of 5 and 10 wt % (Figure 3e,f), showed signals corresponding to PP and TiO_2 resin, and in 443 cm^{-1} a signal that corresponds to Ti-O bond was detected. The other signal that increases intensity in the nanocomposites with greater charge percentages (5 and 10%) is the one corresponding to CH_3, with values in the order of 1373 cm^{-1}.

This coincides with results obtained by Hashing et al. in 2004, when preparing nanocomposites with PP and CNT by mechanical pulverization, where they obtained variation in the intensities of the signals at 1373 cm^{-1}, attributing this result to that some PP chains are strongly bonded to CNT walls, in consequence to strong actions of cut, compression, and friction of mixing [37].

The FT-IR spectrum of neat PP processed at same conditions as nanocomposites shows a small signal in 1745 cm^{-1}, which is related to its thermo-oxidative degradation and the formation of carbonyl groups. The presence of these carbonyl groups into range of 1810 to 1660 cm^{-1} [38] is reported in the literature. On the other hand, the FT-IR spectra nanocomposites do not show the signal in 1745 cm^{-1} which suggests that the NPs mix of CNT and TiO_2 inhibits or reduces the thermo-oxidative degradation effects during processing of PP.

In this regard, it has been reported that the use of TiO_2 as an additive reduces the thermal degradation of PP. When there is an increase of TiO_2 concentration, the signals attributed to asymmetric and symmetric C-H stretching vibration (3000–2800 cm^{-1}), exhibit an intensity rise, which can suggest an increase in thermal stability of PP. The absence of bands in the wave number range 3600–3200 cm^{-1} indicates that formation of hydroperoxides was not favored [39].

3.3. *Evaluation of Melt Flow Index (MFI)*

The results of the evaluation of melt flow index (MFI) for different samples are listed in Table 2. The MFI values suggest that the addition of nanoparticles increase the viscosity, as a result of the materials becoming less fluid after increasing the CNT and TiO_2 content. The PP-TiO_2/CNT-10 nanocomposite showed greater viscosity increase (0.24 MFI g/10 min), in comparison with the value shown by PP without charge (0.76 g/10 min). The nanocharges have a high aspect ratio, which favors strong intermolecular interaction with the polymeric matrix; the adsorption of PP chains in the surface of nanocharges increases viscosity. These interactions increase the deformation resistance to hinder polymer flow in melted state, thereby making difficult or limiting the polymer flow in melted state [40].

Table 2. Flow index of the samples analyzed.

Sample	MFI (g/10 min)
PP	0.76
PP-TiO_2/CNT-1	0.56
PP-TiO_2/CNT-5	0.40
PP-TiO_2/CNT-10	0.24

3.4. *Thermogravimetric Analysis (TGA)*

To define thermal stability of nanocomposites, thermogravimetric analysis was performed. Figure 4 shows these results and it can be noted that materials have similar thermal behavior; the samples analyzed do not exhibited weight loss related to water adsorbed in surface materials. The first weight loss is in the range 358–500 °C, this is attributed to breaking of chains existing in polypropylene structure [41].

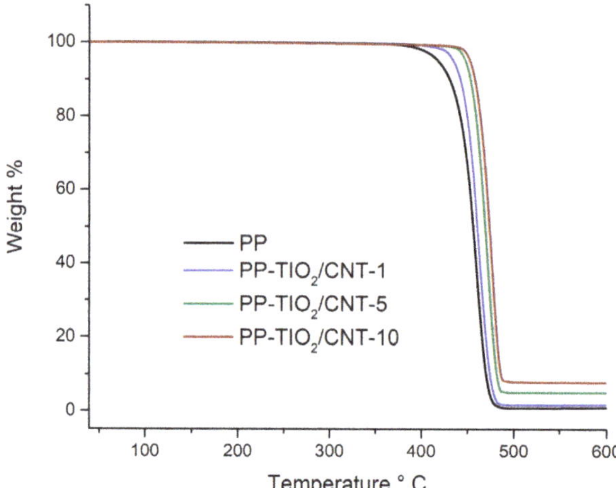

Figure 4. Thermogravimetric analysis of polypropylene (PP) and nanocomposites PP-TiO_2/CNT (carbon nanotubes).

The nanocomposites thermograms show that mass losses at 5 wt % and 50 wt % are detected at different temperatures in each nanocomposite (Table 3). In general, the temperatures $T_{5\%}$ and $T_{50\%}$ increased with the charges content in the nanocomposites. The PP-TiO_2/CNT-10 nanocomposite showed the highest temperature compared to nanocomposites with less nanocharge content. The 5 wt % decompositions for neat PP in PP-TiO_2/CNT-1, PP-TiO_2/CNT-5, and PP-TiO_2/CNT-10 occurred at 435, 448, and 454 °C, respectively. This means that the addition of charges increases the decomposition temperature in 18, 31, and 37 °C. Similar behavior was shown when the nanocomposites reached a 50 wt % of decomposition.

Table 3. Thermal properties of nanocomposite of polypropylene-carbon-nanotubes (PP-TiO_2/CNT).

Sample	$T_{5\%}$ (°C)	$T_{50\%}$ (°C)	T_{max} (°C)	Residue at 550 °C (%)
PP	417	454	459	0
PP-TiO_2/CNT-1	435	460	462	1.00
PP-TiO_2/CNT-5	448	469	472.	4.92
PP-TiO_2/CNT-10	454	474	475	8.02

These results indicated that using TiO_2 and CNT blend improves PP thermal stability; therefore, these combinations are a viable alternative to obtain flame retardant materials. Similar results have been reported when using TiO_2 and CNT nanoparticles individually in polypropylene [42,43].

To define a comparison of maximum decomposition temperature (T_{max}) the second derivative was used in TGA analysis. The values obtained in several formulations corroborate the difference of thermal stability between nanocomposites and PP without charge. The nanocomposites showed greater T_{max}, the PP-TiO_2/CNT-10 sample increased its T_{max} in 16 °C above pure PP.

The increase in starting decomposition temperature can be attributed to the increase in the addition strength in the PP interface and the CNT and TiO_2 nanoparticles. When proper interface interaction exists, the particles are able to restrain the movement of polymer chain, making it more difficult that the breaking of the polymer chains occurs at lower temperature. In consequence, the degradation temperature of nanocomposite is shifted to a higher temperature [44].

Other element to consider when improving thermal stability of nanocomposites is the transport barrier effect of the mass of the CNT hollow structure, and these structures can trap the free radicals generated during PP thermal [45,46].

3.5. Differential Scan Calorimetry (DSC) Analysis

By DSC analysis, the melting temperature T_m, crystallization temperature Tc and degree of crystallinity Xc, were derived from endothermic and exothermic peak temperatures. These thermograms are shown in Figures 5 and 6.

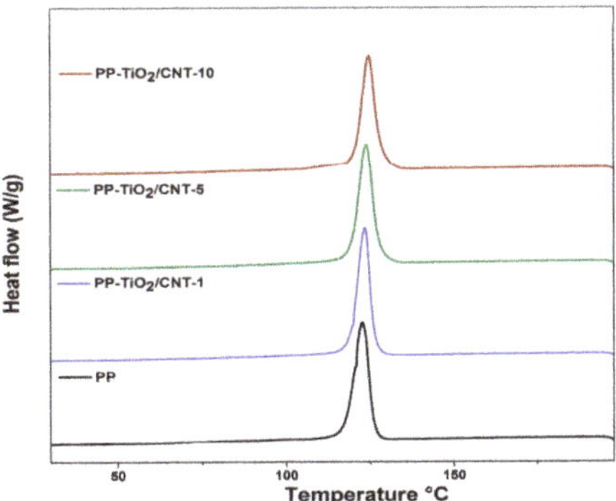

Figure 5. Differential scanning calorimetry (DSC) crystallization exotherms of pure polypropylene (PP) and PP-TiO_2/CNT (carbon nanotubes) (1, 5, 10%).

The DSC analyses reported that Tc and Tm increase gradually with the increase of nanoparticles (TiO_2 and CNT) in the PP, reaching temperatures above 122 °C and 157 °C, respectively.

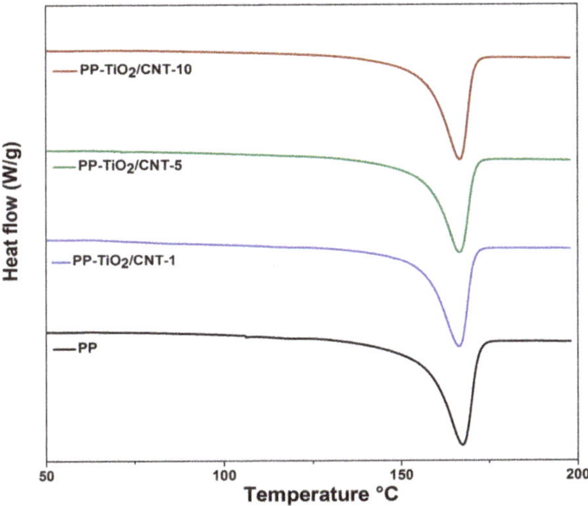

Figure 6. Differential scanning calorimetry (DSC) fusion endotherms of pure polypropylene (PP) and PP-TiO_2/CNT (carbon nanotubes) (1, 5, 10%).

The crystallization process can be noted in the DSC curves; in Table 4 is the summary of fusion enthalpies, crystallization, and degree of crystallinity calculated with Equation (1), for each of the studied materials. The presence of nanocharges affects PP crystallinity, the fusion peaks are narrow in comparison to peak from neat PP, Xc decreases in the extent the nanocharges content (CNT and TiO_2) increases, for nanocomposite with greater content of charges Xc decreased 3.38 wt %, and this behavior can be related to the agglomeration of particles. Some reports indicate that formation of nanocomposites with high content of CNT can induce effects of topologic confinement that can eventually result in reduction of nucleation kinetics and crystallization [47–50].

Table 4. Differential scanning calorimetry (DSC) date for polypropylene (PP) and nanocomposites PP-TiO_2/CNT (carbon nanotubes) (1, 5 y 10%).

Nanocomposite	T_m (°C)	Enthalpy of Fusion (J/g)	Enthalpy of Crystallization (J/g)	X_c (%)
PP	156.93	95.82	93.72	45.84
PP-TiO_2/CNT-1	156.16	93.45	95.35	44.71
PP-TiO_2/CNT-5	157.14	91.06	92.82	43.56
PP-TiO_2/CNT-10	157.07	88.75	93.63	42.46

In 2015, Zhang et al. [51], in their research with PP and CNT nanocomposites, suggest that variation of crystallization temperature is strongly related to different functions that charges have during PP crystallization.

The effect caused by CNT and TiO_2 nanoparticles in the PP matrix is known. Some reports are contradictory because the physicochemical characteristics of the nanoparticles are not always the same, so the aggregation and dispersion in semicrystalline polymers is different. On the other hand, the final properties of most semicrystalline polymers depend on the microstructures, which are mainly affected by crystallization [52]. The polymer crystallization may be intimately related to the type of nanoparticle and concentration, dispersion state, aspect ratio, crystallization conditions, and so on. [53–57] All these conditions can explain the contradictory results that have been reported; in the PP-TiO_2/CNT nanocomposites, the presence of two nanocharges type of blend increases the adverse effects.

The results obtained for the PP-TiO_2/CNT nanocomposites are expected and they coincide with what is reported in literature, where the effect of CNT and TiO_2 nanoparticles individually has been studied. For example, several reports show that CNT accelerates PP crystallization only at contents less than 0.5 wt % due to strong heterogeneous nucleation effect of the CNTs in the PP. When the concentration increases, the formation of aggregates is favored, and the heterogeneous nucleating efficiency of the individual CNTs is lowered [58].

The PP/TiO_2 nanocomposites have similar behavior independent of size and shape of nanoparticles, Xc increases with contents of TiO_2 of 0.5–2 wt %, when the content increases between 3–4 wt %, decreases it, but the value is similar to that shown for PP [22].

In summary, the high CNT content decreases crystallinity, whereas TiO_2 increases the crystallinity, even at high contents of ~2%. The nanocomposites of PP/TiO_2/CNT have a ratio of CNT/TiO_2 of 2:1. Therefore, the expected effect for CNT on the PP matrix is more predominant. According to Table 4, a decrease was detected in Xc and it decreases in the extent of CNT content increases.

3.6. Electron Scan Microscopy

Figure 7 shows SEM micrographs of PP-TiO_2/CNT-1, PP-TiO_2/CNT-5, and PP-TiO_2/CNT-10 nanocomposites. In these SEM images, the presence of CNT and TiO_2 nanoparticles can be verified. All micrographs exhibited two zones with different nanoparticles agglomeration, in zone 1 (Supplementary Materials) agglomerates can be observed containing both nanoparticles and their contact between them. Also, the amount of agglomerates increases with the increase in the CNT and TiO_2 content. The zone 2 showed less agglomeration and it was noted that both nanoparticles are embedded within PP matrix (Figure 7a–c). In these zones, it is difficult to define if CNT and TiO_2 nanoparticles are in contact. In the micrography of PP-TiO_2/CNT-10 nanocomposite with greater content of nanoparticles (Figure 7c), small agglomerates were detected where there is contact between TiO_2 nanoparticles. It is worth to mention that processing conditions of nanocomposites causes coalescence between TiO_2 nanoparticles and in some cases the spherical shape it is not well defined; similar results were previously reported [59]. The coalescence is not exclusive to TiO_2 NPs. The spherical copper nanoparticles (Cu NPs), with an average size of 21 nm when they are processed to obtain Nylon 6/Cu nanocomposites, increased their size and formed aggregates in form of wire [60].

The increase in viscosity caused coalescence of the TiO_2 NPs, the PP-TiO_2/CNT-10 had an average particle size of 200 nm, and the nanocomposites with 1 and 5 wt % of NPs had an average particle size of 557 nm and 286 nm, respectively (Supplementary Materials Figures S1–S3).

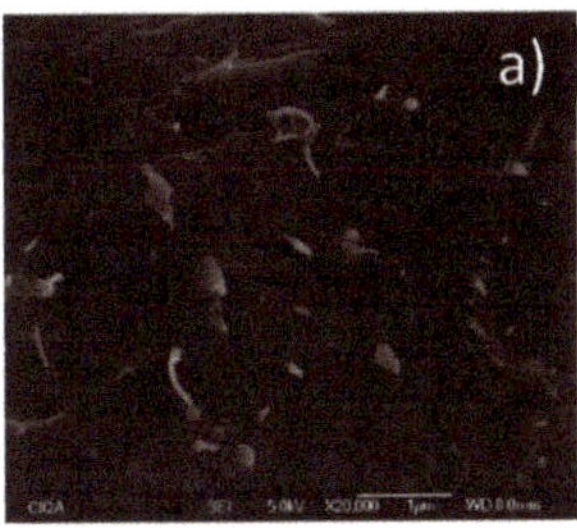

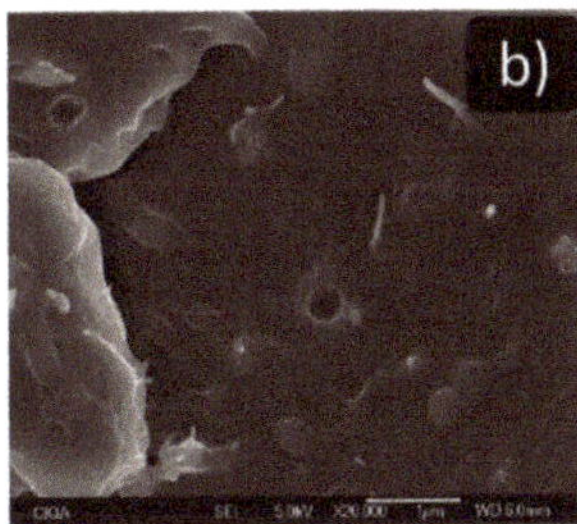

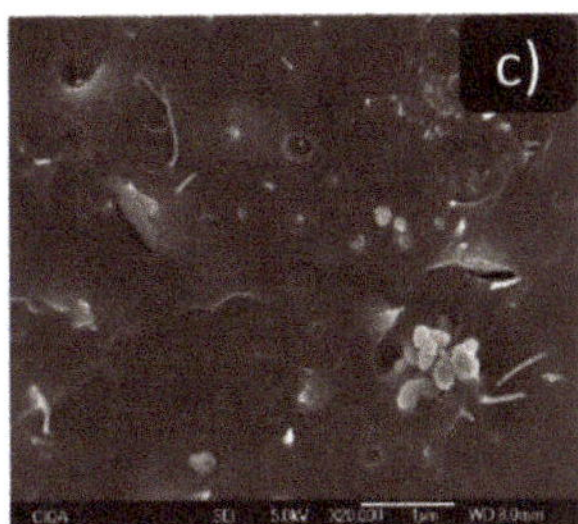

Figure 7. SEM micrographs of nanocomposites. (**a**) polypropylene-carbon-nanotubes (PP-TiO_2/CNT-1), (**b**) PP-TiO_2/CNT-5 and (**c**) PP-TiO_2/CNT-10.

In some of the micrographs, a series of cavities or voids can be observed in the polymeric matrix. This can be due to the fact that the extrusion process of the particles of greater size is sent against the next material, impacting and forming a track or trail [61].

3.7. Electrical Conductivity

The electric resistance of nanocomposites decreases to the extent the nanoparticles content increases; this effect was only significant in the PP-TiO_2/CNT-10 nanocomposite. With the addition of 10% of nanoparticles, the electrical conductivity increased 6 times with regards to pure polypropylene. Table 5 lists the obtained values of electrical resistance and calculated electrical conductivity. The lack of electrical conductivity in nanocomposites with 1 and 5 wt % can be explained by an inefficient dispersion of nanoparticles, and this coincides with SEM analysis. Another possible explanation can be found in the amount of PP that can be bonded to nanoparticles. The adherence of polymer chains to nanoparticles surface prevents electrons flow, and it is known that thermal treatments can improve the electrical conductivity, due to destruction of crystalline phase of existent polymer in the surface [62].

The electrical conductivity of PP/CNT nanocomposites has been deeply studied and it is known that nanocomposites with content of CNT of 10 wt % can have resistivity of up to 10^2 Ω/sq [63].

There are few studies about electrical properties of PP/TiO_2 nanocomposites. To our knowledge, there is only one study about this, and is reported that with the increase in volume content of titanium-dioxide nanoparticles, the value of dielectric permittivity of nanocomposites also increases, and after some point it starts to decrease. The specific resistance of nanocomposites depends on temperature and it was detected that at 116.9 °C the electrical properties of nanocomposite show significant improvement [64].

Table 5. Electrical properties of polypropylene (PP) and PP-TiO_2/CNT (carbon-nanotubes) nanocomposites.

Sample	Surface Resistance Ω/sq	Volumetric Resistance Ω cm	Electric Conductivity S/m
PP	2.35×10^{13}	1×10^{18}	1×10^{-18}
PP-TiO_2/CNT-1	6×10^{16}	3×10^{17}	3.0×10^{-18}
PP-TiO_2/CNT-5	3×10^{12}	5×10^{16}	2.0×10^{-17}
PP-TiO_2/CNT-10	6.5×10^{9}	7.2×10^{9}	1.4×10^{-10}

3.8. Mechanical Properties

The results of mechanical tests are listed in Table 6, where tensile strength, percentage of nominal deformation to rupture, and Young's modulus values for nanocomposites can be noted. In this analysis, it was observed PP without charge presented a tensile strength of 23.93 MPa, while the PP-TiO_2/CNT nanocomposites showed low values, which decreases in the extent the nanoparticles content is increased. These values coincide with the degree of crystallinity (Table 4); in consequence, the decrease of rigidity of matrix of PP is favored.

Table 6. Tensile properties of polypropylene (PP) and PP-TiO_2/CNT (carbon nanotubes) nanocomposites.

Sample	Xc	Tensile Strength (MPa)	Nominal Strain at Break (%)	Young's Modulus (MPa)
PP	56.36	23.93 ± 0.12	53.37	971.57 ± 29.1
PP-TiO_2/CNT-1	54.97	23.93 ± 0.19	45.78	894.98 ± 27.1
PP-TiO_2/CNT-5	53.56	23.32 ± 0.08	33.24	961.28 ± 31.0
PP-TiO_2/CNT-10	52.20	22.75 ± 0.37	22.28	1077.44 ± 26.2

For the nominal percentage at the breaking the observed behavior is similar, such effect is due to the fact that nanotubes particles obstruct the movement of PP chains, reducing rigidity. Young's modulus of PP-TiO_2/CNT-10 nanocomposite showed an increase of 18.3% with regards PP without charges. This was the only case where it was observed that nanoparticles cause higher rigidity and hardness in the PP matrix. It has been reported that adding CNT in polymers will strengthen material, if there is an efficient dispersion [65]

The decrease of Young's modulus when increasing charges content is explained by inefficient dispersion of charges, since the agglomerates present in poorly dispersed composite cause cracks to initiate and easily propagate in polymeric matrix [66]. As noted previously, the TiO_2 nanoparticles tend to coalesce, increasing their size and changing shape. This process seems to be more significant when charges concentrates are lower (1 and 5%). This behavior can be attributed to the increase of viscosity with the nanoparticles' concentration. For low viscosity, the TiO_2 NPs are able to move in the matrix and aggregates, whereas at high concentrations the viscosity is higher and the aggregation is limited by slightly mobility of the nanoparticles.

3.9. Calorimetric Cone

To analyze combustion processes, in real time assessment of nanocomposites the cone calorimetry method was used. Heat release rate (HRR) and its maximum value (PHRR) were some of the parameters obtained in this study.

The heat release rate curves of the PP-TiO_2-CNT nanocomposite are shown in Figure 8. These results showed essentially a similar behavior, the maximum value of heat release rate (PHRR) for PP without charge and PP-TiO_2/CNT-1 nanocomposite were of 1529.53 and 1593.83 kW/m^2 respectively, which means that there is no difference between pure PP and nanocomposite with low load percentages (1%). In contrast, for PP-TiO_2/CNT-5 and TiO_2/CNT-10 nanocomposites, a decrease was noted in PHHR and total heat release rate (THR) values, which indicates that materials have less flame propagation and better resistance to fire. The minimum values of PHHR that could be obtained were of 1058.49 kW/m^2 for PP-TiO_2/CNT-5 and of 1079.94 kW/m^2 for PP-TiO_2/CNT-10.

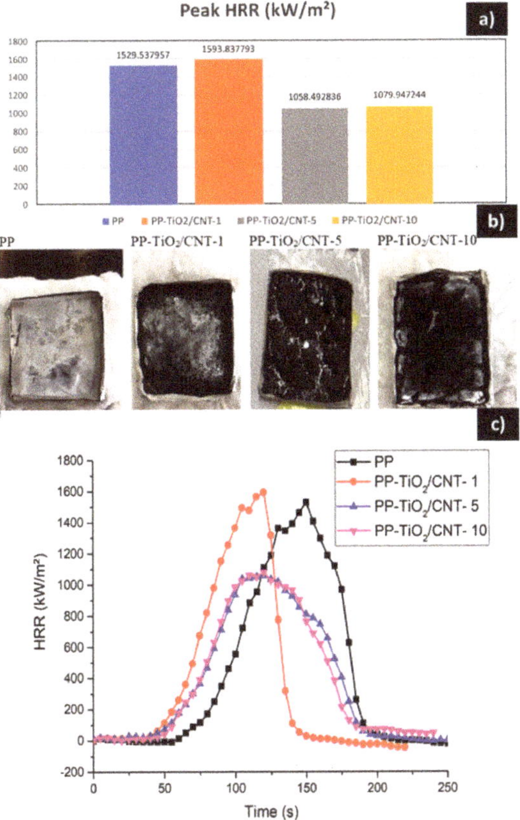

Figure 8. Calorimetric measurements (**a**) Comparison of the peaks of the heat release rate (HRR), (**b**) Photographs of residual material after the cone calorimetric test (**c**) HRR curves for the results.

Table 7 lists data of calorimetric cone, where it was noted that the total heat release rate (THR) reduced when increasing the particles content to 5 and 10% wt. Also, there are listed the wt % of residues obtained after flame-retardant assessment. The number of residues increases dramatically when the nanoparticles content increases. For PP-TiO_2/CNT-10 nanocomposite, the obtained residues were 91.9%. In Figure 8b) the residues appearance can be observed. For the case of PP-TiO_2/CNT-1, it was noted that there was a small amount of scattered dust, while for PP-TiO_2/CNT-5 y TiO_2/CNT-10 a semi-continuous phase that suggests a good distribution of nanocharges in polypropylene matrix was noted. In addition, during the ignition a network forms between nanoparticles and degradation product of PP, and the new materials can have enough resistance to flame in order to avoid complete ignition. A similar explanation was obtained by studying the flammability properties of the PP/zeolite/CNT nanocomposite, where it was proposed to form a protective layer with a continuous network structure that provides the flame-retardant characteristic of polymeric nanocomposites [67].

Table 7. Data of the calorimetric cone test of the samples analyzed.

Sample	Peak HRR (kW/m^2)	THR (MJ/m^2)	Residue (%)
PP	1529.53	66.42	0.09
PP-TiO_2/CNT-1	1593.83	65.91	34.3
PP-TiO_2/CNT-5	1058.49	43.49	88.4
PP-TiO_2/CNT-10	1079.94	44.06	91.9

Results coincide with other studies, for example, PP/CNT nanocomposites are considered to be flame-retardant materials and even more effective than PP/clay nanocomposites. The combination of TiO_2 and CNT to obtain anti-flame additives can be a good idea, because TiO_2 has a high decomposition temperature (700 °C–800 °C) and an oxygen index of 29 and has also been widely used as an anti-flame additive alone or in combination with other additives [5]. Despite the above, its usage, like flame-retardant in polyamides, has been technically questionable [68].

In our study, the CNT and TiO_2 blends lead to obtaining nanocomposites with acceptable flame-retardant properties, but more detailed studies can be required to optimize formulation and to explain its possible mechanism.

4. Conclusions

The PP-TiO_2/CNT nanocomposites was obtained by the melt mixing method using a mixture of nanoparticles TiO_2 and CNT, with contents of 1, 5, and 10 wt %. The thermal stability of the nanocomposites increased when increasing the content of the NPs, for example, for PP-TiO_2/CNT-10, its maximum degradation temperature increased 16 °C with respect to the pure PP. In addition, the thermo-oxidative stability of the material was improved and confirmed by the lack of signals of the carbonyl group in FTIR spectrum. The degree of crystallinity decreased with a high content of NPs. This effect was also reflected in a slight decrease in mechanical properties, and only an in increase in the Young's modulus of 10% to PP-TiO_2/CNT-10 was observed.

The electrical conductivity of PP-TiO_2/CNT-10 nanocomposite was improved by eight orders of magnitude with respect to the pure PP; nanocomposites with low content do not have significant changes.

The melt flow index (MFI) of the nanocomposites decreased with the number of NPs, and the nanocomposites with 1, 5, and 10 wt % gave an MFI of 0.56, 0.40, and 0.24 (g/10 min) respectively. The conditions of the melt processing and the increase in the viscosity caused coalescence of the TiO_2 NPs. This conducted to different average particle size in each nanocomposite.

Finally, the effect of flame retardancy was confirmed by a significant decrease of the peak HRR in the nanocomposites PP-TiO_2/CNT-5 and PP-TiO_2/CNT-10, and besides this, PP-TiO_2/CNT-10 presented a content of residual carbon of 91.9% after ignition.

Supplementary Materials: The following are available online at http://www.mdpi.com/2073-4360/11/7/1204/s1. Figure S1. Morphology nanocomposite PP-TiO_2/CNT-1 analyzed by SEM, histogram of TiO_2 particle. Figure S2. Morphology nanocomposite PP-TiO_2/CNT-5 analyzed by SEM, histogram of TiO_2 particle. Figure S3. Morphology nanocomposite PP-TiO_2/CNT-10 analyzed by SEM, histogram of TiO_2 particle.

Author Contributions: C.C.A. performed experiments on nanocomposite PP-TiO_2/CNT. P.R.R. analyzed the data and contributed to the discussions on DRX and TGA, M.A.G. analyzed the data and contributed to the discussions on FT-IR and Electrical conductivity, V.C.D. and L.M.L. performed experiments and analyzed calorimetric cone, C.A.O. and G.C.P. provided technical discussions and reviewed and contributed to the final revised manuscript. M.P.A. performed the mechanical properties, Z.Q.J. SEM analysis.

Funding: Financial support from project Tlax-2018-01-01-43129 by CONACYT.

Acknowledgments: The authors kindly acknowledge the scholarship postdoctoral (387368) provided by CONACYT-Mexico. The authors are also grateful Jesús Angel Cepeda Garza, Jesús Alejandro Espinosa Muñoz, Ma. Guadalupe Méndez Padilla and Adán Herrera Guerrero, for their technical support.

Conflicts of Interest: The authors declare no conflict of interest.

References

1. Panzino, F.; Piza, A.; Pociello, N.; García, J.J.; Luaces, C.; Pou, J. Estudio multicéntrico sobre factores de riesgo de lesiones en accidentes de automóvil. In *Anales de Pediatría*; Elsevier Doyma: Amsterdam, The Netherlands, 2009; Volume 71, pp. 25–30. [CrossRef]
2. Adanur, S. *Polymers and Fiber. Wellington Sears Handbook of Industrial Textiles*; Routledge: New York, NY, USA, 2017; Volume 2, pp. 12–37. [CrossRef]
3. Kiliaris, P.; Papaspyrides, C.D. Polymer/layered silicate (clay) nanocomposites: An overview of flame retardancy. *Prog. Polym. Sci.* **2010**, *35*, 902–958. [CrossRef]
4. Thomas, N.L. Zinc compounds as flame retardants and smoke suppressants for rigid PVC. *Plast. Rubber Compos.* **2003**, *32*, 413–419. [CrossRef]
5. Lam, Y.L.; Kan, C.W.; Yuen, C.W.M. Effect of titanium dioxide on the flame-retardant finishing of cotton fabric. *J. Appl. Polym. Sci.* **2011**, *121*, 267–278. [CrossRef]
6. Dittrich, B.; Wartig, K.; Hofmann, D.; Mülhaupt, R.; Schartel, B. Carbon black, multiwall carbon nanotubes, expanded graphite and functionalized graphene flame retarded polypropylene nanocomposites. *Polym. Adv. Technol.* **2013**, *24*, 916–926. [CrossRef]
7. Mubarak, Y.A.; Abbadi, F.O.; Tobgy, A.H. Effect of iron oxide nanoparticles on the morphological properties of isotactic polypropylene. *J. Appl. Polym. Sci.* **2010**, *115*, 3423–3433. [CrossRef]
8. Hufenbach, W.; Böhm, R.; Thieme, M.; Winkler, A.; Mäder, E.; Rausch, J.; Schade, M. Polypropylene/glass fibre 3D-textile reinforced composites for automotive applications. *Mater. Des.* **2011**, *32*, 1468–1476. [CrossRef]
9. Ayrilmis, N.; Jarusombuti, S.; Fueangvivat, V.; Bauchongkol, P.; White, R.H. Coir fiber reinforced polypropylene composite panel for automotive interior applications. *Fibers Polym.* **2011**, *12*, 919. [CrossRef]
10. Saujanya, C.; Radhakrishnan, S. Structure development and crystallization behaviour of PP/nanoparticulate composite. *Polymer* **2001**, *42*, 6723–6731. [CrossRef]
11. Mishra, S.; Sonawane, S.H.; Singh, R.P.; Bendale, A.; Patil, K. Effect of nano-$Mg(OH)_2$ on the mechanical and flame-Retarding properties of polypropylene composites. *J. Appl. Polym. Sci.* **2004**, *94*, 116–122. [CrossRef]
12. Alexandre, M.; Dubois, P. Polymer-layered silicate nanocomposites: Preparation, properties and uses of a new class of materials. *Mater. Sci. Eng. R* **2000**, *28*, 1–63. [CrossRef]
13. Demir, H.; Arkis, E.; Balköse, D.; Ülkü, S. Synergistic effect of natural zeolites on flame retardant additives. *Polym. Degrad. Stab.* **2005**, *89*, 478–483. [CrossRef]
14. Ke, C.H.; Li, J.; Fang, K.; Zhu, K.; Zhu, J.; Yan, Q.; Wang, Y. Synergistic effect between a novel hyperbranched charring agent and ammonium polyphosphate on the flame retardant and anti-dripping properties of polylactide. *Polym. Degrad. Stab.* **2010**, *95*, 763–770. [CrossRef]
15. Kausar, A.; Rafique, I.; Muhammad, B. Review of Applications of Polymer/Carbon nanotube and Epoxy/CNT Composites. *Polym. Plast. Technol. Eng.* **2016**, *55*, 1167–1191. [CrossRef]
16. Lau, A.K.T.; Hui, D. The revolutionary creation of new advanced materials-carbon nanotube composites. *Compos. Part B Eng.* **2002**, *33*, 263–277. [CrossRef]
17. Kashiwagi, T.; Grulke, E.; Hilding, J.; Groth, K.; Harris, R.; Butler, K.; Douglas, J. Thermal and flammability properties of polypropylene/carbon nanotube nanocomposites. *Polymer* **2004**, *45*, 4227–4239. [CrossRef]
18. Antunes, M.; Mudarra, M.; Velasco, J.I. Broad-band electrical conductivity of carbon nanofibre-reinforced polypropylene foams. *Carbon* **2011**, *49*, 708–717. [CrossRef]
19. Seo, M.K.; Park, S. Electrical resistivity and rheological behaviors of carbon nanotubes-filled polypropylene composites. *Chem. Phys. Lett.* **2004**, *395*, 44–48. [CrossRef]
20. Avalos, F.; Ramos, L.F.; Ramirez, E.; Sanchez, S.; Mendez, J.; Zitzumbo, R. Nucleating effect of carbón nanoparticles and their influence on the thermal and chemical stability of polypropylene. *J. Nanomater.* **2012**, *2012*, 104. [CrossRef]
21. Kurahatti, R.V.; Surendranathan, A.O.; Kori, S.A.; Singh, N.; Kumar, A.R.; Srivastava, S. Defence applications of polymer nanocomposites. *Def. Sci. J.* **2010**, *60*, 551–563. [CrossRef]
22. Aydemir, D.; Uzun, G.; Gumus, H.; Yildiz, S.; Gumus, S.; Bardak, T.; Gunduz, G. Nanocomposites of polypropylene/nano titanium dioxide: Effect of loading rates of nano titanium dioxide. *Mater. Sci.* **2016**, *22*, 364–369. [CrossRef]
23. Esthappan, S.K.; Kuttappan, S.K.; Joseph, R. Thermal and mechanical properties of polypropylene/titanium dioxide nanocomposite fibers. *Mater. Des.* **2012**, *37*, 537–542. [CrossRef]

24. El-dessouky, H.M.; Lawrence, C.A. Nanoparticles dispersion in processing functionalized PP/TiO_2 nanocomposite: Distribution and properties. *J. Nanopart. Res.* **2011**, *13*, 1115–1124. [CrossRef]
25. Cassagnau, P.; Legare, V.; Fenouillot, F. Reactive processing of thermoplastic polymer: A review of the fundamental aspect. *Int. Polym. Process.* **2007**, *22*, 218–258. [CrossRef]
26. Chiamori, H.C.; Brown, J.W.; Adhiprakasha, E.V.; Hantsoo, E.T.; Straalsund, J.B.; Melosh, N.A.; Pruitt, B.L. Suspension of nanoparticles in SU-8: Processing and characterization of nanocomposite polymers. *Microelectron. J.* **2008**, *39*, 228–236. [CrossRef]
27. Logakis, E.; Pollatos, E.; Pandis, C.; Peoglos, V.; Zuburtikis, I.; Delides, C.G.; Vatalis, A.; Gjoka, M.; Syskakis, E.; Viras, K.; et al. Structure–property relationships in isotactic polypropylene/multi-walled carbon nanotubes nanocomposites. *Compos. Sci. Technol.* **2010**, *70*, 328–335. [CrossRef]
28. Crossman, R.A. Conductive composites past, present, and future. *Polym. Eng. Sci.* **1985**, *25*, 507–513. [CrossRef]
29. Wang, S.; Abdellah, A.; Shaoyun, G.; Chuanxi, X. Preparation of microporous polypropylene/titanium dioxide composite membranes with enhanced electrolyte uptake capability via melt extruding and stretching. *Polymers* **2017**, *9*, 110. [CrossRef]
30. Bahloul, W.; Walid, B.; Flavien, M.; Veronique, B.L.; Philippe, C. Structural characterisation and antibacterial activity of PP/TiO_2 nanocomposites prepared by an insitu sol–gel method. *Mater. Chem. Phys.* **2012**, *134*, 399–406. [CrossRef]
31. Wang, S.; Zhang, J. Effect of nucleating agent on the crystallization behavior, crystal form and solar reflectance of polypropylene. *Sol. Energy Mater. Sol. Cells* **2013**, *117*, 577–584. [CrossRef]
32. Zohrevand, A.; Ajji, A.; Mighri, F. Morphology and properties of highly filled iPP/TiO_2 nanocomposites. *Polym. Eng. Sci.* **2014**, *54*, 874–886. [CrossRef]
33. Liu, Z.; Jian, Z.; Fang, J.; Xu, X.; Zhu, X.; Wu, S. Low-temperature reverse microemulsion synthesis, characterization, and photocatalytic performance of nanocrystalline titanium dioxide. *Int. J. Photoenergy* **2011**, *2012*, 702503. [CrossRef]
34. Leon, A.; Reuquen, P.; Garin, C.; Segura, R.; Vargas, P.; Zapata, P.; Orihuela, P. FTIR and Raman characterization of TiO_2 nanoparticles coated with polyethylene glycol as carrier for 2-methoxyestradiol. *Appl. Sci.* **2017**, *7*, 49. [CrossRef]
35. Nguyen, V.H.; Shim, J.J. Green synthesis and characterization of carbon nanotubes/polyaniline nanocomposites. *J. Spectrosc.* **2015**, *2015*, 297804. [CrossRef]
36. Yildrim, A.; Seckin, T. In situ preparation of polyether amine functionalized MWCNT nanofiller as reinforcing agents. *Adv. Mater. Sci. Eng.* **2014**, *2014*, 356920. [CrossRef]
37. Xia, H.; Wang, Q.; Li, K.; Hu, G.H. Preparation of polypropylene/carbon nanotube composite powder with a solid-state mechanochemical pulverization process. *J. Appl. Polym. Sci.* **2004**, *93*, 378–386. [CrossRef]
38. Qian, Z.; Qian, J.; Lerou, F.; Tanga, P.; Li, D. Antioxidant intercalated hydrocalumite as multifunction nanofiller for Poly (propylene): Synthesis, thermal stability, light stability, and anti-migration property. *Polym. Degrad. Stab.* **2017**, *140*, 9–16. [CrossRef]
39. Esthappan, S.K.; Kuttappan, S.K.; Joseph, R. Effect of titanium dioxide on the thermal ageing of polypropylene. *Polym. Degrad. Stab.* **2012**, *97*, 615–620. [CrossRef]
40. Nurul, M.S.; Mariatti, M. Effect of thermal conductive fillers on the properties of polypropylene composites. *J. Thermoplast. Compos.* **2013**, *26*, 627–639. [CrossRef]
41. Bhagat, N.A.; Shrivastava, N.K.; Suin, S.; Maiti, S.; Khatua, B.B. Development of electrical conductivity in PP/HDPE/MWCNT nanocomposite by melt mixing at very low loading of MWCNT. *Polym. Compos.* **2013**, *34*, 787–798. [CrossRef]
42. Zhou, T.Y.; Tsui, G.C.P.; Liang, J.Z.; Zou, S.Y.; Tangm, C.Y. Thermal properties and thermal stability of PP/MWCNT. *Compos. Part B Eng.* **2016**, *90*, 107–114. [CrossRef]
43. Hapuarachchi, T.D.; Peijs, T.; Bilotti, E. Thermal degradation and flammability behavior of polypropylene/clay/carbon nanotube composite systems. *Polym. Adv. Technol.* **2013**, *24*, 331–338. [CrossRef]
44. Yang, F.; Nelson, G.L. Polymer/silica nanocomposites prepared via extrusion. *Polym. Adv. Technol.* **2006**, *17*, 320–326. [CrossRef]
45. Chu, C.C.; White, K.L.; Liun, P.; Zhangn, X.; Suen, H.J. Electrical conductivity and thermal stability of polypropylene containing well-dispersed multi-walled carbon nanotubes disentangled with exfoliated nanoplatelets. *Carbon* **2012**, *50*, 4711–4721. [CrossRef]

46. Al-Shere, S.Z.; Al-Amshany, Z.M.; Al Sulami, Q.A.; Tashkandi, N.Y.; Hussein, M.A.; El-Shishtawy, R. The preparation of carbon nanofillers and their role on the performance of variable polymer nancomposites. *Des. Monomers Polym.* **2019**, *22*, 8–53. [CrossRef]
47. Vega, J.; Martinez, J.; Trujillo, M.; Arnal, M.; Muller, A.; Bredeau, S.; Dubois, P. Rheology, Processing, Tensile Properties, and Crystallization of Polyethylene/Carbon Nanotube Nanocomposites. *Macromolecules* **2009**, *42*, 4719–4727. [CrossRef]
48. Vega, J.F.; da Silva, Y.; Vicente Alique, E.; Nuńez Ramírez, R.; Trujillo, M.; Arnal, M.L.; Müller, A.J.; Dubois, P.; Martínez Salazar, J. Influence of Chain Branching and Molecular Weight on Melt Rheology and Crystallization of Polyethylene/Carbon Nanotube Nanocomposites. *Macromolecules* **2014**, *47*, 5668–5681. [CrossRef]
49. Trujillo, M.; Arnal, M.; Müller, A.J.; Bredeau, S.; Bonduel, D.; Dubois, P.; Hamley, I.; Castelletto, V. Thermal Fractionation and Isothermal Crystallization of Polyethylene Nanocomposites Prepared by in Situ Polymerization. *Macromolecules* **2008**, *41*, 2087–2095. [CrossRef]
50. Trujillo, M.; Arnal, M.; Müller, A.J.; Laredo, E.; Bredeau, S.; Bonduel, D.; Dubois, P. Thermal and Morphological Characterization of Nanocomposites Prepared by in Situ Polymerization of High- Density Polyethylene on Carbon Nanotubes. *Macromolecules* **2007**, *40*, 6268–6276. [CrossRef]
51. Zhang, X.; Yan, X.; He, Q.; Wei, H.; Long, J.; Guo, J.; Gu, H.; Yu, J.; Liu, J.; Ding, D.; et al. Electrically conductive polypropylene nanocomposites with negative permittivity at low carbón nanotube loading levels. *ACS Appl. Mater. Interfaces* **2015**, *7*, 6125–6138. [CrossRef]
52. Laird, E.D.; Li, C.Y. Structure and Morphology Control in Crystalline Polymer-Carbon Nanotube Nanocomposites. *Macromolecules* **2013**, *46*, 2877–2891. [CrossRef]
53. Grady, B.P.; Pompeo, F.; Shambaugh, R.L.; Resasco, D.E. Nucleation of polypropylene crystallization by single-walled carbon nanotubes. *J. Phys. Chem. B* **2002**, *106*, 5852–5858. [CrossRef]
54. Bhattacharyya, A.R.; Sreekumar, T.V.; Liu, T.; Kumar, S.; Ericson, L.M.; Hauge, R.H.; Smalley, R.E. Crystallization and orientation studies in polypropylene/single wall carbon nanotube composite. *Polymer* **2003**, *44*, 2373–2377. [CrossRef]
55. Xu, D.H.; Wang, Z.G. Role of multi-wall carbon nanotube network in composites to crystallization of isotactic polypropylene matrix. *Polymer* **2008**, *49*, 330–338. [CrossRef]
56. Haggenmueller, R.; Fischer, J.E.; Winey, K.I. Single wall carbon nanotube/polyethylene nanocomposites: Nucleating and templating polyethylene crystallites. *Macromolecules* **2006**, *39*, 2964–2971. [CrossRef]
57. Funck, A.; Kaminsky, W. Polypropylene carbon nanotube composites by in situ polymerization. *Compos. Sci. Technol.* **2007**, *67*, 906–915. [CrossRef]
58. Wang, J.; Yang, J.; Deng, L.; Fang, H.; Zhang, Y.; Wang, Z. More dominant shear flow effect assisted by added carbon nanotubes on crystallization kinetics of isotactic polypropylene in nanocomposites. *ACS Appl. Mater. Interfaces* **2015**, *7*, 1364–1375. [CrossRef]
59. Maharramov, A.M.; Ramazanov, M.A.; Ahmadova, A.B.; Hajiyeva, F.V.; Hasanova, U.A. Structure and dielectric properties of nanocomposites based on isotactic polypropylene and titanium nanoparticles. *Dig. J. Nanomater. Bios.* **2016**, *11*, 781–786.
60. Sierra, R.; Pérez, M.; Valdez, J.; Ávila, C.; Jimenez, E.J.; Mata, J.; Soto, E.; Cadenas, G. Synthesis and Thermomechanical Characterization of Nylon 6/Cu Nanocomposites Produced by an Ultrasound-Assisted Extrusion Method. *Adv. Mater. Sci. Eng.* **2018**, *2018*, 4792735. [CrossRef]
61. Pan, Y.; Feng, H.K.; Li, L.; Chan, S.; Zhao, J.; Kay, Y. Effects of hybrid fillers on the electrical conductivity and EMI shielding efficiency of polypropylene/conductive filler composites. *Macromol. Res.* **2013**, *21*, 905–910. [CrossRef]
62. Pan, Y.; Cheng, H.K.F.; Li, L.; Chan, S.H.; Zhao, J.; Juay, Y.K. Annealing induced electrical conductivity jump of multi-walled carbon nanotube/polypropylene composites and influence of molecular weight of polypropylene. *J. Polym. Sci. Polym. Phys.* **2010**, *48*, 2238–2247. [CrossRef]
63. Steinmann, W.; Vad, T.; Weise, B.; Wulfhorst, J.; Seide, G.; Gries, T.; Heidelmann, M.; Weirich, T. Extrusion of CNT-modified polymers with low viscosity-influence of crystallization and CNT orientation on the electrical properties. *Polym. Polym. Compos.* **2013**, *21*, 473–482. [CrossRef]
64. Ramazanov, MA.; Hajiyeva, F.V.; Maharramov, A.M. Structure and properties of PP/TiO_2 based polymer nanocomposites. *Integr. Ferroelectr.* **2018**, *192*, 103–112. [CrossRef]
65. Rafiee, R. Fabrication of carbon nanotube/polymer nanocomposite. In *Carbon Nanotube-Reinforced Polymers: From Nanoscale to Macroscale*; Elsevier: Amsterdam, The Netherlands, 2017; Volume 1, pp. 61–76.

66. Song, Y.S.; Youn, J.R. Influence of dispersion states of carbon nanotubes on physical properties of epoxy nanocomposites. *Carbon* **2005**, *43*, 1378–1385. [CrossRef]
67. Zhao, Q.; Hu, Y.; Wang, X. Mechanical performance and flame retardancy of polypropylene composites containing zeolite and multiwalled carbon nanotubes. *J. Appl. Polym. Sci.* **2016**, *133*, 42875–42879. [CrossRef]
68. Apaydin, K.; Laachachi, A.; Ball, V.; Jimenez, M.; Bourbigot, S.; Toniazzo, V.; Ruch, D. Intumescent coating of (polyallylamine-polyphosphates) deposited on polyamide fabrics via layer-by-layer technique. *Polym Degrad Stab.* **2014**, *106*, 158–164. [CrossRef]

Article

Direct Writing Supercapacitors Using a Carbon Nanotube/Ag Nanoparticle-Based Ink on Cellulose Acetate Membrane Paper

Xipeng Guan [1,2], Lin Cao [2], Qin Huang [2], Debin Kong [2], Peng Zhang [2], Huaijun Lin [2], Wei Li [2], Zhidan Lin [2,*] and Hong Yuan [1,*]

1 MOE Key Laboratory of Disaster Forecast and Control in Engineering, School of Mechanics and Construction Engineering, Jinan University, Guangzhou 510632, China; guanxipeng@jnu.edu.cn

2 Institute of Advances Wear & Corrosion Resistant and Functional Materials, Jinan University, Guangzhou 510632, China; linc19993@163.com (L.C.); hq2502@126.com (Q.H.); 13580457530@163.com (D.K.); tzhangpeng@jnu.edu.cn (P.Z.); hjlin@jnu.edu.cn (H.L.); liweijnu@126.com (W.L.)

* Correspondence: linzd@jnu.edu.cn (Z.L.); tyuanhong@jnu.edu.cn (H.Y.); Tel.: +86-020-8522-0890 (Z.L.)

Received: 22 April 2019; Accepted: 17 May 2019; Published: 3 June 2019

Abstract: In this work, we present a cellulose acetate membrane flexible supercapacitor prepared through a direct writing method. A carbon nanotube (CNT) and silver (Ag) nanoparticle were prepared into ink for direct writing. The composite electrode displayed excellent electrochemical and mechanical electrochemical performance. Furthermore, the CNT-Ag displayed the highest areal capacity of 72.8 F/cm^3. The assembled device delivered a high areal capacity (17.68 F/cm^3) at a current density of 0.5 mA/cm^2, a high areal energy (9.08–5.87 mWh/cm^3) at a power density of 1.18–0.22 W/cm^3, and showed no significant decrease in performance with a bending angle of 180°. The as-fabricated CNT/Ag electrodes exhibited good long-term cycling stability after 1000 time cycles with 75.92% capacitance retention. The direct writing was a simple, cost-effective, fast, and non-contact deposition method. This method has been used in current printed electronic devices and has potential applications in energy storage.

Keywords: Ag; CNT; flexible supercapacitor electrode; polymer conductive film; cellulose acetate membrane

---

1. Introduction

With the development of the social economy, human beings are paying more and more attention to green energy and the ecological environment [1–4]. As a new type of energy storage device, the supercapacitor is generally environmentally friendly and has an irreplaceable superiority [5,6]. The two most important points in the design and fabrication of flexible supercapacitors are electrode materials and structure. In the past few years, the number of scientific articles in the fields of laser, direct writing, and printing has increased significantly [7–9]. The direct writing exhibits advantages such as low temperature, being environmentally friendly, shorter reaction times, energy saving, and excellent control over experimental parameters. In addition, the direct writing method allows the active material to undergo catalyst-free growth on insulating substrates.

Carbon-based materials have become attractive nanomaterials in various applications due to their unique properties [10]. A carbon-based nanotube (CNT) has a unique hollow structure, a high specific surface area, mechanical flexibility, a high stability, and a low cost, and it is suitable for electrolyte ion migration pores. Its use as an electrode material can significantly improve the power characteristics and frequency response characteristics of supercapacitors. Additionally, supercapacitors using carbon material-based materials mainly rely on the electrical double layer capacitor (EDLC), exhibiting fast charge-discharge times and surface-area-dependent capacities [11–13]. At present, many studies are devoted to the study of conductive patterns in carbon nanotube printing [14–16]. However, the printed carbon nanotubes have an overly high resistance on flexible substrates and cannot be widely used in optoelectronic devices [17,18].

Both Ag nanowires and sintered Ag nanoparticles have a high conductivity [19–22]. However, nanoparticles are easier to print without clogging the nozzle. The mixing of Ag nanoparticles with graphene can significantly reduce the sheet resistance [23]. Therefore, the mixing of Ag nanoparticles with carbon nanotubes can also reduce the sheet resistance without changing the mechanical flexibility. Combining metal nanoparticles with CNTs can combine the excellent properties of these two types of nanomaterials, and the obtained nanoassembled materials have great application potential in the fields of optics, electricity, and electrocatalysis. The micro/nanostructured metals and metal oxides, such as Ag, Au, Ru_xO_y, Mn_xO_y, and Co_xO_y, as well as their hybrids, can facilitate electron charge transfer induced by electrosorption, redox reactions, and intercalation processes at the interfaces, thereby remarkably improving the energy densities [24–29].

The cellulose acetate membrane paper is usually composed of cellulose fibers arranged in *N*-dimensional form, and the surface of the cellulose acetate membrane paper is not only very rough but also has a highly porous structure as compared with a conventional flexible substrate. In particular, we can make full use of the porous structure of the cellulose film [30–32] to prepare high-performance flexible electrodes.

In this work, we report a low-temperature, fast, controlled, and straightforward approach for the fabrication of CNT/Ag electrodes on a flexible cellulose acetate membrane by the direct writing method. In the new energy system of CNT/Ag, CNT contribution of specific capacitance serves as carbon support and Ag promotion of electrical conductivity as a conducting agent, while the cellulose acetate membrane provides support and a porous structure as a flexible substrate. Furthermore, the capacitive behaviors of the CNT/Ag electrode were tested mainly including cyclic voltammetry (CV), galvanostatic charge-discharge (GCD), and Nyquist, as well as the derivative specific capacitance. This work shows the potential application of flexible direct writing electrode materials for the operation of various flexible devices in the near future.

2. Materials and Methods

2.1. Materials

All the chemicals used in this study were of analytical reagent grade. H_2SO_4 and Na_2SO_4 were obtained from Guangzhou Chemical Reagent Factory (Guangzhou, China). Poly(vinyl alcohol) (PVA) was purchased from Shanghai Macklin Biochemical Co., Ltd. (Shanghai, China). Sodium dodecylbenzene sulfonate (SDBS) was provided by the Tianjin Damao Chemical Reagent Factory (Tianjin, China). Cellulose acetate membrane was provided by Tianjin Jinteng Experimental Equipment Co., Ltd. (Tianjin, China) CNT was purchased from Korea Kumho Corporation (Seoul, Korea). Ag nanoparticles (Ag) were obtained from the Shanghai Chaowei Nanotechnology Co., Ltd. (Shanghai, China).

2.2. Ag and CNT Ink Production

To prepare CNT ink, 0.20 g of SDBS was dissolved in 100 mL deionised water, after which 0.15 g of CNT was slowly added to the solution, and this was followed by ultrasonication for 0.5 h (ultrasonic power of 100 W, ultrasonic frequency of 80 kHz). For the Ag ink production, the 0.04, 0.12, and 0.20 g Ag nanoparticles were dissolved with 0.20 g SDBS in 100 mL deionised water, then added into a glass bottle, and the next procedure was similar to the preparation of the CNT ink.

2.3. Ag and MWCNT Electrode

The process for rapid direct-write of the composite electrode consisted of three steps, as illustrated in Figure 1. During the entire direct-writing process, the cellulose acetate membrane paper was fixed on the experimental bench. First, CNT and Ag ink were printed onto the cellulose acetate membrane paper with a ballpoint pen core with an inner diameter of 1 mm. By using the venous infusion apparatus to provide liquid pressure and control, the writing speed was 500 mm/s by using a digital plotter. With a vacuum oven temperature of 60 °C, the ink was fully dried.

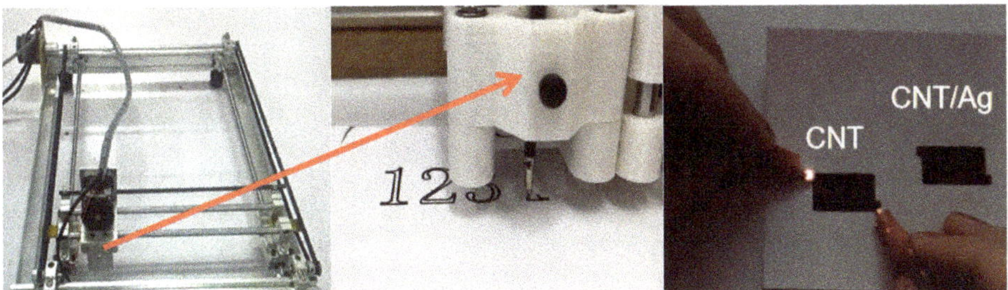

Figure 1. Overview of the fabrication process for direct-writing: design using a personal computer and direct-writing using a Plotter and a photo of patterned electrodes.

2.4. Characterization

Morphological observations were carried out using a field-emission scanning electron microscope (FE-SEM, Zeiss, Oberkochen, Germany). The resistance was tested using a KDB-1 digital four-probe resistance tester (Guangzhou Kunde Technology Companies (Guangzhou, China)). Galvanostatic charge-discharge (GCD), cyclic voltammetry (CV), and electrochemical impedance spectroscopy (EIS) were performed on the flexible electrode using a Princeton (PARSTAT 4000) electrochemical station. The capacitances (C in $F \cdot cm^{-2}$) of each device at different current densities were calculated from the discharge curves obtained from GCD tests using the following formula:

$$C = 2I\Delta t/S\Delta U, \quad (1)$$

where I is the applied discharge current (A), Δt is the discharge time (seconds), and ΔU (V) is the discharge voltage after the iR drop (ohmic voltage) is removed. S (cm^2) is the volume of the active materials of all electrodes. Here, the surface areas of the active materials are 1 cm^2.

The volumetric energy density and power density can provide more reliable performance metrics for porous, nanomaterial-based, thin-film devices compared to gravimetric capacitance. As a result, the volumetric energy density ($Wh \cdot cm^{-2}$) of each device was calculated using the following formula [33,34]:

$$E = 0.5C\Delta U^2/3.6, \quad (2)$$

The volumetric power density ($W \cdot cm^{-2}$) of the device was calculated from the following formula:

$$P = 3600E/\Delta t, \quad (3)$$

We investigated the electrochemical properties of all electrodes on a three-electrode system using a Pt plate as the counter electrode and a saturated calomel reference electrode (SCE) in 1 M Na_2SO_4 aqueous electrolyte solution.

3. Results and Discussion

The fabrication steps of the electrodes and direct-write devices are illustrated in Figure 1. A personal computer was used to design the pattern of the electrodes, which were direct-written using a purchased plotter. The ink concentration was the most important factor for direct-writing, because a concentration that was too high clogged the top of the pen, and a concentration that was too low decreased the active substance content. Therefore, 5 wt % CNT and 1–5 wt % Ag were chosen for the printed ink.

To further prepare the paper-based capacitor from a cellulose acetate membrane, we prepared a CNT and CNT/Ag ink for direct writing and prepared a paper-based capacitor with a cellulose acetate membrane by superposing the direct writing of CNT and Ag. As shown in Figure 2a, cellulose acetate membrane paper is usually composed of cellulose fibres arranged in an N-dimensional hierarchy, and the surface of the cellulose acetate membrane paper was not only very rough but also had a highly porous structure, in comparison to ordinary flexible substrates. We could also make full use of the porous structure of cellulose acetate membrane paper; when soaked, the active materials in the porous structure increased the contact area of the active material with the electrolyte. It could be used to improve the performance of electrochemical energy storage devices such as supercapacitors. The SEM image in Figure 2b shows that the CNT electrodes were composed of CNT with a diameter of about 50 nm. Then, different wt % Ag values were used to dope the CNT ink, and Ag particles were uniformly dispersed between carbon nanotubes (Figure 2c).

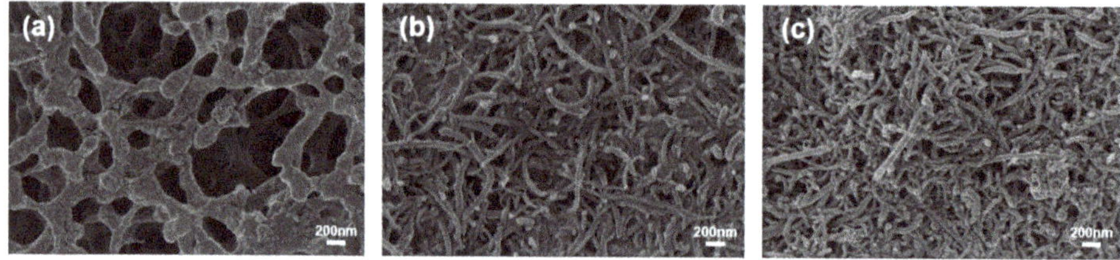

Figure 2. SEM images of (**a**) cellulose acetate membrane paper; (**b**) carbon nanotube (CNT); (**c**) CNT/Ag-5.

The CV, GCD, and EIS measurements were conducted in a three-electrode configuration in 1 M Na_2SO_4 aqueous solution. The results about the electrochemical performance of the CNT electrode before and after the doping of Ag nanoparticles are shown in Figure 3a–h. As expected, in Figure 3a, the rectangular CV curves of the electrode showed an ideal capacitive behavior. The GCD performance of the CNT electrode was further tested in Figure 3b. The volumetric capacitance for the CNT electrode was 25.3 F/cm^3 at a discharge current density of 0.5 mA/cm^2.

Moreover, the electrodes doping Ag exhibited a larger volume specific capacitance than CNT electrodes under the same voltage sweep speed (Figure 3a,c). The GCD curves of the CNT/Ag electrodes with different Ag doping doses at a current density of 0.5 mA/cm^2 are shown in Figure 3d. It can be observed that as the Ag doping dose increased from 1 to 5 wt %, the charge and discharge time of CNT/Ag electrodes increased, and the iR drop decreased. This appearance indicates an increase in the capacitance of the CNT electrode after Ag doping. The volumetric capacitances for the 1, 3, and 5 wt % Ag doping CNT/Ag electrode were 25.7, 37.4, and 72.8 F/cm^3, respectively, at a current density of 0.5 mA/cm^2, which were much larger than the value of the CNT electrode (25.3 F/cm^3). The high volumetric capacitance of CNT/Ag electrodes could be ascribed to the fact that the Ag particles could promote the contact between the electrolyte ions and the electrode surfaces, as well as the enhanced conductivity of the composite electrodes. It was also observed in Figure 3e that the addition of Ag nanoparticles significantly improved the conductivity. The experimental results showed that by doping of Ag conductive phase, the resistivity of the electrode could be 5.1×10^{-4} Ω/cm, which was 98.2% less than the CNT electrode. The main reason for the decrease in resistivity was the excellent conductivity of Ag nanoparticles and the good connection between Ag and CNT. It can be seen from Figure 3f that the equivalent series resistance (ESR) for the CNT/Ag-5 electrode was about 13.4 Ω/cm^2, and CNT electrode ESR was 51.7 Ω/cm^2. Due to the high conductivity of Ag particles, the iR dropped and was much lower than the CNT electrode. The CV curve of the CNT/Ag-5 electrode under different sweep speeds is shown in Figure 3g. The CNT/Ag-5 electrodes had very symmetrical and nearly rectangular CV curves in a potential window of 0–1 V, indicating that the device also had a excellent capacitance performance. Good linearity of GCD curves with different current densities (Figure 3h) further confirms the good electrochemical behavior of the device.

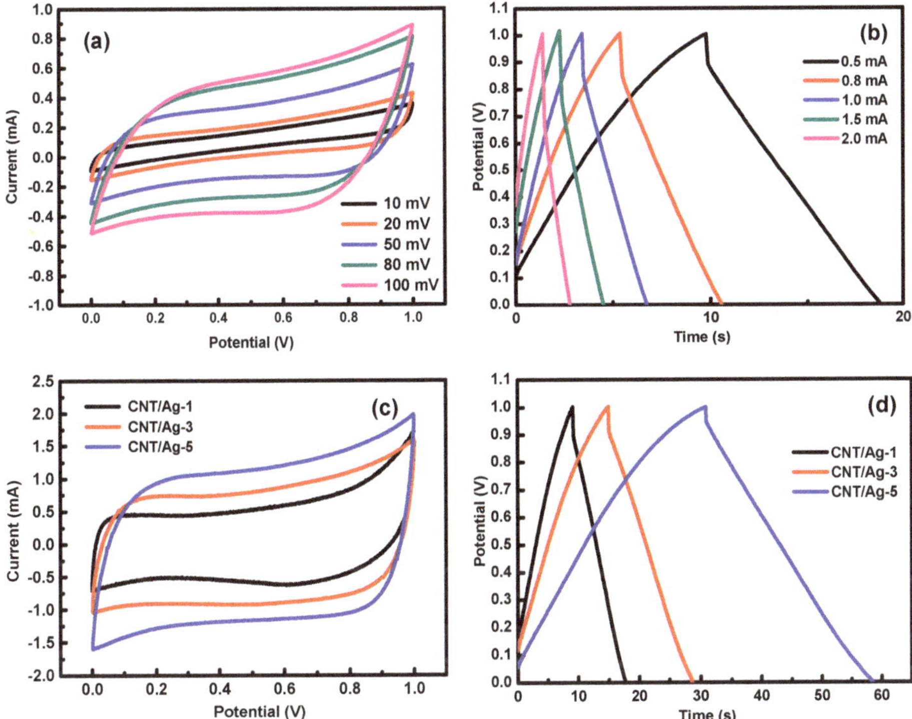

Figure 3. *Cont.*

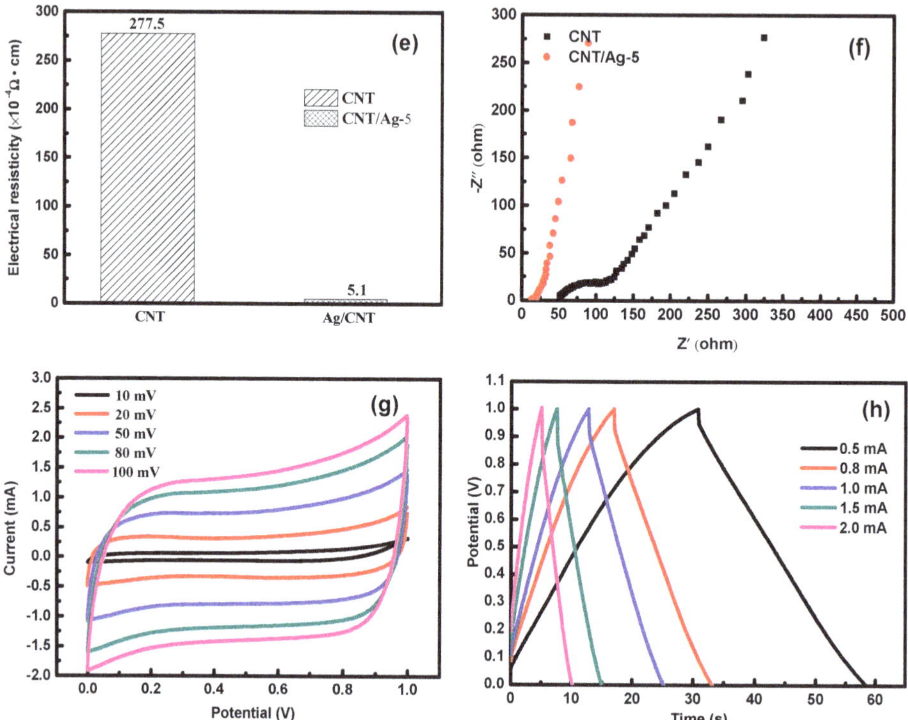

Figure 3. (**a**) Cyclic voltammetry (CVs) of the CNT electrodes with scan rate (from 10.0 to 100.0 mV/s). (**b**) Galvanostatic charge-discharge (GCD) of the CNT electrodes with 0.5 to 2.0 mA/cm^2. (**c**) CV of CNT/Ag electrodes with different Ag. (**d**) GCD of CNT/Ag electrodes with different Ag. (**e**) Resistivity of the electrodes. (**f**) Nyquist plot of the electrodes. (**g**) CV of the CNT/Ag-5 electrodes with scan rate (from 10.0 to 100.0 mV/s). (**h**) GCD of the CNT/Ag-5 electrodes with 0.5 to 2.0 mA/cm^2.

The CNT/Ag composite papers could be directly used as electrodes for a flexible solid-state supercapacitor (FSC) because they possess flexibility and low sheet resistance. As a proof of concept, we used two identical CNT/Ag-5 electrodes for the assembled supercapacitors, and the performance is depicted in Figure 4a. Figure 4a shows that the CV curves displayed no obvious distortion in the shape at a high scan rate of 100 mV/s. From the GCD curves (Figure 4b), we can determine that the single volumetric capacitance was 17.68 F/cm^3 at a current density of 0.5 mA/cm^2.

The CV curves under different bending angles of solid-state symmetric supercapacitors (at the scanning rate of 100 mV/s) is shown in Figure 4c. All of the CV curves almost overlapped under various bending conditions. This confirmed that no structure failure and capacitance loss occur upon bending up to 180°, which is the requirement for wearable applications. The specific energy and specific power were the two sticking points for evaluating the practical application in FSC.

Figure 4d presents the Ragone plot comparing the energy density and the power density of the flexible energy storage devices with the CNT/Ag-5 electrodes and CNT electrodes. It was worth noting that the assembly in this work had shown an outstanding energy density and power density; the energy density was 9.08–5.87 mWh/cm^3 with a power density of 1.18–0.22 W/cm^3. These powerful mechanical features allowed the device to be connected in serial geometry that provided high power to maintain a steady red LED of 1.7 V (Figure 4d). Figure 4e shows the excellent cycling stability of CNT/Ag-5 flexible solid-state symmetric supercapacitors after 1000 cycles at a current density of

0.5 mA/cm^2. The CNT/Ag-5 sample retained up to 75.92% of its initial capacitance after the first 1000 cycles. GCD cycling stability of the first ten cycles and last ten cycles is shown in Figure 4f.

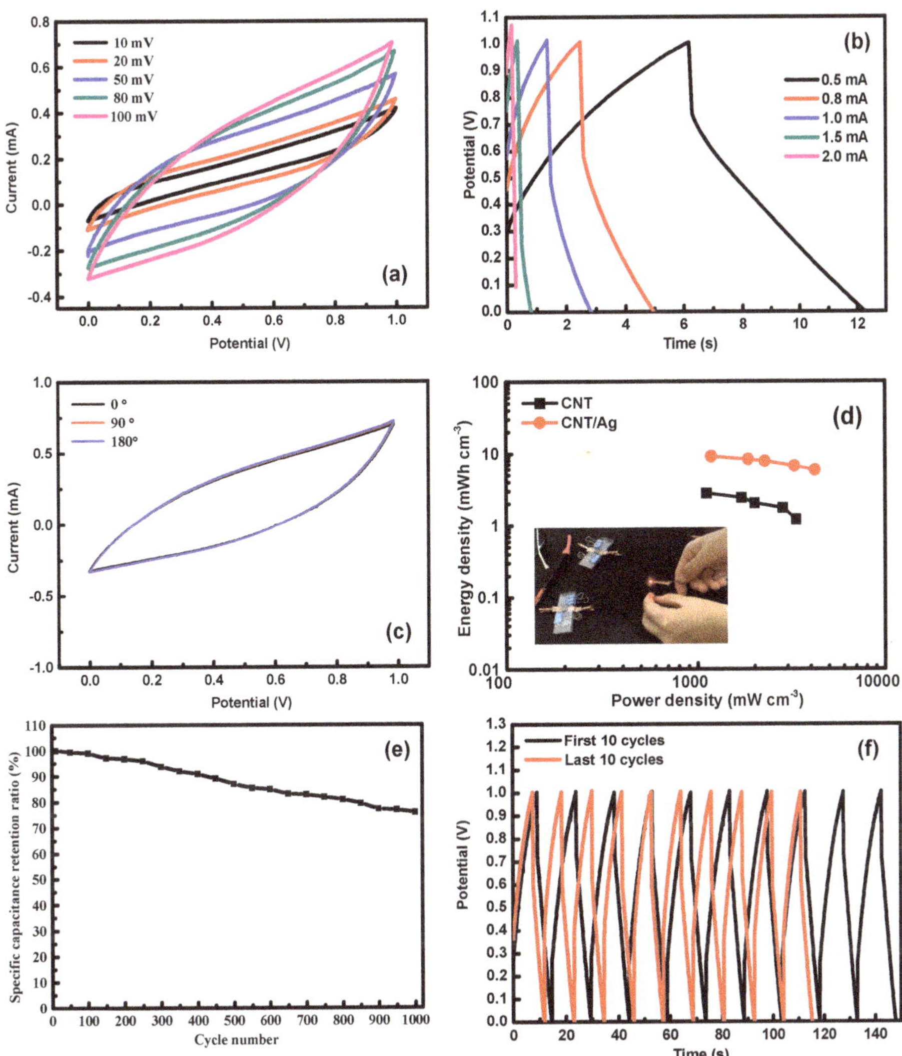

Figure 4. (**a**) CV and (**b**) discharge curves at different scan rates and current densities. (**c**) CV curves under different bending angles of solid-state symmetric supercapacitors. (**d**) Ragone plots of the areal energy density and power density and digital photographs of the assembled flexible solid-state symmetric supercapacitors (two in series) lighting LEDs. (**e**) Cycling stability of the flexible solid-state symmetric supercapacitors. (**f**) GCD cycling stability of the first ten cycles and last ten cycles

4. Conclusions

In summary, we reported a process for the design and fabrication of a CNT/Ag cellulose acetate membrane. The good solubility of CNT in SDBS solution, the way in which carbon nanotubes combined well with the Ag particles, and the application of the direct writing method all increased the binding force between the CNT/Ag and the cellulose acetate membrane. The incorporation of Ag into CNT largely modulated the conductivity, as well as the resulting capacitive behaviors, including the enhancement of specific capacitance, rate capability, coulombic efficiency, and cycling stability. The as-fabricated CNT/Ag-5 exhibited excellent cycling stability, with 75.92% capacitance retention. The experimental results show that by doping Ag conductive phase, the resistivity of the electrode can be 5.1×10^{-4} Ω/cm, which was 98.2% less than the CNT electrode. In addition, the doping of Ag nanoparticles not only improved the conductivity of the composite electrode, but also enhanced the specific volumetric capacitance up to 72.8 F/cm^3. The as-fabricated FSC showed an outstanding flexibility for wearable applications, without showing any decline in performance under bending conditions. It also had a high energy density of 9.08–5.87 mWh/cm^3 at a power density of 1.18–0.22 W/cm^3.The present doping method in this work can provide more kinds of alternative electrode materials for high-performance supercapacitors. The conducting agent doping in carbon materials could be easily extended to other substances and have promising prospects in future developments.

Author Contributions: Conceptualization, L.C.; Data curation, X.G., L.C., Q.H., and D.K.; Formal analysis, X.G., L.C., and P.Z.; Methodology, H.L., Z.L., and H.Y.; Project administration, X.G., Z.L., and H.Y.; Resources, P.Z., H.L., W.L., Z.L., and H.Y.; Supervision, Z.L. and H.Y.; Writing–original draft, X.G. and L.C.

Funding: This research was funded by [Jinan University Science and Technology Key Platform Construction Special Project] grant number [21617422], [Guangdong Science and Technology Project Fund] grant number [2015A030310488] and [Scientific cultivation and innovation fund project of Jinan University] grant number [21617427].

Conflicts of Interest: The authors declare no conflict of interest.

References

1. Kouchachvili, L.; Yaïci, W.; Entchev, E. Hybrid battery/supercapacitor energy storage system for the electric vehicles. *J. Power Sources* **2018**, *374*, 237–248. [CrossRef]
2. Guo, H.; Yeh, M.-H.; Zi, Y.; Wen, Z.; Chen, J.; Liu, G.; Hu, C.; Wang, Z.L. Ultralight cut-paper-based self-charging power unit for self-powered portable electronic and medical systems. *ACS Nano* **2017**, *11*, 4475–4482. [CrossRef] [PubMed]
3. Niu, Z.; Zhang, L.; Liu, L.; Zhu, B.; Dong, H.; Chen, X. All-solid-state flexible ultrathin micro-supercapacitors based on graphene. *Adv. Mater.* **2013**, *25*, 4035–4042. [CrossRef] [PubMed]
4. Guo, R.; Chen, J.; Yang, B.; Liu, L.; Su, L.; Shen, B.; Yan, X. In-plane micro-supercapacitors for an integrated device on one piece of paper. *Adv. Funct. Mater.* **2017**, *27*, 1702394. [CrossRef]
5. Beidaghi, M.; Gogotsi, Y. Capacitive energy storage in micro-scale devices: Recent advances in design and fabrication of micro-supercapacitors. *Energy Environ. Sci.* **2014**, *7*, 867–884. [CrossRef]
6. González, A.; Goikolea, E.; Barrena, J.A.; Mysyk, R. Review on supercapacitors: Technologies and materials. *Renew. Sustain. Energy Rev.* **2016**, *58*, 1189–1206. [CrossRef]
7. Kumar, R.; Singh, R.K.; Singh, D.P.; Joanni, E.; Yadav, R.M.; Moshkalev, S.A. Laser-assisted synthesis, reduction and micro-patterning of graphene: Recent progress and applications. *Coord. Chem. Rev.* **2017**, *342*, 34–79. [CrossRef]
8. Kumar, R.; Savu, R.; Joanni, E.; Vaz, A.R.; Canesqui, M.A.; Singh, R.K.; Timm, R.A.; Kubota, L.T.; Moshkalev, S.A. Fabrication of interdigitated micro-supercapacitor devices by direct laser writing onto ultra-thin, flexible and free-standing graphite oxide films. *RSC Adv.* **2016**, *6*, 84769–84776. [CrossRef]
9. Kumar, R.; Joanni, E.; Singh, R.K.; da Silva, E.T.; Savu, R.; Kubota, L.T.; Moshkalev, S.A. Direct laser writing of micro-supercapacitors on thick graphite oxide films and their electrochemical properties in different liquid inorganic electrolytes. *J. Colloid Interface Sci.* **2017**, *507*, 271–278. [CrossRef] [PubMed]
10. Kumar, R.; Joanni, E.; Singh, R.K.; Singh, D.P.; Moshkalev, S.A. Recent advances in the synthesis and modification of carbon-based 2d materials for application in energy conversion and storage. *Prog. Energy Combust. Sci.* **2018**, *67*, 115–157. [CrossRef]
11. Zhang, L.L.; Zhao, X. Carbon-based materials as supercapacitor electrodes. *Chem. Soc. Rev.* **2009**, *38*, 2520–2531. [CrossRef] [PubMed]

12. Zang, Z.; Zeng, X.; Wang, M.; Hu, W.; Liu, C.; Tang, X. Tunable photoluminescence of water-soluble aginzns–graphene oxide (go) nanocomposites and their application in-vivo bioimaging. *Sens. Actuators B Chem.* **2017**, *252*, 1179–1186. [CrossRef]
13. Zhai, Y.; Dou, Y.; Zhao, D.; Fulvio, P.F.; Mayes, R.T.; Dai, S. Carbon materials for chemical capacitive energy storage. *Adv. Mater.* **2011**, *23*, 4828–4850. [CrossRef]
14. Kim, T.; Song, H.; Ha, J.; Kim, S.; Kim, D.; Chung, S.; Lee, J.; Hong, Y. Inkjet-printed stretchable single-walled carbon nanotube electrodes with excellent mechanical properties. *Appl. Phys. Lett.* **2014**, *104*, 113103. [CrossRef]
15. Tortorich, R.P.; Song, E.; Choi, J.-W. Inkjet-printed carbon nanotube electrodes with low sheet resistance for electrochemical sensor applications. *J. Electrochem. Soc.* **2014**, *161*, B3044–B3048. [CrossRef]
16. Shimoni, A.; Azoubel, S.; Magdassi, S. Inkjet printing of flexible high-performance carbon nanotube transparent conductive films by "coffee ring effect". *Nanoscale* **2014**, *6*, 11084–11089. [CrossRef]
17. El-Kady, M.F.; Strong, V.; Dubin, S.; Kaner, R.B. Laser scribing of high-performance and flexible graphene-based electrochemical capacitors. *Science* **2012**, *335*, 1326–1330. [CrossRef] [PubMed]
18. Frackowiak, E.; Beguin, F. Carbon materials for the electrochemical storage of energy in capacitors. *Carbon* **2001**, *39*, 937–950. [CrossRef]
19. Hu, L.; Kim, H.S.; Lee, J.-Y.; Peumans, P.; Cui, Y. Scalable coating and properties of transparent, flexible, silver nanowire electrodes. *ACS Nano* **2010**, *4*, 2955–2963. [CrossRef]
20. Chiolerio, A.; Maccioni, G.; Martino, P.; Cotto, M.; Pandolfi, P.; Rivolo, P.; Ferrero, S.; Scaltrito, L. Inkjet printing and low power laser annealing of silver nanoparticle traces for the realization of low resistivity lines for flexible electronics. *Microelectron. Eng.* **2011**, *88*, 2481–2483. [CrossRef]
21. Sawangphruk, M.; Suksomboon, M.; Kongsupornsak, K.; Khuntilo, J.; Srimuk, P.; Sanguansak, Y.; Klunbud, P.; Suktha, P.; Chiochan, P. High-performance supercapacitors based on silver nanoparticle–polyaniline–graphene nanocomposites coated on flexible carbon fiber paper. *J. Mater. Chem. A* **2013**, *1*, 9630–9636. [CrossRef]
22. Kalambate, P.K.; Dar, R.A.; Karna, S.P.; Srivastava, A.K. High performance supercapacitor based on graphene-silver nanoparticles-polypyrrole nanocomposite coated on glassy carbon electrode. *J. Power Sources* **2015**, *276*, 262–270. [CrossRef]
23. Li, L.; Guo, Y.; Zhang, X.; Song, Y. Inkjet-printed highly conductive transparent patterns with water based ag-doped graphene. *J. Mater. Chem. A* **2014**, *2*, 19095–19101. [CrossRef]
24. Moon, H.; Lee, H.; Kwon, J.; Suh, Y.D.; Kim, D.K.; Ha, I.; Yeo, J.; Hong, S.; Ko, S.H. Ag/au/polypyrrole core-shell nanowire network for transparent, stretchable and flexible supercapacitor in wearable energy devices. *Sci. Rep.* **2017**, *7*, 41981. [CrossRef]
25. Usman, M.; Pan, L.; Sohail, A.; Mahmood, Z.; Cui, R. Fabrication of 3D vertically aligned silver nanoplates on nickel foam-graphene substrate by a novel electrodeposition with sonication for efficient supercapacitors. *Chem. Eng. J.* **2017**, *311*, 359–366. [CrossRef]
26. Shin, D.; Shin, J.; Yeo, T.; Hwang, H.; Park, S.; Choi, W. Scalable synthesis of triple-core–shell nanostructures of TiO_2@ MnO_2@ c for high performance supercapacitors using structure-guided combustion waves. *Small* **2018**, *14*, 1703755. [CrossRef] [PubMed]
27. Hu, Y.; Wu, Y.; Wang, J. Manganese-oxide-based electrode materials for energy storage applications: How close are we to the theoretical capacitance? *Adv. Mater.* **2018**, *30*, 1802569. [CrossRef] [PubMed]
28. Liu, Y.-H.; Xu, J.-L.; Gao, X.; Sun, Y.-L.; Lv, J.-J.; Shen, S.; Chen, L.-S.; Wang, S.-D. Freestanding transparent metallic network based ultrathin, foldable and designable supercapacitors. *Energy Environ. Sci.* **2017**, *10*, 2534–2543. [CrossRef]
29. Wei, J.; Zang, Z.; Zhang, Y.; Wang, M.; Du, J.; Tang, X. Enhanced performance of light-controlled conductive switching in hybrid cuprous oxide/reduced graphene oxide (Cu_2 o/rGO) nanocomposites. *Opt. Lett.* **2017**, *42*, 911–914. [CrossRef]
30. Sawitri, R.A.; Suryanti, L.; Zuhri, F.U.; Diantoro, M. *Dielectric Properties of Dirt Sugarcane Sediment (Dss) Extract-Batio3 for Organic Supercapacitors*; IOP Conference Series: Materials Science and Engineering; IOP Publishing: Bristol, UK, 2019; p. 012062.
31. Liu, W.; Cui, M.; Shen, Y.; Zhu, G.; Luo, L.; Li, M.; Li, J. Waste cigarette filter as nanofibrous membranes for on-demand immiscible oil/water mixtures and emulsions separation. *J. Colloid Interface Sci.* **2019**, *549*, 114–122. [CrossRef] [PubMed]

32. Miao, Y.-E.; Liu, T. Electrospun nanofiber electrodes: A promising platform for supercapacitor applications. In *Electrospinning: Nanofabrication and Applications*; Elsevier: Amsterdam, The Netherlands, 2019; pp. 641–669.
33. Li, K.; Liu, J.; Huang, Y.; Bu, F.; Xu, Y. Integration of ultrathin graphene/polyaniline composite nanosheets with a robust 3d graphene framework for highly flexible all-solid-state supercapacitors with superior energy density and exceptional cycling stability. *J. Mater. Chem. A* **2017**, *5*, 5466–5474. [CrossRef]
34. Chen, B.; Jiang, Y.; Tang, X.; Pan, Y.; Hu, S. Fully packaged carbon nanotube supercapacitors by direct ink writing on flexible substrates. *ACS Appl. Mater. Interfaces* **2017**, *9*, 28433–28440. [CrossRef] [PubMed]

MDPI
St. Alban-Anlage 66
4052 Basel
Switzerland
Tel. +41 61 683 77 34
Fax +41 61 302 89 18
www.mdpi.com

Polymers Editorial Office
E-mail: polymers@mdpi.com
www.mdpi.com/journal/polymers

www.ingramcontent.com/pod-product-compliance
Lightning Source LLC
LaVergne TN
LVHW070122170726
843515LV00007B/1732
9783039289912